Lecture Notes in Business Information Processing 543

LNBIP reports state-of-the-art results in areas related to business information systems and industrial application software development – timely, at a high level, and in both printed and electronic form.

The type of material published includes

- Proceedings (published in time for the respective event)
- Postproceedings (consisting of thoroughly revised and/or extended final papers)
- Other edited monographs (such as, for example, project reports or invited volumes)
- Tutorials (coherently integrated collections of lectures given at advanced courses, seminars, schools, etc.)
- Award-winning or exceptional theses

LNBIP is abstracted/indexed in DBLP, EI and Scopus. LNBIP volumes are also submitted for the inclusion in ISI Proceedings.

Massimo Marchiori · Francisco García Peñalvo
Editors

Web Information Systems and Technologies

19th International Conference, WEBIST 2023
Rome, Italy, November 15–17, 2023
Revised Selected Papers

Springer

Editors
Massimo Marchiori
University of Padua
Padua, Italy

Francisco García Peñalvo
University of Salamanca
Salamanca, Spain

ISSN 1865-1348 ISSN 1865-1356 (electronic)
Lecture Notes in Business Information Processing
ISBN 978-3-031-89620-0 ISBN 978-3-031-89621-7 (eBook)
https://doi.org/10.1007/978-3-031-89621-7

This Springer imprint is published by the registered company Springer Nature Switzerland AG
The registered company address is: Gewerbestrasse 11, 6330 Cham, Switzerland

Preface

The present book includes extended and revised versions of a set of selected papers from the 19th International Conference on Web Information Systems and Technologies (WEBIST 2023), held in Rome, Italy, from 15–17 November.

The purpose of the International Conference on Web Information Systems and Technologies (WEBIST) is to bring together researchers, engineers and practitioners interested in technological advances and business applications of web-based information systems. The conference has four main tracks, covering different aspects of Web Information Systems, namely Internet Technology, Web Intelligence and Semantic Web, Social Network Analytics, and HCI in Mobile Systems and Web Interfaces. WEBIST focuses on real-world applications; therefore authors should highlight the benefits of Web Information Systems and Technologies for industry and services, in addition to academic applications. Ideas on how to solve business problems using web-based information systems and technologies are discussed at the conference.

WEBIST 2023 received 77 paper submissions from 25 countries, of which 17% were included in this book.

The papers were selected by the event chairs and their selection was based on a number of criteria that included the classifications and comments provided by the program committee members, the session chairs' assessment and also the program chairs' global view of all papers included in the technical program. The authors of selected papers were then invited to submit revised and extended versions of their papers having at least 30% innovative material.

The papers selected to be included in this book contribute to the understanding of relevant trends of current research on Web Information Systems and Technologies, including: Human Computer Interaction, Research Projects and Internet Technology, UX and User Centric Systems, Web Programming, Web Tools and Languages, Applications, Research Projects and Web Intelligence, Natural Language Processing, Human Factors, Web Information Filtering and Retrieval, and Web Interfaces and Applications.

We would like to thank all the authors for their contributions and also the reviewers who helped to ensure the quality of this publication.

November 2023

Massimo Marchiori
Francisco García Peñalvo

Organization

Conference Chair

Massimo Marchiori	University of Padua, Italy

Program Chair

Francisco García Peñalvo	University of Salamanca, Spain

Program Committee

Alsayed Algergawy	University of Jena, Germany
Elarbi Badidi	United Arab Emirates University, UAE
Demis Ballis	University of Udine, Italy
Faiza Belala	Constantine II University, Algeria
Fernando Bobillo	University of Zaragoza, Spain
Pasquina Campanella	University of Bari "Aldo Moro", Italy
Dickson Chiu	University of Hong Kong, China
Pedro J. Clemente	University of Extremadura, Spain
Liviu-Adrian Cotfas	Bucharest University of Economic Studies, Romania
Christophe Cruz	University of Burgundy, France
Daniel Cunliffe	University of South Wales, UK
Guglielmo De Angelis	IASI-CNR, Italy
Toon De Pessemier	Ghent University - Imec, Belgium
Enrico Denti	Alma Mater Studiorum - Università di Bologna, Italy
Martin Drlik	Constantine the Philosopher University in Nitra, Slovak Republic
Jon Dron	Athabasca University, Canada
Karim El Guemhioui	Université du Québec en Outaouais, Canada
Larbi Esmahi	Athabasca University, Canada
Przemyslaw Falkowski-Gilski	Gdańsk University of Technology, Poland
Luís Ferreira Pires	University of Twente, Netherlands
Karla Fook	Instituto Tecnológico de Aeronáutica, Brazil
Pasi Fränti	University of Eastern Finland, Finland

Piero Fraternali	Politecnico di Milano, Italy
Xiang Fu	Hofstra University, USA
Ombretta Gaggi	Università di Padova, Italy
Francisco García-Sánchez	University of Murcia, Spain
Massimiliano Garda	Università degli Studi di Brescia, Italy
John Garofalakis	University of Patras, Greece
Ilche Georgievski	University of Stuttgart, Germany
Jose Gonzalez	University of Seville, Spain
Francesco Guerra	University of Modena and Reggio Emilia, Italy
Shanmugasundaram Hariharan	Vardhaman College of Engineering, India
Ioannis Hatzilygeroudis	University of Patras, Greece
Sebastian Heil	Technische Universität Chemnitz, Germany
Andreas Hinderks	University of Applied Sciences and Arts Hannover, Germany
Ronghuai Huang	Beijing Normal University, China
Sergio Ilarri	University of Zaragoza, Spain
Edmond Jajaga	UBT - Higher Education Institution, Kosovo
Monique Janneck	Lübeck University of Applied Sciences, Germany
Karel Jezek	University of West Bohemia, Czech Republic
Mohsen Kahani	Ferdowsi University of Mashhad, Iran
Georgia Kapitsaki	University of Cyprus, Cyprus
Sokratis Katsikas	Norwegian University of Science and Technology, Norway
Ilan Kirsh	ObjectDB Software, UK
Andreas Klein	University of Seville, Spain
Matthias Klusch	German Research Center for Artificial Intelligence (DFKI) GmbH, Germany
Hiroshi Koide	Kyushu University, Japan
Fotios Kokkoras	University of Thessaly, Greece
Maxime Lefrançois	École des Mines de Saint-Étienne, France
Isaac Lera	Universitat de les Illes Balears, Spain
Martin Llamas-Nistal	University of Vigo, Spain
Angelica Lo Duca	IIT-CNR, Pisa, Italy
Francisco J. Lopez-Pellicer	Universidad de Zaragoza, Spain
Miroslaw Mazurek	Rzeszów University of Technology, Poland
Inmaculada Medina-Bulo	Universidad de Cádiz, Spain
Michele Melchiori	University of Brescia, Italy
Santiago Meliá	Universidad de Alicante, Spain
Ingo Melzer	Daimler Truck North America, USA
Cédric Mouza	CNAM, France
Martin Musicante	Universidade Federal do Rio Grande do Norte, Brazil

Akiyo Nadamoto	Konan University, Japan
Anand Nayyar	Duy Tan University, Vietnam
Martin Necaský	Charles University in Prague, Czech Republic
Alex Norta	Tallinn University of Technology, Estonia
Daniel O'Leary	University of Southern California, USA
Panagiotis Papadopoulos	iProov Ltd., UK
Kalpdrum Passi	Laurentian University, Canada
David Paul	University of New England, Australia
Bhanu Prasad	Florida A&M University, USA
Jim Prentzas	Democritus University of Thrace, Greece
Birgit Pröll	Johannes Kepler University Linz, Austria
Muhammad Asif Razzaq	Data Science Institute, Insight SFI Research Centre for Data Analytics, University of Galway, Ireland
Marina Ribaudo	Università di Genova, Italy
Thomas Risse	University Library Johann Christian Senckenberg, Germany
Riccardo Rosati	Sapienza University of Rome, Italy
Lloyd Rutledge	Open University of the Netherlands, Netherlands
Victor Saquicela	University of Cuenca, Ecuador
Comai Sara	Politecnico di Milano, Italy
Claudio Schifanella	Università degli Studi di Torino, Italy
Georg Schneider	Trier University of Applied Sciences, Germany
Wieland Schwinger	Johannes Kepler University Linz, Austria
Álvaro Sicilia	Ramon Llull University, Spain
Marianna Sigala	University of South Australia Business School, Australia
Selver Softic	CAMPUS 02 University of Applied Sciences, Austria
Eliza Stefanova	Sofia University, Bulgaria
Ilias Tachmazidis	University of Huddersfield, UK
Giorgio Terracina	University of Calabria, Italy
Christos Troussas	University of West Attica, Greece
Ananthanarayana V. S.	Independent Researcher, India
Liliana Patricia Valencia	King Juan Carlos University, Spain
Costas Vassilakis	University of the Peloponnese, Greece
Jari Veijalainen	University of Jyvaskyla, Finland
Tony Wasserman	Software Methods and Tools, USA
Jason Whalley	Northumbria University, UK

Invited Speakers

Christian Bizer	University of Mannheim, Germany
Yannis Manolopoulos	Open University of Cyprus, Cyprus
Dimitris Kiritsis	EPFL, Switzerland

Contents

Impact of Item Polarity on Semantic Differential Scales in User Experience Measurement

Martin Schrepp[1], Jessica Kollmorgen[2(✉)], and Jörg Thomaschewski[2]

[1] SAP SE, Walldorf, Germany
martin.schrepp@sap.com
[2] University of Applied Sciences Emden/Leer, Emden, Germany
{jessica.kollmorgen,joerg.thomaschewski}@hs-emden-leer.de

Abstract. Different UX questionnaires utilize various types of items, e.g., rating scales or semantic differentials. Another difference in UX questionnaires is item polarity. Some questionnaires maintain a consistent polarity with positive terms on the right, while others randomly change this position. Studies have shown that item polarity does not significantly impact the results if statements are used as item format. However, such investigations are currently lacking for semantic differentials. We investigate in two studies whether changing to a one-sided polarity of semantic differential items influences the effort required to answer the items and the measured scale scores. In the first study, we used the User Experience Questionnaire (UEQ) to evaluate Microsoft PowerPoint, WhatsApp and Google Maps with the original UEQ and a modified version with one-sided polarity. In the second study, we investigated again WhatsApp with the HED/UT and Trustdiff questionnaires. Results showed that the effort to complete the questionnaire was lower for the one-sided polarity version, but differences were small and practically not relevant. The impact of item polarity on scale scores was low, only for one of the UEQ scales there is a significant difference. Scale consistency was slightly higher with one-sided version, likely due to a small answer tendency introduced by the one-sided polarity. Overall, the impact of item polarity on the results of semantic differential scales is low, but differences can arise depending on the scale's semantics and the investigated product. Comparisons of questionnaire versions that do not have the same item polarity should be done carefully.

Keywords: User experience · UX measurement · UX survey · UX · questionnaire · Semantic differential · Item polarity

1 Introduction

Modern products are expected to deliver a high level of user experience (UX). If a product fails to meet that expectation, this will reduce its ability to retain its user base in the face of competition, especially in the long run. Thus, measuring the user experience of interactive products is important to verify that their UX quality is good enough to ensure their long-term market success [1] and to defend the company's position towards competitors. Standardized UX questionnaires such as the User Experience Questionnaire

M. Marchiori and F. García Peñalvo (Eds.): WEBIST 2023, LNBIP 543, pp. 1–29, 2025.
https://doi.org/10.1007/978-3-031-89621-7_1

(UEQ; [2]), the System Usability Scale (SUS; [3] or [4]), the Usability Metric for User Experience (UMUX, [5]) or the Purdue Usability Testing Questionnaire (PUTQ, [6]) are popular instruments for this purpose.

The available standardized UX questionnaires measure different UX aspects with their scales. In addition, different item formats are used. Existing UX questionnaires differ, for example, in the number of response options for the items (most frequent are 5- or 7-point response scales) or the item type (statements to which a respondent can agree or disagree or semantic differentials). A detailed analysis of different item formats in UX questionnaires can be found in [7]. A special option that is investigated in this paper is the polarity of the items in the questionnaire.

On the one hand, a distinction can be made between unipolar and bipolar polarities. Bipolar questions measure opposite constructs (e.g. easy/difficult), which are often represented by numerical ratings, for example −3 to 3, in the scale. Unipolar questions measure the (non-)presence of a particular item (e.g., not easy at all/very easy), often in the form of numerical ratings, for example 1 to 7 [8].

Another important difference is the arrangement of the endpoints within an item.

For example, semantic differentials describe a semantic continuum by a positive and negative term with semantically opposite meaning, for example:

unattractive o o o o o o o attractive

Another typical item format are statements to which participants can agree or disagree. For example, the following items from the SUS [3]:

I thought that the system is easy to use.

Strongly disagree o o o o o Strongly agree

I Found the System Unnecessarily Complex.

Strongly disagree o o o o o Strongly agree

Agreement to the first of these two items represents a positive impression, agreement to the second item a negative impression.

A questionnaire usually contains a higher number of items. For semantic differentials it is possible to place the positive term (in the example above "attractive") always in the right position or to alter the position of the positive and negative term between different items.

Thus, when constructing a UX questionnaire that contains semantic differentials as items, a decision must be made either to place all positive endpoints (e.g., good, easy, efficient) continuously to the right-hand or left-hand side of the response scale (so-called one-sided polarity), or to change polarity between items by changing the position of positive and negative endpoints between different items (so called mixed polarity). If mixed polarity is chosen the position of the positive endpoints is usually randomized.

This does not only apply to questionnaires with semantic differential items. For example, the System Usability Scale (SUS, see [3]) uses short statements to which respondents can agree or disagree. For half of the SUS items (for example, "I thought the system was easy to use"), agreement to the statement describes a positive evaluation;

for the other half (for example, "I found the system unnecessary complex") a negative evaluation.

In a strictly positive worded version of the SUS [9], items are reformulated so that agreement always represents a positive evaluation (for example "I found the system unnecessarily complex" was changed to "I found the system to be simple"). Empirical studies [9, 10] compared these two SUS versions and found no major deviations in the measured scores of various products. Thus, for questionnaires that use statements to which participants can disagree or agree as item format there seems to be no difference between one-sided and mixed polarity.

For UX questionnaires that use semantic differentials as item format similar studies do not exist, thus it is unclear if polarity has an impact here.

Several established UX questionnaires use semantic differentials. The UEQ [2] and its variants measure several pragmatic and hedonic UX aspects. The AttrakDiff [11] has a similar focus, but with a very strong focus on hedonic UX aspects. The Trustdiff questionnaire [12] measures trust of users in web services and eCommerce applications. The HED/UT questionnaire [13] is an established method from consumer research (not restricted to interactive products) to measure consumers' hedonic and utilitarian motivations to purchase a product. The INTUI questionnaire measures the perception that a product can be used intuitively [14].

Most of the UX questionnaires mentioned above use a mixed polarity and randomly change the position of the positive and negative term between items. The main reason seems to be the wish to avoid answer tendencies like the agreement bias (but please note that it is questionable if such a bias exist in UX questionnaires, see [15]). But there are examples for questionnaires using semantic differential with a one-sided polarity.

The original UEQ [2] is implemented with mixed polarity. The modular extension of the UEQ, the UEQ+ [16], uses a one-sided polarity. The reason for this change in item format compared to the original UEQ was that this allows, together with the grouping of items per scale, a flexible and suitable composition of a concrete questionnaire from various individual scales. The short version of the UEQ, the UEQ-Short [17], also uses one-sided polarity to facilitate answering the questionnaire. Thus, it is an interesting research question which effect a similar change to a one-sided polarity would have on the results of an UEQ.

As we described above existing studies [9, 10] imply that the polarity of items has not much impact if the items are statements to which participants can disagree respectively agree. Why can we not simply transform these findings to semantic differentials?

Semantic differentials have advantages and disadvantages compared to the classic statement format. They allow survey participants to directly express feelings or subjective impressions concerning the UX of a product. Response time is shorter than for items in the statement format [18], since the participant needs to read only two terms and not a complete sentence.

On the other hand, semantic differentials do not allow to formulate very detailed questions, for example "Is the cursor placement consistent?" (item from PUTQ, [6]). They also require a certain level of language skills and abstract thinking from the participants of a study [19].

What is relevant for our research question is that semantic differentials encourage a faster and more intuitive response behavior. On the one hand, the time interval between the rating of two items is typically shorter. On the other hand, if many semantic differential items must be answered in a row, the likelihood of a one-sided item polarity introducing an answer bias is higher compared to questionnaires that use statements as item format.

Building on this, the following research questions emerged:

- RQ1: *Are UX questionnaires using semantic differential items easier to answer if a consistent one-sided polarity is used?*
- RQ2: *Does a change in item polarity for a UX questionnaires using semantic differential items impact the measured scale scores?*

The empirical studies described in this paper try to answer these two questions. In Sect. 2 we describe the UX questionnaires we use in our studies. Section 3 describes our first study, which consists of a total of six product evaluations. Our first study examines a polarity change from the original (mixed polarity) to a one-sided UEQ in the evaluation of the three well-known products Microsoft PowerPoint, WhatsApp and Google Maps. We investigate if there will be advantages in shortening the completion time of the questionnaires as well as in reducing inconsistencies, and if there will be a good comparability of the scales. Section 4 describes a second study that replicates the design of our first study but uses the HED/UT and Trustdiff questionnaires. Section 5 summarizes the results of both studies and relates them to our two research questions. An outlook to further research and limitations of our studies are discussed in Sect. 6.

The work presented in this paper is an extension of results presented in [20]. It contains new data, especially study 2, which allow to extend the findings obtained for the UEQ to a wider range of questionnaires.

2 Semantic Differentials in UX Measurement

We describe in this section the UX questionnaires used in our studies.

2.1 User Experience Questionnaire

The User Experience Questionnaire (UEQ, see [2]) is a widely utilized instrument for measuring the subjective experiences of individuals interacting with a product. This tool facilitates the evaluation of UX across six distinct dimensions: *Attractiveness*, *Efficiency*, *Perspicuity*, *Dependability*, *Stimulation*, and *Novelty*.

Each of these dimensions carries a specific semantic interpretation, as follows:

- *Attractiveness:* Do users like or dislike the product (valence towards the product)?
- *Efficiency:* Do users think they can perform tasks in the product fast and without unnecessary effort?
- *Perspicuity:* Do users have the impression that the usage of the product is easy to understand and easy to learn?
- *Dependability:* Do users feel in control of the interaction?
- *Stimulation:* Do users feel that it is exciting and motivating to use the product?

- *Novelty:* Is the design of the product perceived as inventive and original and thus catches the interest of users?

The 26 items (Attractiveness is measured with 6 items, the five other scales with 4 items) are semantic differentials with a 7-point answer scale. An item consists of a pair of adjectives with opposite meaning that spans a semantic dimension. Figure 1 shows a subset of the original UEQ items. The full list of items is shown in the Appendix.

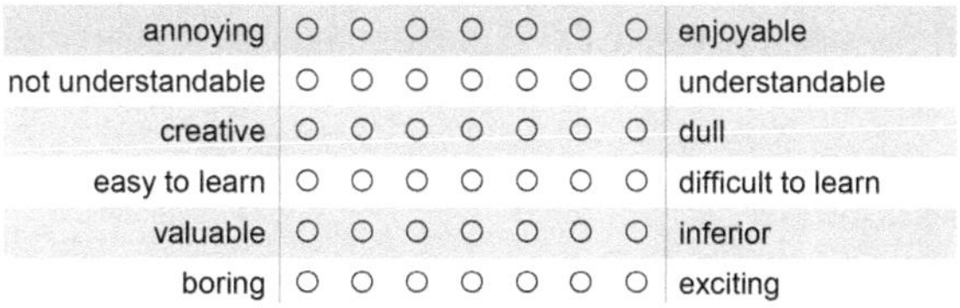

Fig. 1. Excerpt of original UEQ items (English language version).

As can be seen in Fig. 1, half of the items have the positive term on the right side (e.g., "enjoyable"), while the other half have the positive term on the left side (e.g., "valuable"). This means that rows 1, 2 and 6 from Fig. 1 show a right-sided polarity, while rows 3, 4 and 5 show a left-sided polarity.

Schrepp & Thomaschewski [16] describe a modular extension of the UEQ, called UEQ +. The UEQ+ contains scales for 27 different aspects of user experience (the 6 UEQ scales are included). This modular extension allows researchers to select the scales that are most relevant to their specific research question, enabling them to create a customized questionnaire. This allows studies to be conducted that are tailored to their specific needs and the product being studied. Thus, the UEQ+ is not a questionnaire with a fixed scale structure, but a framework for the creation of concrete UX questionnaires. One notable difference between the original UEQ and the UEQ+ is the item format. In the UEQ+, all items are arranged with one-sided polarity. The positive term is always in the right position. Additionally, items belonging to the same scale are grouped together, providing a clear separation of scales and the ability to combine them in any order. Figure 2 shows the items of the efficiency scale of the UEQ+.

Fig. 2. Efficiency scale of the UEQ+ (English translation).

There is also a short version of the UEQ, the UEQ-Short (UEQ-S). In this version all items are arranged with one-sided polarity (positive term on the right side). In contrast to the original UEQ which consists of 26 items, the UEQ-S contains only 8 items (4 from the hedonic quality scales *Stimulation* and *Novelty*, and 4 from the pragmatic quality scales *Efficiency*, *Perspicuity*, and *Dependability*). The goal of the UEQ-S is to reduce the required completion time. The choice of a one-sided polarity was based on the assumption that this reduces cognitive complexity and thus response time compared to mixed polarity [17].

Therefore, both the modular extension UEQ+ and the short version UEQ-S employ a one-sided item polarity. This raises questions about the potential advantages and disadvantages of consistently modifying the original UEQ to have a one-sided item polarity. Additionally, it is important to determine whether the scale values obtained through this modification would be comparable to the values obtained using the original measurement approach. These questions can be answered by the studies described below.

2.2 Hed/Ut Questionnaire

The HED/UT (the abbreviations stand for *Hedonic* and *Utilitarian*) questionnaire [13, 21] is an established tool in consumer research to measure consumers' hedonic and utilitarian motivations to purchase a product. It helps to understand whether buying decisions are driven by the desire for pleasure and enjoyment (hedonic) or by practical and functional considerations (utilitarian).

This questionnaire is not restricted to interactive products, it can be used to evaluate any type of product. The scales of the questionnaire can be interpreted as follows:

- *Hedonic quality*: Measures the extent to which purchase decisions of consumers are motivated by pleasure, enjoyment, and emotional gratification.
- *Utilitarian quality*: Measures the extent to which purchase decisions of consumers are motivated by practicality, functionality, and usefulness.

Each scale consists of 5 items in the form of a semantic differential with a 7-point response scale. Examples for utilitarian items are *effective/ineffective* and *practical/impractical*. Examples for hedonic items are *not fun/fun* or *enjoyable/unenjoyable*. All items are listed in the Appendix in the description of Study 2.

2.3 Trustdiff

Trust of users in systems is vital for many online scenarios, for example eCommerce or online banking apps. The Trustdiff questionnaire [12] measures trust of users in such systems. The questionnaire contains three scales, which can be interpreted as follows (see [12]):

- *Benevolence:* Is the provider of a service interested in the user's welfare and motivated to establish a fair relationship?
- *Integrity:* Is the provider of a service honest and willing to fulfil given promises.
- *Competence:* Is the provider of a service able to successfully complete transactions and sustain the customer relation?

The questionnaire contains 10 items in the form of a semantic differential with a 7-point answer scale. Examples for items are ignoring/caring (Benevolence), dishonest/honest (Integrity) and incompetent/competent (Competence). The complete list of items of the Trustdiff questionnaire can be found in the description of Study 2 in the Appendix.

3 Study 1: UEQ

In the following, the structure of the study, product-related results of the corresponding UEQ versions, and possible impacts are analyzed.

3.1 Study Design

In order to answer the research questions, it was first necessary to collect data to determine the scale scores and influencing factors. For the first study, the UEQ was selected as questionnaire. As explained, it consists of semantic differentials and uses a mixed polarity. Half of the items show the positive end point at the left position of the answer scale and the other half shows the positive end point at the right position. As part of the initial study, two versions of the UEQ were created, one featuring a one-sided arrangement and the other maintaining the original, mixed setup of semantic differentials. These two questionnaire versions were used to evaluate the three well-known and widely used products Microsoft PowerPoint (PPT), WhatsApp (WA) and GoogleMaps (GM), which each support a different use case and can therefore be compared independently of each other.

Setup of the Questionnaire. The initial segment of the survey remained consistent across both questionnaire formats (mixed and one-sided UEQ). In this section, demographic information on age and gender was collected.

In the second part of the questionnaire, the 26 UEQ items were assessed based on the product. One questionnaire utilized the mixed item polarity, while the other employed a one-sided polarity.

The third section of the questionnaire remained the same for both the mixed and one-sided UEQ. Within this section, an additional set of 6 items were included as part of the Key Performance Indicator (KPI) extension. These items aim to gauge the relative importance of different scales, facilitating an overall assessment of the respective product [22]. The complete survey is described in detail in the Appendix.

Study Implementation. Participants were enlisted through a survey portal and received compensation for their involvement. Six distinct samples were assembled, meaning each participant evaluated only one product using either of the two UEQ versions.

The data collection occurred over three days in June and November 2022. Accordingly, a total of 438 completed questionnaires, distributed across the 6 surveys, were gathered (see Table 1). The comprehensive findings of this study are available in the research protocol [23].

3.2 Results

In the following results, information on the age and gender of the participants and on the scale and importance ratings of the individual products is discussed in order to investigate possible impacts of the change in polarity.

Demographic Variables. Table 1 provides a summary of the collected data, with a specific emphasis on the initial section of the survey, which includes demographic information. The table displays the results of the six studies conducted on Microsoft PowerPoint, WhatsApp, and Google Maps. It includes the participant counts for both the mixed and one-sided UEQ versions. Additionally, the average age of participants and the distribution based on gender (male M, female F, no answer NA) are presented.

The participant counts (N) per product show similarity between the mixed and one-sided UEQ.

Table 1. UEQ format (mix = mixed, one = one-sided), product (PPT = Microsoft PowerPoint, WA = WhatsApp, GM = Google Maps), number of participants (N), age and gender information (M = male, F = female, NA = no answer) for the studies.

Study	Prod	UEQ	N	Age	Gender
1	PPT	mix	64	34.9	18M, 45F, 1NA
2	PPT	one	81	36.4	28M, 53F, 0NA
3	WA	mix	63	35.6	16M, 46F, 1NA
4	WA	one	80	36.9	27M, 52F, 1NA
5	GM	mix	60	33.6	18M, 41F, 1NA
6	GM	one	90	37.8	34M, 56F, 0NA

Moreover, variations in the average age distribution are minimal, with a maximum difference of 4.2 years observed across all questionnaires. Similarly, the distribution of participants by gender remains consistent across the surveys. It is noteworthy, however, that female participants are evidently overrepresented in all samples.

Microsoft PowerPoint. To assess the UX quality of an examined product, the most important information derived from a UX questionnaire is the scale scores. Figure 3 displays the scale scores measured for Microsoft PowerPoint. It is evident that, across all scales, the one-sided UEQ shows higher values for Microsoft PowerPoint compared to the mixed UEQ.

Nevertheless, statistically significant differences were identified only in the mean scores for the scale *Novelty*, as indicated by a two-sample t-test ($p < 0.05$). This aligns with the observations in Fig. 3, where the disparity in scale means between the mixed and one-sided UEQ is highest for the *Novelty* scale. No significant effects were observed for the other scales.

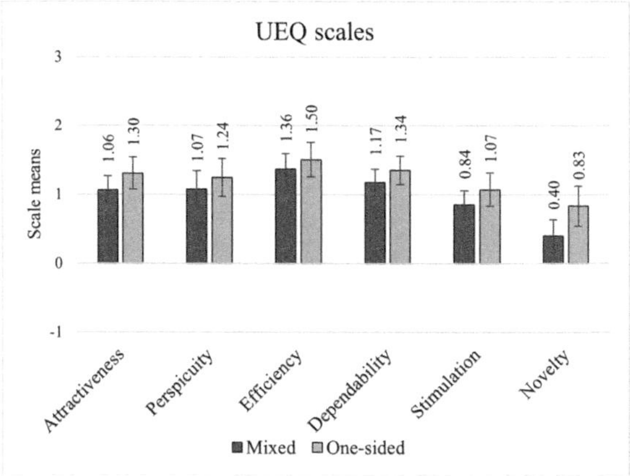

Fig. 3. Scale means of the mixed (N = 64) vs. one-sided (N = 81) UEQ for Microsoft PowerPoint (section from −1 to 3; the original scale ranges from −3 to 3).

Additionally, to assess the reliability of the scales, Cronbach's alpha [24, 25] serves as a commonly employed metric. This metric relies on the correlations among the items within each scale. Table 2 presents the Cronbach Alpha values for the UEQ scales related to Microsoft PowerPoint.

Table 2. Values of the Cronbach's alpha coefficient for the mixed (mix) vs. one-sided (one) UEQ scales for Microsoft PowerPoint.

Scale	Cronbach's α mix. UEQ N = 64	Cronbach's α one. UEQ N = 81
Attractiveness	0.83	0.95
Perspicuity	0.85	0.89
Efficiency	0.75	0.93
Dependability	0.61	0.76
Stimulation	0.75	0.88
Novelty	0.77	0.88

From the information presented in Table 2, it is evident that all scales have a good reliability (> 0.7), except for *Dependability* in the mixed UEQ. Additionally, the Cronbach's alpha coefficient values consistently prove higher in the one-sided UEQ compared to the mixed UEQ across all scales. This indicates that the correlations among items within a scale are stronger for the one-sided UEQ.

However, it is important to note that the sample sizes are moderate, and correlations can be quite unstable with small sample sizes [26]. While the Alpha coefficient is more stable than a single correlation as it is based on the intercorrelations of all items in a scale, it remains sensitive to sampling effects [27]. Consequently, caution should be exercised in overinterpreting the results shown in Table 2.

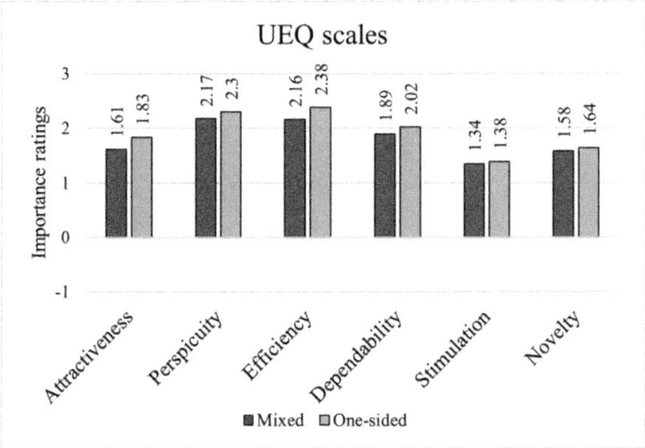

Fig. 4. Importance ratings of the scales of the mixed (N = 64) vs. one-sided (N = 81) UEQ for Microsoft PowerPoint (section from −1 to 3; the original scale ranges from −3 to 3).

With regard to the importance ratings (see Fig. 4), it is apparent that the scales in the one-sided UEQ received minimally higher ratings for Microsoft PowerPoint. However, these differences are not statistically significant (two-sample t-test, $p < 0.05$). As previously explained, this outcome is expected at this stage, given that no polarity change was implemented in the importance rating scales.

WhatsApp. The same key figures are examined for WhatsApp. As shown in Fig. 5, unlike Microsoft PowerPoint, there is no clear trend in the scale scores for WhatsApp. While the mixed UEQ scores slightly higher for *Attractiveness* and *Perspicuity*, the opposite holds true for the other four scales in the one-sided UEQ. The highest difference is shown for *Novelty*, and the difference is also significant (two-sample t-test, $p < 0.05$). However, no significant difference between the mixed and one-sided UEQ is demonstrated for the other five scales, although a trend is noticeable for *Stimulation*.

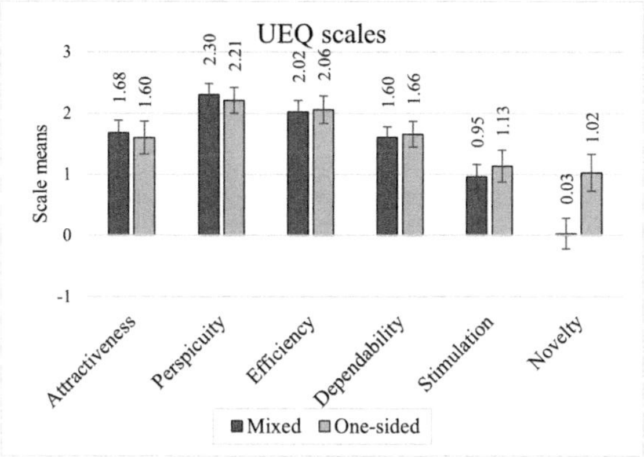

Fig. 5. Scale means of the mixed (N = 63) vs. one-sided (N = 80) UEQ for WhatsApp (section from −1 to 3; the original scale ranges from −3 to 3).

Similar to Microsoft PowerPoint, the Cronbach's alpha values for the one-sided UEQ are higher than those for the original, mixed version (see Table 3).

Table 3. Values of the Cronbach's alpha coefficient for the mixed (mix) vs. one-sided (one) UEQ scales for WhatsApp.

Scale	Cronbach's α mix. UEQ N = 63	Cronbach's α one. UEQ N = 80
Attractiveness	0.90	0.94
Perspicuity	0.83	0.88
Efficiency	0.82	0.95
Dependability	0.57	0.75
Stimulation	0.79	0.88
Novelty	0.74	0.92

Nonetheless, given that all values (except for *Dependability* in the mixed UEQ) fall within the very good range, and considering the moderate sample sizes, further interpretations in this context are not advisable.

This is also applicable to the importance ratings, shown in Fig. 6. The ratings for the mixed and one-sided UEQ are very close. Nevertheless, this aligns with expectations, as no polarity change has been implemented for the importance rating scales.

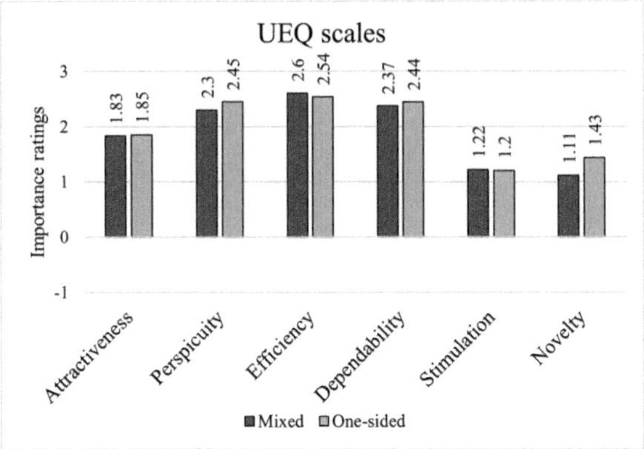

Fig. 6. Importance ratings of the scales of the mixed (N = 63) vs. one-sided (N = 80) UEQ for WhatsApp (section from −1 to 3; the original scale ranges from −3 to 3).

Google Maps. Evaluations are also conducted for the third product, Google Maps. Similar to WhatsApp, there is no apparent trend in the scale means (refer to Fig. 7). A significant difference between the mixed and one-sided UEQ is evident for *Novelty*, and only a trend is observed for *Stimulation*. Hence, it is once again evident that the four pragmatic scales do not show discernible differences, in contrast to the two hedonic scales.

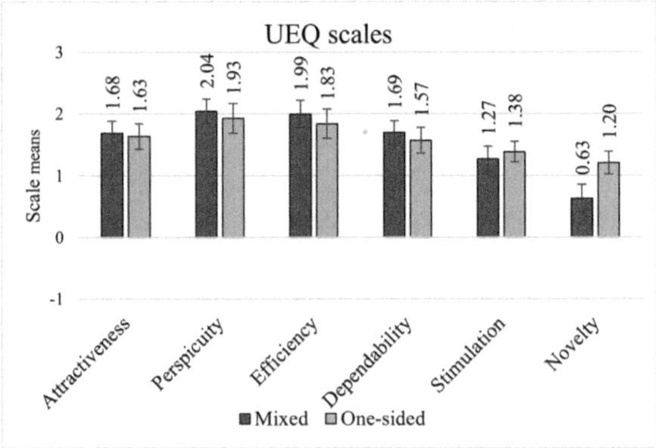

Fig. 7. Scale means of the mixed (N = 60) vs. one-sided (N = 90) UEQ for Google Maps (section from −1 to 3; the original scale ranges from −3 to 3).

However, similar to Microsoft PowerPoint and WhatsApp, a clear trend emerges indicating that the scales measured with the one-sided UEQ show higher reliability compared to the mixed UEQ.

Table 4 also shows that this time, however, two alpha values generally fall below the threshold value of 0.7, partly due to the moderate sample sizes. Therefore, it is advisable to refrain from further interpretations.

Table 4. Values of the Cronbach's alpha coefficient for the mixed (mix) vs. one-sided (one) UEQ scales for Google Maps.

Scale	Cronbach's α mix. UEQ N = 60	Cronbach's α one. UEQ N = 90
Attractiveness	0.82	0.90
Perspicuity	0.77	0.88
Efficiency	0.82	0.94
Dependability	0.62	0.77
Stimulation	0.70	0.73
Novelty	0.49	0.72

Finally, Fig. 8 offers insight into the importance ratings for Google Maps. Similarly, no clear trend is discernible or statistically significant. The variations between the ratings of both UEQ formats are small (< 0.2 differences). Once more, the expectation of no differences was confirmed, as no polarity change was implemented in the importance rating scales.

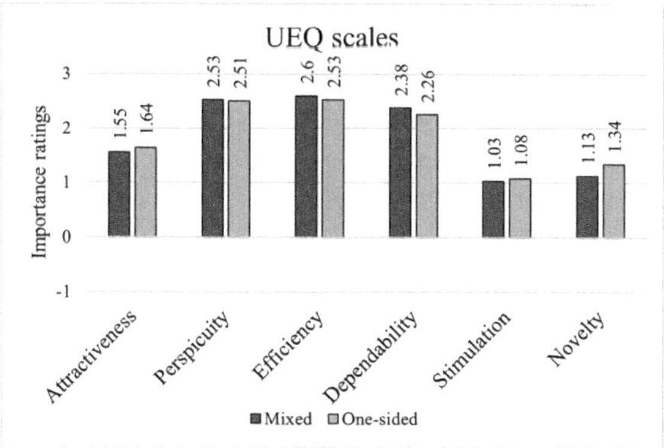

Fig. 8. Importance ratings of the scales of the mixed (N = 60) vs. one-sided (N = 90) UEQ for Google Maps (section from −1 to 3; the original scale ranges from −3 to 3).

Time Savings. One of the anticipated positive outcomes of the transition from the mixed UEQ to the one-sided version is the expectation that the absence of polarity change should expedite participants' completion of the questionnaire. Consequently, the general metrics of time and clicks will be explored below to assess this hypothesis.

Table 5 presents the median and mean duration between the survey initiation and clicking the submit button. Additionally, it displays the number of clicks made during the survey process. In cases where a participant accidentally selected the wrong answer category and recognized the mistake, an additional click was necessary to correct the error. Therefore, the number of clicks serves as an indicator for the correction of erroneous decisions.

The number of clicks required to complete the survey is relatively similar for both versions of the UEQ, with an exception observed in studies related to Microsoft PowerPoint.

Table 5. UEQ format (mix = mixed, one = one-sided), product (PPT = Microsoft PowerPoint, WA = WhatsApp, GM = Google Maps), and information on time and clicks.

Product	UEQ Version	Time		Clicks
		Median	Mean	
PPT	mix	84	111	34.92
PPT	one	80	93	44.30
WA	mix	94	126	41.10
WA	one	85	115	42.20
GM	mix	85	102	41.82
GM	one	85	99	44.48

In terms of the time taken for processing, it is advisable to consider the median rather than the mean due to the presence of certain participants with unusually high and unrealistic response times. Since the surveys were executed online, the researchers had no control over the general conditions and potential interruptions faced by respondents. These disruptions, especially when prolonged, can significantly influence the mean, underscoring the stability of the median against such influences.

It is evident from the data presented in Table 5 that, for all products, the median processing time is shorter when using the one-sided UEQ compared to the mixed UEQ. However, the observed differences in processing time and the count of clicks do not show statistical significance (t-test, $p < 0.05$).

Consequently, it can be concluded that the transition to a one-sided polarity for all UEQ items had only a minimal and practically irrelevant impact on the effort required to complete the questionnaire.

Inconsistency Reductions. With the modification from the mixed UEQ to the one-sided UEQ, a second assumption emerged, aiming to reduce potential inconsistencies in participants' responses.

All items within a UEQ scale measure the same quality aspect. Consequently, it is expected that responses to these items should generally only show little variation. This is illustrated in the following example for the *Efficiency* scale:

slow o o o o o x o fast

efficient o o o o o o x inefficient

impractical o o o o x o o practical

organized o o o o o x o cluttered

This response behavior is right-sided. Yet, when the item polarity is changed to a one-sided scale (ranging from negative (1) to positive (7)), responses fall between 1 and 6, showing a 5-point difference within items of the same scale. Such a notable difference (exceeding 3 points between the least and most favorable responses within a scale) implies inconsistent or random answering behavior.

However, it is important to ensure that these inconsistencies appear more frequently, and not, for example, simply because only one item within a scale was misunderstood. A simple heuristic to differentiate between irregularities is that a dataset is deemed suspicious if there is a notable disparity in the item ratings across 2 or 3 scales [28].

Table 6 displays the quantity of observed discrepancies for the six studies. The columns >2 and >3 indicate the counts of participants who provided more than two and more than three inconsistent responses. The column labeled "All" contains the total number of participants with inconsistent answers, and the column "All %" contains the relative frequency of participants with inconsistent answers in the sample.

Table 6. UEQ format (mix = mixed, one = one-sided), product (PPT = Microsoft PowerPoint, WA = WhatsApp, GM = Google Maps), and information on observed inconsistencies.

Product	UEQ Version	All	> 2	> 3	All (%)
PPT	mix	34	6	3	53
PPT	one	22	6	1	28
WA	mix	19	2	0	30
WA	one	21	3	1	26
GM	mix	24	4	2	40
GM	one	20	3	0	25

The number of inconsistencies (two-sample t-test, $p < 0.05$) does not significantly differ between the two UEQ versions for WhatsApp and Google Maps. Conversely, for MS PowerPoint, the mixed UEQ resulted in statistically significantly more inconsistencies. In summary, fewer inconsistent responses were observed in the one-sided version compared to the mixed version. However, except for the studies related to Microsoft PowerPoint, the differences between the two versions of UEQ are relatively minor.

Item Comparison. After the considerations of the scale level, also the comparability on the item level will now be examined two answer the research questions.

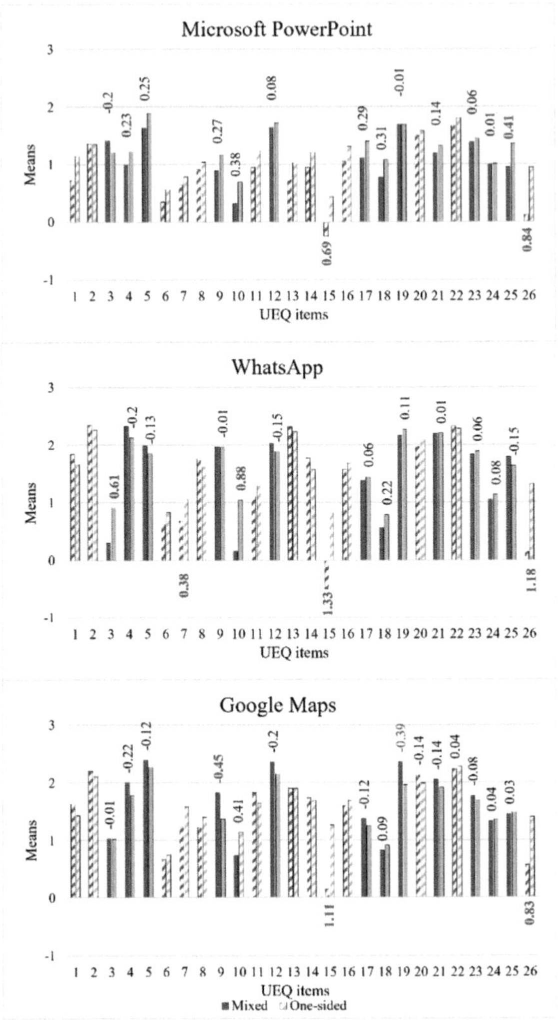

Fig. 9. Summary of mean UEQ item differences for Microsoft PowerPoint, WhatsApp and Google Maps (section from −1 to 3; original scale ranges from −3 to 3). Unmodified UEQ items are shown patterned. Statistically significant differences are marked in red and bold font. Full texts for items 1–26 can be found in the Appendix.

Figure 9 shows the item scores of the 26 UEQ items for the three products Microsoft PowerPoint, WhatsApp and Google Maps for the mixed and one-sided UEQ respectively. The concrete terms for the items 1 to 26 can be found in the Appendix.

The bars of the items whose polarity was not modified are shown patterned. Also shown are the differences between the mixed and one-sided UEQ (numbers above the bars), and all statistically significant changes are printed in red and bold font. Statistically significant changes in unmodified items are printed in red and bold font below the bars.

The size of the difference is only shown for items that have a different polarity in both surveys, or that show a significant difference.

As shown in Fig. 9, not only the items whose polarity was reversed were affected, but also items that remained unchanged. Therefore, this influence of the modification is not an item specific effect, but a general effect that has an impact on all items.

Furthermore, when looking at the items, a total of 9 of the 12 identified significant differences come from the scales *Novelty* and *Stimulation*, which is consistent with the observations from Sect. 3.2.

Due to these influences, a full comparability of the mixed with the one-sided UEQ values is not given. Thus, this implies that a comparison of the UEQ scales with those of the UEQ+ or UEQ-S can also be problematical.

4 Study 2: HED/UT and Trustdiff

In order to make the results comparable and to answer the research questions, the structure of the second study and the associated results are presented again in the following.

4.1 Study Design

To rule out that the findings in Study 1 are caused by some special properties of the UEQ, we replicate this study with two other UX questionnaires that use semantic differentials. We investigate again the product WhatsApp with the HED/UT and Trustdiff questionnaires to establish comparability. For this reason, we again created two surveys, one with one-sided and one with mixed polarity.

Setup of the Questionnaire. The first part was identical for both surveys. Here, the two demographic variables age and gender were queried.

In the second part of the survey, the 10 HED/UT items were shown, with one-sided or mixed item polarity depending on the questionnaire version.

In the third part of the survey, the 10 items from Trustdiff were shown, also with different polarity depending on the survey. The complete survey is described in detail in the Appendix.

Study Implementation. The participants were recruited via a survey portal and compensated for their participation. Two different samples were recruited, thus participants rated WhatsApp only once with the respective survey version randomly assigned to them.

The data was acquired on January 22, 2024. As a result, a total of 222 completed questionnaires (for both surveys together) could be collected (see Table 7).

4.2 Results

The relevant demographic and product-related results and possible impacts for the polarity change for the second study are also presented in the following.

Demographic Variables. The demographic data collected in the first part of the two surveys is shown in Table 7. The same number of participants was recorded for each survey. In contrast to study 1, the gender ratio is more balanced here. Both target groups are highly similar concerning age and gender distribution. The average age of the participants is slightly higher than in Study 1.

Table 7. Questionnaire format (mix = mixed, one = one-sided), number of participants, age and gender information (M = male, F = female, NA = no answer) for the studies.

Study	Format	N	Age	Gender
1	mix	111	39.4	56M, 54F, 1NA
2	one	111	41.2	56M, 55F, 0NA

WhatsApp and HED/UT. First, we compare the scale scores for the HED/UT questionnaire. As we can see in Fig. 10, the measured scale scores are nearly identical. A t-test (two-sided, $p = 0.05$) shows no significant difference between the scales with mixed and one-sided polarity.

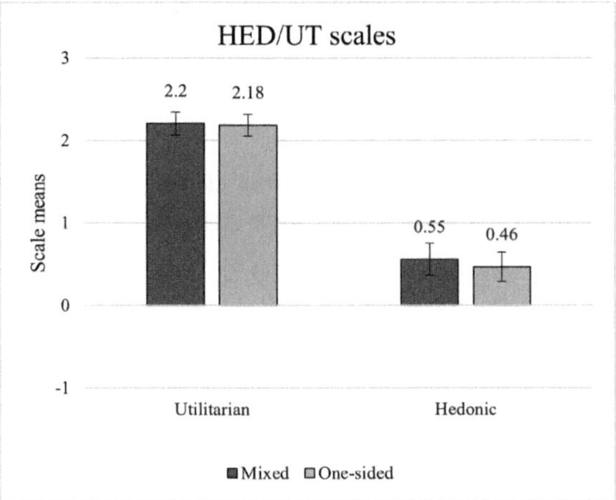

Fig. 10. Scale means of the mixed (N = 111) vs. one-sided (N = 111) HED/UT for WhatsApp (section from −1 to 3; the original scale ranges from −3 to 3).

The observed Alpha values (see Table 8) are high for both the mixed polarity version as well as the one-sided version. The result is in line with the results we found in all six studies with the UEQ. The scales measured with the one-sided polarity have a higher reliability than the scales using a mixed polarity. The differences are a bit lower than the ones observed in the UEQ studies, but this is hard to interpret, since the number of items per scale differs (and this number together with the mean correlations of the items determines the value of the Alpha coefficient).

Table 8. Values of the Cronbach's alpha coefficient for the mixed (mix) vs. one-sided (one) HED/UT scales for WhatsApp.

Scale	Cronbach's α mix. HED/UT N = 111	Cronbach's α one. HED/UT N = 111
Utilitarian	0.81	0.83

WhatsApp and Trustdiff. Now we focus on the scale scores for the Trustdiff questionnaire. Again, the scale means for the scales with mixed or one-sided polarity are very similar (see Fig. 11). The is no significant difference for any of the scales between both conditions (t-test, two-sided, p = 0.05).

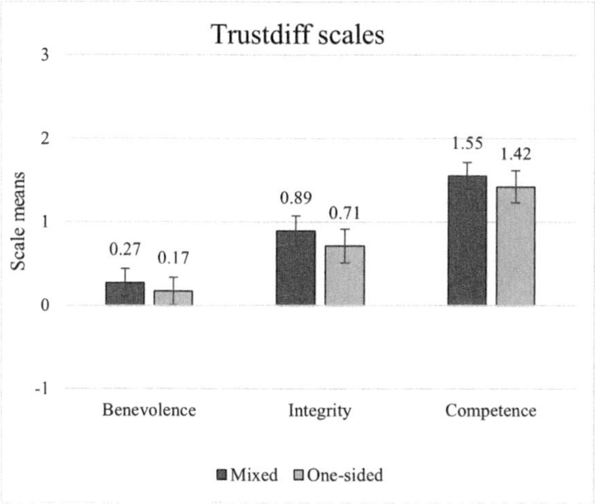

Fig. 11. Scale means of the mixed (N = 111) vs. one-sided (N = 111) Trustdiff for WhatsApp (section from −1 to 3; the original scale ranges from −3 to 3).

Table 9 shows the observed values of the Alpha coefficient, i.e., the reliability of the scales.

Table 9. Values of the Cronbach's alpha coefficient for the mixed (mix) vs. one-sided (one) Trustdiff scales for WhatsApp.

Scale	Cronbach's α mix. Trustdiff N = 111	Cronbach's α one. Trustdiff N = 111
Benelovence	0.85	0.91
Integrity	0.78	0.91
Competence	0.80	0.92

We see the same pattern as for the other scales. The scales measured with the one-sided polarity have a higher reliability than the scales with a mixed polarity. Differences are higher than the ones shown in Table 9, but again the number of items differs.

Time Savings. As already discussed in our first study one of the main reasons for the usage of a one-sided polarity in questionnaire design is the hope that this will decrease the time participants need to answer the items compared to a mixed polarity.

Regarding the influence of the questionnaire format on the processing time (see Table 10), both surveys were combined. The reason for this is that both questionnaires were shown in the same survey as part of the data collection for Study 2. The median is used for this, as there were again some very high outliers (which are most likely caused by pausing the questionnaire, leaving the browser window open and submitting at a later time). As expected, the survey with the one-sided polarity was answered a bit faster than the survey with the mixed polarity. But again (see Table 10), a 5 s difference for a survey that takes approximately a minute to complete is practically not relevant. In addition, the difference in processing time is not statistically significant (two-sample t-test, $p < 0.05$).

Table 10. Questionnaire format of HED/UT and Trustdiff (mix = mixed, one = one-sided), product (WA = WhatsApp), and information on completion time.

Product	Version	Time	
		Median	Mean
WA	mix	65	106
WA	one	60	63

For Study 2, a different survey tool was used, and the recoding of clicks would have required a larger effort to implement. Study 1 showed that the number of clicks required to finish a survey was not significantly different for a mixed versus one-sided polarity, so this part of the survey was not repeated.

Inconsistency Reduction. As in Study 1, we consider a response of a participant as inconsistent if there is a notable difference between the ratings for the items within a scale. Thus, if the difference of the ratings between the least and most favourable item within a scale exceeds 3 points. Table 11 shows the number of inconsistent responses for both questionnaires and both item formats. The chance to produce an inconsistency depends on the number of items in a scale. Since the scales in HED/UT contain 5 items and the scales in Trustdiff 3 respectively 4 items it is not possible to compare the inconsistencies between these two questionnaires or the UEQ investigated in Study 1.

Table 11. Questionnaire version (mix = mixed, one = one-sided) for HED/UT and Trustdiff, and information on observed inconsistencies.

Questionnaire	Version	All	All (%)
HED/UT	mix	20	18.0
HED/UT	one	13	11.7
Trustdiff	mix	8	7.2
Trustdiff	one	1	0.9

The difference in the number if inconsistencies is statistically not significant for HED/UT (t-test, two-sided, $p = 0.05$). For Trustdiff there is a significant difference ($t = 2.40$, $df = 220$, $p = 0.0171$). The differences are even for Trustdiff small compared to the overall number of responses, so from a practical perspective not really relevant.

Item Comparison. In order to answer the research questions, the differences between the individual items of the questionnaire versions are examined in more detail, similar to Study 1. The scale means for UEQ, HED/UT and Trustdiff in relation to the product WhatsApp are therefore considered in Fig. 12 to enable a comparison. Statistically significant differences are highlighted in red and bold font.

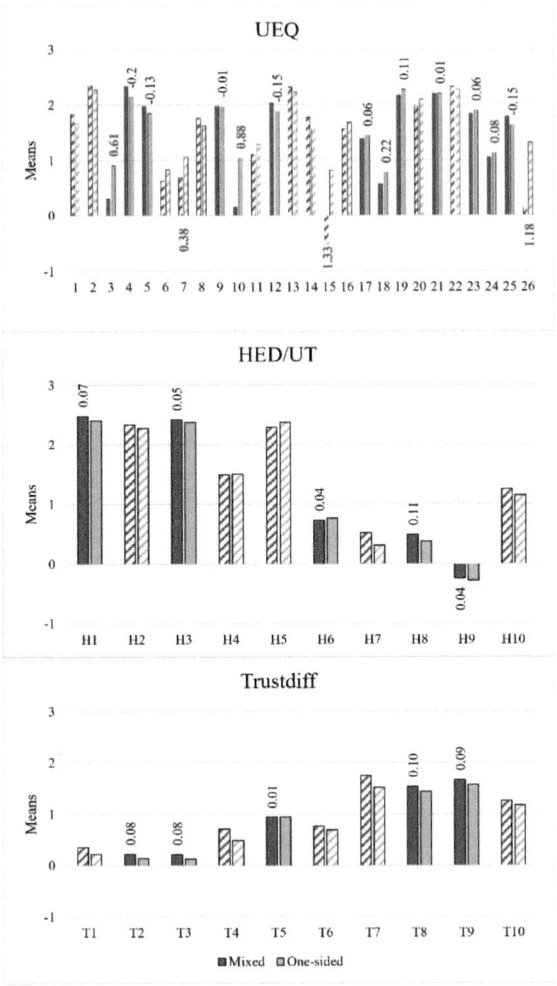

Fig. 12. Summary of mean item differences for WhatsApp, measured with the UEQ, HED/UT and Trustdiff (section from −1 to 3; original scale ranges from −3 to 3; N = 111). Unmodified items are shown patterned. Statistically significant differences are marked in red and bold font. Full texts for all individual items can be found in the Appendix.

As can be seen in Fig. 12, the UEQ items 3, 7, 10, 15 and 26 are rated significantly lower in the mixed compared to the one-sided version. All these items belong to the hedonic scales (see Sect. 3.2).

For the HED/UT items, there were no significant differences in the one-sided and mixed versions of the questionnaire, which is consistent with the results from Sect. 4.2. This is also true for the Trustdiff items.

Thus, a direct comparison of the item polarity in relation to the same product with different questionnaires clearly shows that only for 5 of the 46 items from the 3 questionnaires show a statistically significant change.

5 Conclusion

We investigated how item polarity in UX questionnaires using semantic differentials as item format influences the results. Therefore, we used three established UX questionnaires (UEQ, HED/UT and Trustdiff) that use this item format to evaluate three different products (Microsoft PowerPoint, WhatsApp, Google Maps) in two studies.

In summary, the results of our two studies indicate that modifying a UX questionnaire using semantic differentials to a constant one-sided version offers only marginal benefits on response times and number of clicks required to complete the survey. The observed impact on response time is small and compared to the overall time required to complete a survey not of practical relevance (see Sect. 4.2).

Cronbach's Alpha shows constantly a higher value in the one-sided version compared to the mixed version of the questionnaire. In addition, the number of inconsistent responses is always lower in the one-sided compared to the mixed format. Of course, these effects are connected. For a mixed polarity, the participant must decide per item on which side the positive respectively negative term resides. If many such items must be answered, then some random response errors are likely. Thus, a respondent may accidentally click on the wrong side of the scale based on the expected position derived from the previous item answered. In a one-sided polarity, the chance of such a mistake is of course reduced. Therefore, these response errors increase the chance of inconsistencies and reduce the correlations of the items within a scale, and thus decrease the value of Cronbach's Alpha.

However, we need to note here two important points. First, a sufficient level of the Alpha coefficient is usually interpreted as a sign for a good reliability of a scale. But an Alpha level that is too high is also not desirable. Assuming a scale with 4 items and an observed maximal value of 1 for Cronbach's Alpha coefficient, this means in fact that the correlations between all items are 1 and thus all items are answered by all participants completely identically. In other words, 3 of the 4 items are completely redundant. Thus, usually we aim for a high but not too high value of Alpha in questionnaire construction. Also, moderate levels of Alpha are sufficient for UX questionnaires, since we interpret always mean scores over all participants of a study and not results from single persons [27]. In the light of this argument, the increase of Alpha by a one-sided polarity is not so important and we can see from our results that the observed Alpha values for the mixed versions were in nearly all cases already quite high.

A reduction of response errors by a one-sided polarity is in fact desirable, but it also comes with a price. Especially in online surveys we cannot assume that all participants of a study spend sufficient time and effort in answering the items. There are often participants that rush over the items without carefully reading them and without much effort to come to a careful decision. Thus, it is a problem to filter these participants out. One common method is to use the overall response time. A second established and relatively save method is to simply exclude participants that show inconsistencies in a number of scales [28]. But this only works with mixed polarity.

Thus, we can now answer our first research question RQ1: *Are UX questionnaires using semantic differential items easier to answer if a consistent one-sided polarity is used?*

As described above there are no convincing and practically relevant benefits in switching an existing UX questionnaire with mixed polarity to a one-sided polarity for all items.

Scale scores can differ between a mixed and a one-sided version of a semantic differential. For the Novelty scale of the UEQ this was found for three different product evaluations. However, for the other five UEQ scales no significant differences were found between the two types of item polarity. For HED/UT and Trustdiff no significant differences for the scale scores could be found between the mixed and the one-sided version. Thus, the effect of the item polarity for UX scales exists but is not massive and does only impact some scales.

That the UEQ scale Novelty is the only scale that is significantly impacted can be explained by the content of the scale and the products evaluated. PowerPoint, WhatsApp and Google Maps are quite established and mature products. They are available since many years and most of our participants have most likely a long usage experience. In addition, there were no drastic changes concerning these products in the last years. The items in Novelty try to measure if a product is seen as innovative, original, and new. Thus, for these three mature products we can expect low ratings for the scale compared to other scales. This can be also seen if we look at the scale means for the Novelty scale in the original mixed version of the UEQ. The significant difference to the scale means in the one-sided version result from the fact that the Novelty ratings were much higher in this version.

Furthermore, a trend towards lower ratings of the products was also visible for the scales Stimulation in the UEQ, Hedonic Quality in the HED/UT and Benelovence in the Trustdiff, even if it could not be proven to be statistically significant. On the one hand, this could be related to the fact that the considered products focus on the fulfilment of pragmatic goals or tasks (e.g., efficient communication for WhatsApp). However, scales such as Stimulation or Novelty belong to the hedonic UX factors (e.g., fun of using the product) and are therefore less relevant for the products under consideration, meaning that more uncertainties in the evaluation are to be expected here.

In addition, we have seen that not only the Novelty items whose polarity was reversed were affected, but also items that remained unchanged. Thus, this influence of the modification is not an item specific effect, but a general effect that has an impact on all items. The most natural explanation for this is that in the one-sided version a small answer tendency is imposed. Items that are not so well fitting to the evaluated product or seen as not so relevant for the overall UX quality are then rated accordingly to the overall impression. Such effects do generally exist [29, 30] but are most likely be stronger in the case of a one-sided polarity. It is also natural to assume that such an effect increases with the length of a questionnaire, which can explain why it was found for the UEQ (26 items) but not for HED/UT and Trustdiff (10 items).

This enables us to answer our second research question RQ2: *Does a change in item polarity for a UX questionnaires using semantic differential items impact the measured scale scores?*

Independently of the reason for the impact of the item polarity on the scale scores the results clearly show that we should not directly compare results obtained with versions of the same questionnaire if the item polarity is not identical for these versions. This has a direct practical implication since it shows that we cannot directly compare scale values obtained with the UEQ (mixed polarity) to the same scales in the UEQ+ (one-sided polarity with the positive term always on the right side). But in general, the impact of the item polarity on the scale scores is low. This is consistent with existing findings concerning questionnaires that use statements as item format [9, 10].

6 Limitations

It is important to note that our studies have certain limitations that should be acknowledged and taken into account when interpreting the results.

Firstly, it is important to note that our sample sizes are moderate and may not accurately represent the overall population in terms of age and gender. In Study 1 we have a higher proportion of female participants. Study 2 shows a balanced gender distribution. But in both studies the average age of our participants is below the average age of the general population. However, it is worth mentioning that existing research shows that demographic variables such as age and gender do not have a significant impact on usability ratings. For instance, Lewis [4, 15] provides a comprehensive summary of studies that have investigated the influence of demographic variables on System Usability Scale scores, and most of these studies have found no significant effects.

Secondly, it is important to consider that the product being investigated can have an impact on the results. User experience (UX) items are always interpreted within the specific context of the evaluated product. Furthermore, the three products that were examined in this study are all task-centric, meaning that users primarily engage with them to accomplish clearly defined tasks, such as creating presentations, communicating with friends or colleagues, or obtaining detailed information about a specific geographical location or planning routes. To enhance the validity and generalizability of the findings, it would be beneficial to replicate the study using products from different application categories, such as games or social networks.

Disclosure of Interests. The authors have no competing interests to declare that are relevant to the content of this article.

Appendix

Survey for Study 1

A transcript of the survey used is shown below, using WhatsApp as an example. As explained, Parts 1 and 3 were used identically in all 6 applications, while Part 2 was divided into mixed and one-sided UEQ depending on the questionnaire. The items whose polarity has changed in the one-sided version are shown in italics.

1. How do you rate WhatsApp?

Please only take part in the survey if you have already used WhatsApp!

Your age.

<number input>

Your gender

<male, female, no answer>

2. Please rate WhatsApp

Decide as spontaneously as possible which of the following contrasting terms better describes Google Maps. There is no "right" or "wrong" answer. Only your personal opinion counts!

Mixed UEQ (7-point rating scale):

Item Left	*Right*
1 annoying	enjoyable
2 not understandable	understandable
3 *creative*	*dull*
4 *easy to learn*	*difficult to learn*
5 *valuable*	*inferior*
6 boring	exciting
7 not interesting	interesting
8 unpredictable	predictable
9 *fast*	*slow*
10 *inventive*	*conventional*
11 obstructive	supportive
12 *good*	*bad*
13 complicated	easy
14 unlikable	pleasing
15 usual	leading edge
16 unpleasant	pleasant
17 *secure*	*not secure*
18 *motivating*	*demotivating*
19 *meets expectations*	*does not meet expectations*
20 inefficient	efficient
21 *clear*	*confusing*
22 impractical	practical
23 *organized*	*cluttered*
24 *attractive*	*unattractive*
25 *friendly*	*unfriendly*
26 conservative	innovative

3. Please rate how important certain properties of the product are for your overall impression of the product!

The product should look attractive, enjoyable, friendly and pleasant.

I should perform my tasks with the product fast, efficient and in a pragmatic way.

The product should be easy to understand, clear, simple and easy to learn.

The interaction with the product should be predictable, secure and meets my expectations.

Using the product should be interesting, exiting and motivating.

The product should be innovative, inventive and creatively designed.

<Completely unimportant, important (7-point rating scale)>

Survey for Study 2

A transcript of the survey on WhatsApp is shown below. As explained, Part 1 was used identically in both applications, while Part 2 and 3 were divided into mixed and one-sided item polarity depending on the questionnaire. The items whose polarity has changed in the one-sided version are shown in italics.

1. How do you rate WhatsApp?

Please only take part in the survey if you have already used WhatsApp!

Your age.

<short input>

Your gender.

<male, female, other, no answer>

2. I evaluate the usage of WhatsApp as:

HED/UT (7-point rating scale):

Item	*Left*	*Right*
1	*effective*	*ineffective*
2	unhelpful	helpful
3	*functional*	*not functional*
4	*necessary*	*unnecessary*
5	impractical	practical
6	*fun*	*not fun*
7	dull	exciting
8	*delightful*	*not delightful*
9	not thrilling	thrilling
10	unenjoyable	enjoyable

3. I perceive WhatsApp as:

Trustdiff (7-point rating scale):

Item	*Left*	*Right*
1	ignoring	caring
2	*sensitive*	*insensitive*
3	*empathic*	*inconsiderate*
4	dishonest	honest
5	*believable*	*unbelievable*
6	untruthful	truthful
7	incompetent	competent
8	*proficient*	*unqualified*
9	*capable*	*incapable*
10	inept	resourceful

References

1. Schrepp, M.: User Experience questionnaires: how to use questionnaires to measure the user experience of your products? Self-published . (2021)
2. Laugwitz, B., Schrepp, M., Held, T.: Construction and Evaluation of a User Experience Questionnaire. In: Holzinger, A. (ed.) USAB 2008, LNCS 5298, pp. 63–76. Springer (2008)
3. Brooke, J.: SUS - A quick and dirty usability scale. Usabil. Eval. Ind. **189**(194), 4–7 (1996)

4. Lewis, J.R.: The system usability scale: past, present, and future. Int. J. Hum.-Comput. Interact. **34**(7), 577–590 (2018)
5. Finstad, K.: The usability metric for user experience. Interact. Comput. **22**(5), 323–327 (2010)
6. Lin, H., Choong, Y., Salvendy, G.: A proposed index of usability: a method for comparing the relative usability of different software systems. Behav. Inf. Technol. **16**(4/5), 267–278 (1997)
7. Schrepp, M.: A comparison of UX Questionnaires - what is their underlying concept of user experience? In: Hansen, C., Nürnberger, A., Preim, B. (eds.) Mensch und Computer 2020 – Workshopband. Gesellschaft für Informatik e.V., Bonn (2020)
8. Jebb, A.T., Ng, V., Tay, L.: A review of key likert scale development advances: 1995–2019. Front. Psychol. **12** (2021)
9. Sauro, J., Lewis, J.R.: When designing usability questionnaires, does it hurt to be positive? In: Proceedings of the SIGCHI Conference on Human Factors in Computing Systems, pp. 2215–2224. Association for Computing Machinery, New York (2011)
10. Kortum, P., Acemyan, C.Z., Oswald, F.L.: Is it time to go positive? Assessing the positively worded system usability scale (SUS). Hum. Factors **63**(6), 987–998 (2021)
11. Hassenzahl, M., Burmester, M., Koller, F.: AttrakDiff: Ein Fragebogen zur Messung wahrgenommener hedonischer und pragmatischer Qualität. In: Ziegler, J., Szwillus, G. (eds.) Mensch & Computer 2003 – Interaktion in Bewegung, pp. 187–196. BG Teubner, Stuttgart, Leipzig (2004)
12. Brühlmann, F., Petralito, S., Rieser, D.C., Aeschbach, L.F., Opwis, K.: TrustDiff: development and validation of a semantic differential for user trust on the web. J. Usability Stud. **16**(1), 29–38 (2020)
13. Voss, K.E., Spangenberg, E.R., Grohmann, B.: Measuring the hedonic and utilitarian dimensions of consumer attitude. J. Mark. Res. **40**(3), 310–320 (2003)
14. Ullrich, D., Diefenbach, S.: INTUI: exploring the facets of intuitive interaction. In: Ziegler, J., Schmidt, A. (eds.) Mensch & Computer 2010, pp. 251–260. Oldebourg, München (2010)
15. Lewis, J.R.: Comparison of item formats: agreement vs item-specific endpoints. J. Usabil. Stud. **11**(1), 48–60 (2018)
16. Schrepp, M., Thomaschewski, J.: Design and validation of a framework for the creation of user experience questionnaires. Int. J. Interact. Multimedia Artif. Intell. **5**(7), 88–95 (2019)
17. Schrepp, M., Hinderks, A., Thomaschewski, J.: Design and evaluation of a short version of the user experience questionnaire (UEQ-S). IJIMAI **4**(6), 103–108 (2017)
18. Chin, W.W., Johnson, N., Schwarz, A.: A fast form approach to measuring technology acceptance and other constructs. MIS Q. **32**(4), 687–703 (2008)
19. Hinderks, A., Schrepp, M., Rauschenberger, M., Olschner, S., Thomaschewski, J.: Konstruktion eines Fragebogens für jugendliche Personen zur Messung der User Experience. In: Brau, H., Lehmann, A., Petrovic, K., Schroeder, M. (eds.) Usability Professionals 2012, pp. 78–83. German UPA e.v, Stuttgart (2012)
20. Schrepp, M.; Kollmorgen, J., Thomaschewski, J.: Impact of item polarity on the scales of the user experience questionnaire (UEQ). In: Proceedings of the 19th International Conference on Web Information Systems and Technologies, vol. 1: WEBIST, INSTICC, pp. 15–25. SciTePress, Rome (2023)
21. Spangenberg, E.R., Voss, K.E., Crowley, A.E.: Measuring the hedonic and utilitarian dimensions of attitude: a generally applicable scale. Adv. Consum. Res. **24**(1), 235–241 (1997)
22. Hinderks, A., Schrepp, M., Domínguez Mayo, F.J., Escalona, M.J., Thomaschewski, J.: Developing a UX KPI based on the user experience questionnaire. Comput. Stand. Interfaces **65**, 38–44 (2019)
23. Schrepp, M., Kollmorgen, J., Thomaschewski, J.: Impact of item polarity on the scales of the User Experience Questionnaire (UEQ) - Research Protocol (2023). https://doi.org/10.13140/RG.2.2.23530.08645

24. Cronbach, L.J.: Coefficient alpha and the internal structure of tests. Psychometrika **16**, 297–334 (1951)
25. Cronbach, L.J.: My current thoughts on coefficient Alpha and successor procedures. Educ. Psychol. Meas. **64**(3), 391–418 (2004)
26. Schönbrodt, F.D., Perugini, M.: At what sample size do correlations stabilize? J. Res. Pers. **47**(5), 609–612 (2013)
27. Schrepp, M.: On the usage of cronbach's alpha to measure reliability of UX scales. J. Usabil. Stud. **15**(4), 247–258 (2020)
28. Schrepp, M.: Datenqualität bei Online-Fragebögen sicherstellen. In: Hess, S., Fischer, H. (eds.) Mensch und Computer 2016 – Usability Professionals. Gesellschaft für Informatik e.V. and German UPA e.V., Aachen (2016)
29. Lance, C.E., LaPointe, J.A., Stewart, A.M.: A test of the context dependency of three causal models of halo rater error. J. Appl. Psychol. **79**(3), 332–340 (1994)
30. Ford, G.T., Smith, R.A.: Inferential beliefs in consumer evaluations: an assessment of alternative processing strategies. J. Cons. Res. **14**, 363–371 (1987)

An End-User Tool for Supporting Frequent Multisource Web Search Workflows

Alex Tacuri[1], Gonzalo Ripa[1,2], Alejandro Fernandez[1], Gustavo Rossi[1,3], Juan Cruz Gutierrez Saravia[1], and Sergio Firmenich[1,2](✉)

[1] LIFIA, CIC, Facultad de Informática, UNLP, La Plata, Argentina
alex.tacuri@lifia.info.unlp.edu.ar
[2] CONICET, Buenos Aires, Argentina
sergio.firmenich@gmail.com
[3] Facultad de Tecnologia Informatica, Universidad Abierta Interamericana, Buenos Aires, Argentina

Abstract. Current web browser support for search tasks often lead to tedious and time-consuming processes especially when information needs span across various domains and when these tasks are done frequently. This paper addresses this challenge by proposing an approach that empowers end users to seamlessly conduct complex web search workflows. By prioritizing end-user programming and controllability, our approach enables users to customize and automate their search processes according to individual preferences and needs. We emphasize the importance of maintaining user control and awareness of information sources. Through our approach, we aim to enhance user experience and efficiency in accessing information across diverse online platforms.

Keywords: Web scraping · Web search · End-user development · Prompt engineering

1 Introduction

To initiate a search, users commonly rely on either a crawler-based search service like Google or Bing, or they utilize the search functionality of a particular website when they need to retrieve a domain-specific information objects [3]. Both methods are crucial for searching in various contexts, and both may entail conducting ancillary searches [4] to acquire supplementary information. However, certain web tasks associated with accessing information require conducting searches across multiple websites. When such tasks are performed frequently, the process can become tedious and time-consuming. In a very interesting study about the user behaviour when searching, previous works have classified users between 'navigators' and 'explorers', and although explorers seem to do more branches when looking for information, it is valid for both to say that users utilize more than just one website when searching for information in order to accomplish a task [17].

For example, researchers in any academic field typically have a set of repositories or web applications from which they access relevant information for their work using the search engines provided by those applications. During a typical workday, researchers

M. Marchiori and F. García Peñalvo (Eds.): WEBIST 2023, LNBIP 543, pp. 30–44, 2025.
https://doi.org/10.1007/978-3-031-89621-7_2

may need to search for papers on a specific topic on the Springer website. Subsequently, for several of the resulting papers, they might wish to find additional information from other sources. This could involve checking the number of citations on Google Scholar, finding more articles by the same authors on DBLP, or examining scientific ranking services for more details about the journal or conference where the paper was published. This process goes beyond a simple web search; it constitutes a search workflow that demands greater interaction and time. Furthermore, additional effort is required if the researchers aim to consolidate the information scattered across different sources due to the absence of a unified user interface that supports the overall task.

Similarly, various professions, occupations, or activities on the web rely on this kind of workflows using specific websites or web applications. While web browsers, whether mainstream or not, typically offer sufficient support for such search workflows when they are conducted infrequently—often managed through tab management by end users—using them on a daily basis presents challenges. In fact, if multiple search workflows are carried out simultaneously, the user experience provided by web browsers is often inadequate.

In the past, various works have addressed this issue through different approaches such as mash-ups [14] and web augmentation tools [4]. Regardless of the final product or outcome, the process of creating these tools has also been the focus of research, resulting in both code-oriented tools and end-user development ones. Despite living in an API-driven world, it remains true that a significant number of websites do not offer APIs, which complicate integration. Furthermore, even if every website were to provide an API, it would still be complex for users to consume them, despite interesting approaches such as ScrAPIr [2] or the significant advancements in service composition. As we transition towards a world with a growing presence of AI agents, it is unclear which role web searches will play in general information retrieval. It becomes important to consider controllability, a term very well-known in user interface adaptation and recommender systems [10]. In the context of the approach, it means in the best of the cases that users may decide which information sources to use, and in the worst, at least to let them known from where the information is obtained. However, it is certain that various user activities will still require searching information, whether on the internet or on intranets. With the fast advances on AI-based tools and agents, it is clear that web searches and search results consumption surely will suffer profound changes in several dimensions. In this work we focus on how the kind of search workflows described before may be supported, how search results are integrated and how these results are presented to users.

This paper introduces an approach aimed at empowering end users with the ability to seamlessly perform complex web search workflows across diverse online platforms. By placing emphasis on end-user programming and controllability, our approach equips users with the tools to effortlessly customize and automate their search processes according to their unique preferences and requirements. In that process, we strongly believe that generative AI models are useful for different construction steps for a final product, but what would be extremely important is that users still maintain control and knowledge on which information is used and from where that information was obtained.

The article is structured as follows. Section 2 provides an overview of the approach by means of a motivating example. Section 3 explains how search APIs are defined with support from generative AI, whereas Sect. 4 discusses how search APIs are composed. Section 5 presents the UI for search workflows through an example. Section 6 discusses related works. Section 7 presents conclusions and future works.

2 The Approach in a Nutshell

This approach focuses on those search workflows that are composed by a sequence of search tasks performed in several websites, combining different information pieces that are provided by each specific website to build more complete search results. To illustrate this concept, we will detail an example on the research domain, where a user needs to perform a very typical and daily research task.

Firstly, the search workflow starts by searching some topic in the Springer repository. From that process, the researcher is interested in several of the resulting articles, and for each one of them requires further information that is not present at Spring. In this regard, the researcher probably opens a new tab to open Google Scholar to search those relevant papers from the first search, in order to look its citations. Also the researcher could be interested in opening another browser tab to search at DBLP for further papers of authors to see if authors have newer publications on the same research line.

This simple search workflow implies to use one main search service (Springer), and for each relevant resulting paper, to open at least two more websites to look further information. It is up to the user to remember the semantic connections among tabs, and to maintain consistency when navigating in any of them.

A workflow like this one could be easily automated by a developer if all of these web applications offered APIs to retrieve information. But in absence of these APIs, and mainly from the point of view of end users without programming skills, a new approach is needed. Such new approach should support end users in defining and triggering complex searches connecting results from multiple sites, and presenting them in a useful and user-friendly mode. Its building blocks are: a) uniform search APIs for any web-site, defined by end users; b) composition of search results from multiple search APIs to obtain navigable data objects; c) straight-forward triggering of searches; d) user-friendly presentation of search results.

In previous work [5], we proposed a mechanism to aid users in creating uniform search APIs for websites or web applications that offer search functionality. This was done by emulating users interaction with websites and applying information extractors that were specified by web annotations, which do not require programming skills. Later, we proposed a visual programming tool to combine search APIs and create web scrapers that integrate information spread in several websites [15]. In this article we extend previous visual programming tool to create search workflows that compose search APIs, and by discussing alternative approaches for creating search APIs based on generative AI. In this regard, we designed and developed a web extension as a toolset implementation, whose components overview is shown in Fig. 1.

This involves:

- Search API tool: this tool allows end users to create search APIs for those websites that do not provide one, and also define an execution engine of these specifications.

In previous works this component was based on an end-user development tool that uses web annotation to define the search API. In this work, we present another approach for search API definitions that is supported in existing generative AI.
- Search API composition tool: it provides a development tool that allows end users without programming skills to define a search API composition using a visual editor. In this way, end users may define the search workflows, i.e. how result's properties values are used as inputs to trigger another search with other search API. This component, similar to the previous one, also provides an execution engine that coordinate the execution of search APIs.
- User Interface: we provide a simple graphical user interface in order to allow users to interact with the results obtained from the execution of a specific search workflow. In this work, this GUI interacts with the Search API composition execution engine to obtain search workflows results.

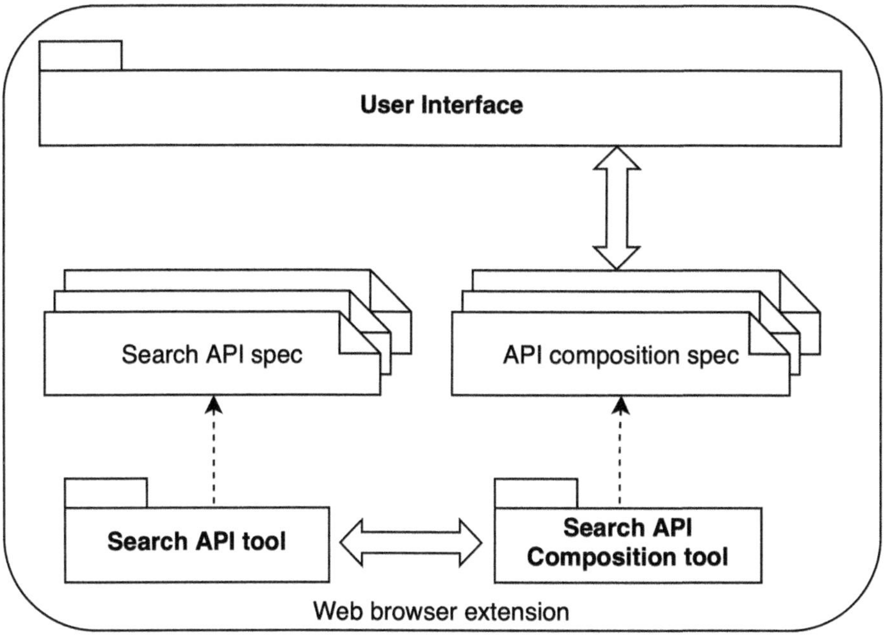

Fig. 1. Web extension overview.

3 Search APIs

3.1 ANDES Search APIs

This work is based on ANDES' concept of Search APIs, which was published before [5]. ANDES offers tools to annotate the UI search components (search input, trigger button, pagination buttons, etc.), but also is the user who defines the type of the resulting object, its properties, and the mapping from the content of the results web page to each object property values. Calling the search operation on the API triggers the corresponding search and scrapping behavior and returns a collection of (JavaScript) objects. In that work, ANDES was mainly focused on provide better support for ancillary searches. In other works, we also worked on a similar concept, not improving the interaction for ancillary searches, but embedding them into a data service layer that helps web extensions developers to don't have harcoded DOM references in their code, but delegating that online information retrieving to this data service layer [15]. Further details about ANDES can be found in those previous works [5].

However, in order to reduce the necessary interaction to define a search API, we developed a prototype using existing generative GPT technologies to solve specific steps in the search API definition process. This approach is presented in the following subsection.

3.2 AI-Assisted Search APIs Creation

In this work we explore the creation of search APIs, APIs similar to the ones proposed in ANDES, by using existing generative AI technologies capabilities. After several tests, we found an efficient way to develop search APIs by dividing the process in small tasks. For solving each of these tasks we based the new approach on prompt engineering, taking advantage of large language model (LLM) technologies currently available, particularly OpenAI's GPT series of models. We developed a tool that guides end users during this short process, giving proper feedback through messages appearing on top of Web sites. This speeds up considerably the process of creation, eliminating the annotation tasks that end users needed to accomplish to build the Search APIs in ANDES.

Creation. This creation process consist in several steps including:

1. Detects a parameterized URL search expression.
2. Identify a common xPath from the results web page, necessary to access the results objects during real time searches.
3. Build a template of semantic properties that describe a result object.

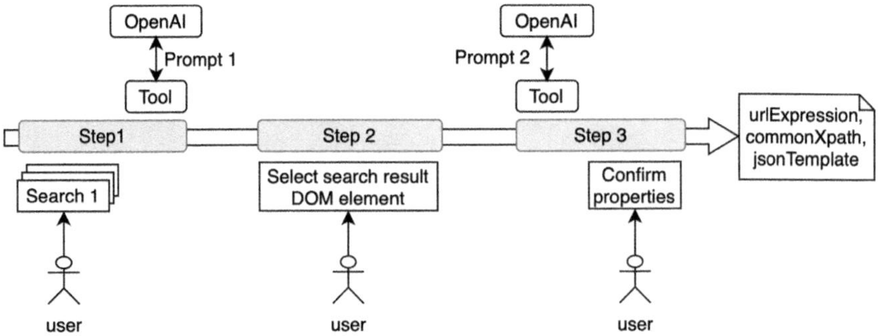

Fig. 2. Search API definition process.

During this process, our tool coordinates the interaction required by part of users and the calls required to OpenAI's API. Figure 2 depicts this process.

Step 1. To accomplish this step, users need to open a website they are interested in to create its Search API. Our tool detects when users are in a new website and ask them to perform a few searches using the corresponding search input elements (step 1 in Fig. 3). The tool also detects the keywords searched by the user giving feedback with proper messages showed on top of the screen during the three searches required (step 2 in Fig. 3).

After three searches are completed, the prompt engineering starts taking action, and when it is finished, user gets feedback (step 3 in Fig. 3). Behind scenes, our tool communicates with the OpenAI's API, sending a request using the gpt-3.5-turbo-0125 model, with the next prompt:

Prompt 1. *"You are a helpful assistant. Based on these examples: {**setOfUrls**}, give me a JSON with searchToken and url properties."*

where {**setOfUrls**} corresponds to an array containing the search urls detected previously. Then, OpenAI's service will respond with a JSON object containing searchToken and url properties, used later to trigger real searches. An example of a url expression returned is: "https://listado.mercadolibre.com.ar/{searchToken}".

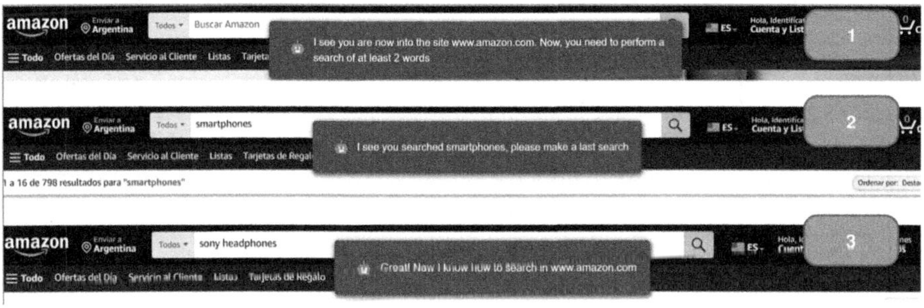

Fig. 3. Website's URL search detection.

Step 2. After URL is obtained, the second step needs to be done, and our tool will highlight with a red color the sections of the web page where the user is pointing with the cursor (Fig. 4). To accomplish 2, the user needs to click on the section that better represents a search result and includes as much information as possible of it. The tool will generate a common xPath expression that will identify all search result elements, necessary at the moment of performing real searches using the search APIs created.

Step 3. Once our tool knows how to obtain DOM elements that represent search results, it is necessary to extract information from them and generate a information object's template. Different to ANDES's approach, where users need to select and configure each property for the selected search result in order to know how to extract information,

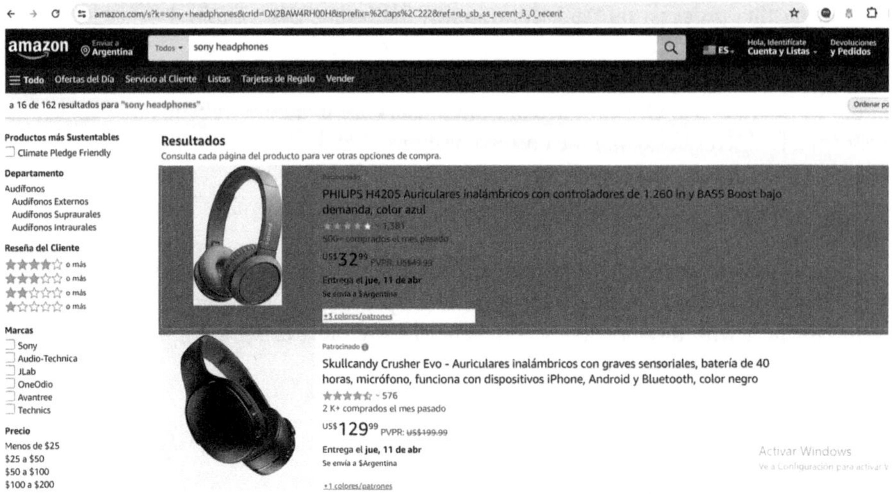

Fig. 4. Result element DOM selection.

this new approach communicates again with OpenAI's service, in order to parse the full text content of a particular search result to construct a JSON object, detecting its semantics and properties. In this case, the prompt defined was the following:

Prompt 2: *"You are a helpful assistant. Generate an object JSON-LD using the schema.org context without aggregation and with each property represented as a string, for the next info:* {***domElementInnerText***}*. Do not nest objects and don't give me a summary. The resulting JSON must have at least a 'name' property, a '@type' property giving a semantic classification, and only have string type properties."*

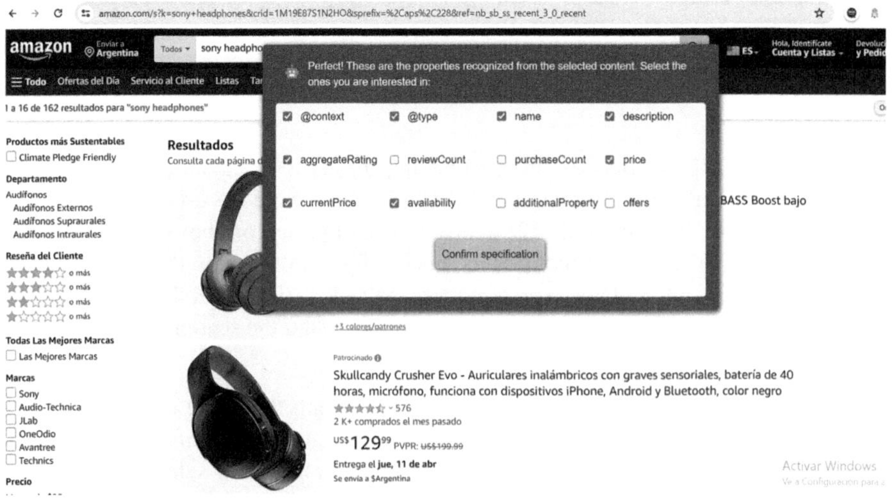

Fig. 5. Properties selection.

where {**domElementInnerText**} corresponds to the innerText obtained by the tool when the user clicked on a search result Web page element. The returned JSON will be a template of semantic properties that describe the result object. Finally, the tool will let users customize the final template with the properties they are interested to show in final search results (Fig. 5), ending with 3.

Search API Execution. To can execute searches using the Search APIs generated through the process described in 3.2.1, the tool will use the common xPath obtained that let to identify each result of the search and get unstructured information through the innerText property of the DOM element, to then communicate with OpenAI's API one more time, sending the next prompt:

Prompt 3: *"You are a helpful assistant. I will provide you with an innerText and you have to generate a JSON-LD object using the schema.org context based on the next JSON template, without aggregation and with each property represented as a string:* {***jsonTemplate***}*. If some property from the template is not found, just complete it with an empty string. Also if you find some new property not present in the template add it to the element. Do not nest objects. The resulting JSON object must have at least a 'name' property, a '@type' property giving a semantic classification and only have string type properties. This is the innerText:* {**domElementInnerText**}*."*

where {**jsonTemplate**} corresponds to the template of semantics properties that is part of the specification created, and {**domElementInnerText**} corresponds to the innerText of a search result obtained using the common xPath expression, both generated during Search API creation process described previously. The search execution will return a final JSON with all the results' information exposed as objects with properties that will be based on {**jsonTemplate**}.

Tests. Regarding the development of search APIs involving Prompt 1 and Prompt 2, the results were very promising, demonstrating excellent detection of search URLs, semantic objects, and template structures across 22 websites spanning four distinct domains.

Using the template derived from the search API definition (crafted by selecting all properties detected by OpenAI services), we conducted tests to parse all results for specific searches on each of the 22 websites, essentially employing Prompt 3. The outcomes are presented in Table 1. The column descriptions are outlined below:

- **Website:** The targeted website.
- **Properties:** The number of properties detected during the creation of the search API.
- **Complete:** The number of search results for which the OpenAI service successfully found values for the given template. For example, for MercadoLibre, the value '5/6' indicates that 5 out of 6 retrieved information objects were complete.
- **Blank AVG (%):** The average percentage of properties for which the value was blank, i.e., where OpenAI could not find a value.
- **Blank STDV (%):** The standard deviation of the percentage of properties for which the value was blank.

Table 1. OpenAI parsing results.

Website	Properties	Complete	Blank AVG (%)	Blank STDV (%)	With new props	New props AVG (%)	New props STDV (%)
Mercadolibre	6	5/6	2,77	6,80	3/6	30,55	35,61
Fravega	5	8/8	0	0	0/8	0	0
Amazon	8	4/9	9,72	12,14	2/9	4,16	8,83
eBay	12	2/9	9,25	6,51	1/9	3,70	11,11
Alibaba	14	1/8	12,5	11,29	0/8	0	0
Biblioteca New York	12	4/8	13,54	16,62	0/8	0	0
Biblioteca UBA	7	7/9	6,34	14,48	0/9	0	0
Biblioteca Australia	7	8/8	0	0	0/8	0	0
Biblioteca España	7	9/9	0	0	0/9	0	0
Biblioteca MIT	13	1/6	29,48	21,43	0/6	0	0
Booking	18	1/9	12,34	7,23	0/9	0	0
Airbnb	6	0/9	24,07	8,78	1/9	1,85	5,55
TripAdvisor	5	4/8	22,5	31,05	2/8	12,5	23,75
IMDb	5	4/4	0	0	0/4	0	0
TheMovieDB	4	9/9	0	0	0/9	0	0
Letterboxd	5	4/4	0	0	0/4	0	0
Metacritic	4	7/9	11,11	22,04	5/9	30,55	32,54
DBLP	9	0/8	11,11	0	0/8	0	0
Google Scholar	10	7/8	1,25	3,53	8/8	10	0
Springer	7	0/8	25	6,61	0/8	0	0
Springer Nature Citations	6	7/9	5,55	11,78	0/9	0	0
ACM	9	8/8	0	0	0/8	0	0

- **With New Props:** The number of information objects returned by OpenAI that have additional properties compared to those specified in the template. For example, for MercadoLibre, the value '3/6' indicates that for 3 out of 6 search results, OpenAI service could extract new properties compared to the original template.
- **New Props AVG (%):** The average percentage of new properties.
- **New Props STDV (%):** The standard deviation of the number of new properties.

Overall, the values for all properties of the information objects were accurate, even when the properties were not explicitly specified in the original template. It is noteworthy that in most cases, the completeness of objects was not 100%; however, considering the average of blank properties, it is very low in most instances.

4 Search API Composition

To start talking about the composition of search APIs, let's recall an example that we have all experienced: when we search for information, we begin with a search engine and then explore other sites to gain a more comprehensive understanding of the topic. With this premise in mind, let's revisit our motivational example from Sect. 2. In this example, we start by searching for a specific topic on Springer, and from the results obtained, we look for the number of citations, or link to the document, or a PDF file on Google Scholar, and additionally, we would like to know what other articles the author has written using the DBLP database.

Thanks to the development of our tools, we can now perform the composition or integration of search APIs even without programming knowledge.

The tool enables the creation of various composition models for search APIs, allowing searches to be conducted across each of them. To set up a composition model, it's necessary to first have the search APIs of the relevant websites. For instance, for our motivational example, we need access to the services of Springer, Google Scholar, and DBLP.

Once we have the search APIs ready, we'll create a model in the tool as shown in Fig. 6. To do this, the extension provides a graphical environment with drag-and-drop components; one of them represents each of the previously generated APIs. Additionally, the tool displays a canvas where we add the APIs we need and then establish their integration or composition. The composition between search services can be one-to-one, meaning to connect to the first object that meets the search criteria, or one-to-N, meaning to return a collection with all the result.

Returning to our motivational example, we need to add the 3 APIs from Springer, Google Scholar, and DBLP to the canvas. Now, let's proceed to establish the connections between each of them. Firstly, let's analyze the connection between Springer and DBLP. We need that, for each element of Springer, a list of articles from DBLP authored by the first author be obtained. We must assign a name to this link, which in the example is "Articles". We must also select which property of the Springer object will serve as input data for the search in the DBLP search API, which in this case is "Author". Finally, we must indicate the number of resulting elements, which in this case will be all. The next link will be between Springer and Google Scholar. The assigned name is "Citations", the property that will serve as search text will be "Title", and the number of resulting elements will be the first element containing the searched item.

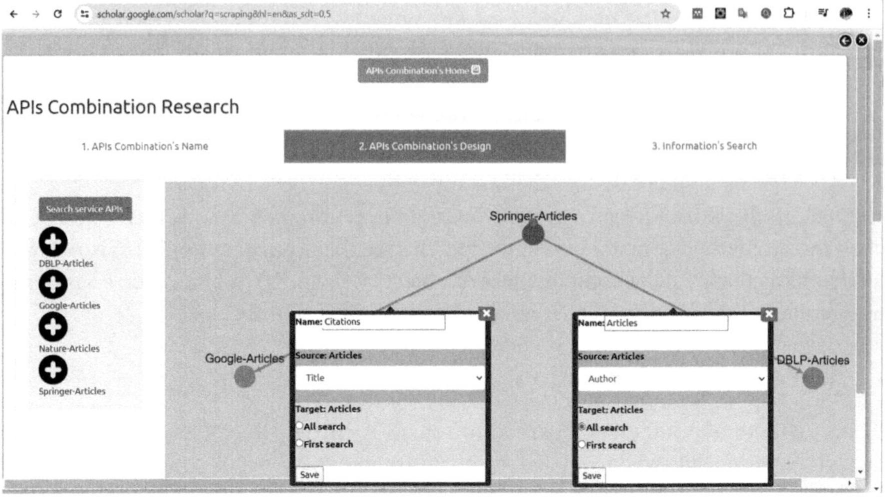

Fig. 6. The API composition tool: Integration between Springer and DBLP search APIs.

5 Generic Workflow User Interface

The interface we have implemented to showcase this approach displays the results obtained from both the 'Search API Tool' and the 'Search API Composition Tool'. These tools provide us with information in JSON format used to create this interface. The example below shows the data obtained from a search performed in the 'Research' model, which was described in the motivational example.

The interface initially presents a form with a textbox for entering the search text and a menu to select the workflow, as depicted in Fig. 7. When a workflow is triggered, and behind the scenes, the tools execute the search APIs composition and integrate their results. For example, we obtain the "title", "abstract", and "authors" of each Springer result, the "number of citations", and the PDF link from Google Scholar. Additionally, clicking on the author of some Springer article, displays a dialog with the search results for the author's articles in DBLP as depicted in Fig. 7.

This generic user interface is quite simple. It lists all the information object's properties, and when one of these properties have been used to trigger the execution of another search API that returns a collection of objects, it is automatically transformed to an anchor widget, which offers an interactive element that opens a modal showing that collection. Also, in order to offer users with feedback about the information's source, the UI adds the favicon of the source application.

We strongly believe that this UI could be improved in several ways, however it is clear that if we compare how long it would take to a user to perform this information search using the traditional method, the time and interaction is quite different. For instance, a search for "artificial intelligence" in our tool takes about 10 s to perform approximately 60 searches on 3 websites, whereas the traditional method would take much longer to perform 60 searches on the three sites and additionally integrate the results.

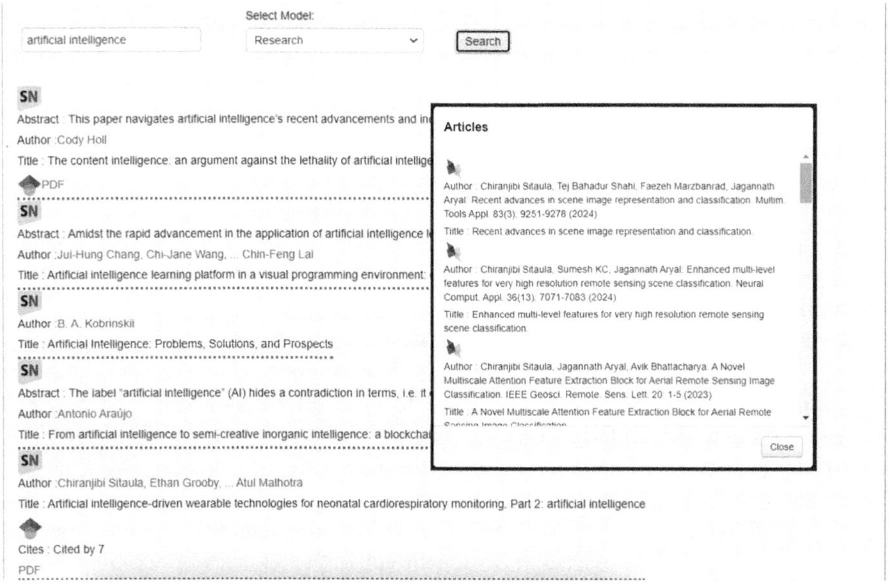

Fig. 7. User interface that utilizes the integration of search services from Springer, DBLP, and Google Scholar.

6 Related Works

There are various dimensions to consider when discussing the approach outlined in this paper within the context of existing literature. To maintain conciseness, we will focus, on one hand, on aspects related to enhancing user experience while navigating the web through external structures [7]. On the other hand, we will delve into techniques associated with scraping online information to construct structured information objects.

Improving user experience without relying on web application owners has been addressed for over 20 years through techniques like web mashups and web augmentation. The former primarily involves integrating content and services to create new applications, while the latter is more focused on adapting existing web pages that fail to meet user requirements.

In the realm of mashups, notable approaches such as Yahoo Pipes [14] and Marmite [18] empower end users without programming skills to define workflows combining services and content from multiple sources to produce new applications. The generic user interface proposed in this article for rendering search workflow results shares similarities with a mashup, but with the added capability of generating a search API for web applications lacking one. Although inspired by mashup approaches, our search API composition tool is exclusively tailored for search workflow specification.

Concerning web augmentation [6], various approaches aim to integrate information with the primary goal of reducing the interaction needed to access it. Some focus on improving information searches, including methods supporting ancillary searches [4], and others closer to our approach, like BEAUD [1], which enhances web page content with information from other sources. Unlike BEAUD, our approach functions as a search engine, orchestrating searches to present a list of results sourced from multiple platforms.

Web browser extensions are commonly used to enhance browsing experiences, with several works proposing end-user programming tools to specify these extensions. Many such approaches emphasize information extraction from websites. Web scraping, the process of obtaining structured information from non-structured or partially structured website data, is widely utilized in web browser extensions [15], and numerous scientific works propose web extensions implementing scraping and web semantic technologies [9], similar to our toolset's methodology.

Several approaches enable users to specify existing content structures to manage relevant information objects more effectively. For example, HayStack [11] offers an extraction tool for populating a semantic-structured information space. Atomate it! [13] provides a reactive platform to set rules for collected objects, alerting users to noteworthy events. [16] facilitates domain-specific application creation working over objects defined in a PIM. Rousillon, an intriguing approach based on end-user programming, allows scraper definition based on hierarchical data [8]. Another approach presented in [12] enables web customization through user-defined web scraping without programming skills. Although our previous work, ANDES, for defining search APIs operates similarly to these approaches, this paper introduces structured information object creation by consuming existing generative AI models. Notably, all these semantics extraction-based approaches do not primarily focus on enhancing search processes.

7 Conclusions and Future Works

In this paper, we have presented an end-user tool designed to enhance the web search experience, particularly for users engaged in frequent multisource search workflows. Our tool addresses several critical aspects of modern web user experience, including the management of searches across multiple web applications and the integration of AI to streamline the search API specifications, reducing the interaction by part of the user in order to specify the semantic structure of the web content.

Through the incorporation of AI technologies, we have aimed to improve the efficiency and effectiveness of scraping specification process and maintaining user's controllability.

Our tests conducted utilizing OpenAI service to create structured information objects from DOM elements' plain text have provided promising results, demonstrating the potential of generative AI in supporting websites scraping. However, we acknowledge that further testing is required to fully explore the capabilities and limitations of generative AI in this context. The emergence of generative AI presents exciting new opportunities for improving information access and search workflows specification. However, it also introduces challenges, particularly regarding the need for structured data and innovative prompt engineering techniques to harness the full potential of AI-based service composition, which is one of the aspects we will study in the future.

References

1. Aldalur, I.: BEAUD: a browser extension to automatize end-user deeds. Softw. Impacts **17**, 100516 (2023). https://doi.org/10.1016/J.SIMPA.2023.100516
2. Alrashed, T., Almahmoud, J., Zhang, A.X., Karger, D.R.: Scrapir: making web data apis accessible to end users. In: Bernhaupt, R., et al. (eds.) CHI '20: CHI Conference on Human Factors in Computing Systems, Honolulu, HI, USA, 25–30 April 2020, pp. 1–12. ACM (2020). https://doi.org/10.1145/3313831.3376691
3. Bhavnani, S.K.: Domain-specific search strategies for the effective retrieval of healthcare and shopping information. In: Terveen, L.G., Wixon, D.R. (eds.) Extended abstracts of the 2002 Conference on Human Factors in Computing Systems, CHI 2002, Minneapolis, Minnesota, USA, 20–25 April 2002, pp. 610–611. ACM (2002). https://doi.org/10.1145/506443.506508
4. Bosetti, G., Firmenich, S., Fernandez, A., Winckler, M., Rossi, G.: From search engines to augmented search services: an end-user development approach. In: Cabot, J., De Virgilio, R., Torlone, R. (eds.) ICWE 2017. LNCS, vol. 10360, pp. 115–133. Springer, Cham (2017). https://doi.org/10.1007/978-3-319-60131-1_7
5. Bosetti, G., et al.: ANDES: an approach to embed search services on the web browser. Comput. Stand. Interfaces **82**, 103633 (2022). https://doi.org/10.1016/j.csi.2022.103633
6. Bouvin, N.O.: Unifying strategies for web augmentation. In: Westbomke, J., Wiil, U.K., Leggett, J.J., Tochtermann, K., Haake, J.M. (eds.) HYPERTEXT '99, Proceedings of the 10th ACM Conference on Hypertext and Hypermedia: Returning to Our Diverse Roots, Darmstadt, Germany, 21–25 February 1999, pp. 91–100. ACM (1999). https://doi.org/10.1145/294469.294493
7. Bouvin, N.O.: From notecards to notebooks: there and back again. In: Atzenbeck, C., Rubart, J., Millard, D.E. (eds.) Proceedings of the 30th ACM Conference on Hypertext and Social Media, HT 2019, Hof, Germany, 17–20 September 2019, pp. 19–28. ACM (2019). https://doi.org/10.1145/3342220.3343666
8. Chasins, S.E., Mueller, M., Bodík, R.: Rousillon: scraping distributed hierarchical web data. In: Baudisch, P., Schmidt, A., Wilson, A. (eds.) The 31st Annual ACM Symposium on User Interface Software and Technology, UIST 2018, Berlin, Germany, 14–17 October 2018, pp. 963–975. ACM (2018). https://doi.org/10.1145/3242587.3242661
9. Huynh, D., Mazzocchi, S., Karger, D.R.: Piggy bank: experience the semantic web inside your web browser. J. Web Semant. **5**(1), 16–27 (2007). https://doi.org/10.1016/J.WEBSEM.2006.12.002
10. Jameson, A., Schwarzkopf, E.: Pros and cons of controllability: an empirical study. In: De Bra, P., Brusilovsky, P., Conejo, R. (eds.) AH 2002. LNCS, vol. 2347, pp. 193–202. Springer, Heidelberg (2002). https://doi.org/10.1007/3-540-47952-X_21
11. Karger, D.R., Bakshi, K., Huynh, D., Quan, D., Sinha, V.: Haystack: a general-purpose information management tool for end users based on semistructured data. In: Second Biennial Conference on Innovative Data Systems Research, CIDR 2005, Asilomar, CA, USA, 4–7 January 2005, Online Proceedings, pp. 13–26 (2005). http://cidrdb.org/cidr2005/papers/P02.pdf
12. Katongo, K., Litt, G., Jackson, D.: Towards end-user web scraping for customization. In: Church, L., Chiba, S., Boix, E.G. (eds.) Programming '21: 5th International Conference on the Art, Science, and Engineering of Programming, Cambridge, United Kingdom, 22–26 March 2021, pp. 49–59. ACM (2021). https://doi.org/10.1145/3464432.3464437
13. Kleek, M.V., Moore, B., Karger, D.R., André, P., m. c. schraefel: atomate it! end-user context-sensitive automation using heterogeneous information sources on the web. In: Rappa, M., Jones, P., Freire, J., Chakrabarti, S. (eds.) Proceedings of the 19th International Conference on World Wide Web, WWW 2010, Raleigh, North Carolina, USA, 26–30 April 2010, pp. 951–960. ACM (2010). https://doi.org/10.1145/1772690.1772787

14. Stolee, K.T., Elbaum, S.G., Sarma, A.: Discovering how end-user programmers and their communities use public repositories: a study on yahoo! pipes. Inf. Softw. Technol. **55**(7), 1289–1303 (2013). https://doi.org/10.1016/J.INFSOF.2012.10.004
15. Tacuri, A., Firmenich, S., Rossi, G., Fernández, A.: A data service layer for web browser extensions. In: García-Peñalvo, F.J., Marchiori, M. (eds.) Proceedings of the 19th International Conference on Web Information Systems and Technologies, WEBIST 2023, Rome, Italy, 15–17 November 2023, pp. 49–58. SCITEPRESS (2023). https://doi.org/10.5220/0012165500003584
16. Van Kleek, M., Smith, D.A., Shadbolt, N., et al.: A decentralized architecture for consolidating personal information ecosystems: the webbox. In: PIM 2012 (2012)
17. White, R.W., Drucker, S.M.: Investigating behavioral variability in web search. In: Williamson, C.L., Zurko, M.E., Patel-Schneider, P.F., Shenoy, P.J. (eds.) Proceedings of the 16th International Conference on World Wide Web, WWW 2007, Banff, Alberta, Canada, 8–12 May 2007, pp. 21–30. ACM (2007). https://doi.org/10.1145/1242572.1242576
18. Wong, J., Hong, J.I.: Making mashups with marmite: towards end-user programming for the web. In: Rosson, M.B., Gilmore, D.J. (eds.) Proceedings of the 2007 Conference on Human Factors in Computing Systems, CHI 2007, San Jose, California, USA, 28 April–3 May 2007, pp. 1435–1444. ACM (2007). https://doi.org/10.1145/1240624.1240842

Overview of Web Application Performance Optimization Techniques

Juho Vepsäläinen(✉), Arto Hellas, and Petri Vuorimaa

Aalto University, Espoo, Finland
juho.vepsalainen@aalto.fi
https://www.aalto.fi/en/department-of-computer-science

Abstract. During its thirty years of existence, the World Wide Web has helped to transform the world and create digital economies. Although it started as a global information exchange, it has become the most significant available application platform on top of its initial target. One of the side effects of this evolution was perhaps suboptimal ways to deliver content over the web, leading to wasted resources and business through lost conversions. Technically speaking, there are many ways to improve the performance of web applications. In this article, we examine the currently available options and the latest trends related to improving web application performance.

Keywords: Optimization · Performance · Web applications · Web development · Web frameworks · Web tooling · World wide web

1 Introduction

The early web was meant as a global information exchange, and it came with critical concepts, such as hyperlinking, allowing documents to be linked to each other [10]. Since then, the web has become the most extensive application platform as predicted by [67], reaching the majority of the population [65]. Growth of the web platform has not been without its problems, as evidenced by the so-called dot-com bubble at the end of the 90 s [44]. In essence, the web has gone through multiple hype cycles [18]. Depending on the viewpoint, the latest iteration of the web could be considered 3.0 or 4.0 [33]. The crux of 4.0 is the seamless integration of physical and digital worlds. It has been estimated that the value of the new space might be up to €800 billion by 2030 [33].

Due to its importance for business, the web is a significant cornerstone of the global economy. However, the size of web pages keeps growing steadily, evidenced by a nearly 600 % increase between 2012 and 2022 [31]. The increase likely has to do with the growing complexity of the sites and user expectations. The growth has a concrete cost regarding wasted resources and lost conversions. To understand what can be done about the situation, we have phrased our research question for this article as follows: What are the currently available ways for improving web application performance?

To achieve this task, we first consider the aspects of web performance in Sect. 2 before reviewing currently available web performance techniques in Sect. 3. Then we

M. Marchiori and F. García Peñalvo (Eds.): WEBIST 2023, LNBIP 543, pp. 45–62, 2025.
https://doi.org/10.1007/978-3-031-89621-7_3

consider the impact of web frameworks on performance in Sect. 4 and discuss our main observations in Sect. 5 before concluding the article with final remarks in Sect. 6.

2 Aspects of Web Application Performance

Traditionally, web services have been built using some variant of the client-server model [51]. This arrangement relies on trust between the parties, and as an alternative, decentralized options, such as peer-to-peer computing, have been proposed [39]. In this review, we will focus on the former as it is the current mainstream model used by the web as we consider the aspects of web application performance.

2.1 What Is Performance

When it comes to performance, it is good to consider what it consists of and what we mean by it. In the ideal case, clients can access a service they want without a noticeable delay. One could even go as far as to say that a good web experience should not differ from an application you might use on a desktop machine or a mobile phone. Web-based technologies such as Electron are commonly used to develop desktop applications these days [59]. React Native and Flutter are comparable examples from the mobile side [84][1]. Compared to these examples, the added requirement of the web has to do with the need for a server.

Web performance has been widely studied, as shown by [5]. Killelea [34] provides an excellent complementary snapshot of the topic and the techniques that were leveraged at the beginning of the millennium. As pointed out by [5], most of the earlier research has focused on landing pages while ignoring internal pages, which is unfortunate as they can have differing performance. To give an example of a set of performance metrics leveraged by many developers, we will focus on Google's Core Web Vitals in this discussion as they capture the main aspects of web application performance.

According to [83], Google evaluates the following six parameters known as Core Web Vitals to rank site speed: Largest Contentful Paint (LCP), First Contentful Paint (FCP), First Input Delay (FID), Interaction to Next Paint (INP), Cumulative Layout Shift (CLS), and Time to First Byte (TTFB). Suffice it to say that there are many ways to look at performance and to make things more complex; the ways to evaluate website speed change over time, as shown by the recent replacement of FID by INP in Google's Core Web Vitals [80]. The reasoning for this change is that INP captures site responsiveness better than FID [80]. Changes like this imply that web developers must know how their websites work against the currently relevant metrics to maintain their visibility in search results, as the metrics also contribute to search engine rankings, as discussed subsequently. Table 1 explains the differences in the metrics in greater detail.

[1] To bridge the gap between native mobile experiences and the web, the concept of Progressive Web Apps (PWAs) was introduced by Google in 2015 [69]. You could say a full parity can likely never be reached, but that is not to say there is no value in trying. PWAs enhance web applications through capabilities, such as push notifications, offline usage, and access via the home screen [69].

Table 1. The table enumerates key differences between Google's Web Vital performance metrics [83]. It is worth noting that each metric looks at performance differently, and doing well in one does not mean you will do well in all of them.

Metric	Description
Largest Contentful Paint (LCP)	LCP measures loading performance, and for good UX, LCP must occur within 2.5 s of when the page starts loading [83]
First Contentful Paint (FCP)	FCP measures the time it takes from when the user navigates to the page to when any part of the page's content is rendered on the screen [81]. Time to Interactive (TTI) is a related metric that helps to identify cases where a page looks interactive but is not [82]
First Input Delay (FID)	FID measures interactivity; ideally, it should be 100 milliseconds or less [83]. Note that INP has replaced FID and FID can be considered as obsolete [80]
Interaction to Next Paint (INP)	INP measures the overall responsiveness of a page by observing the latency of all click, tap, and keyboard interactions [78]
Cumulative Layout Shift (CLS)	CLS measures visual stability, and the value should remain 0.1 or less [83]
Time to First Byte (TTFB)	TTFB measures the time between the request and when the first byte of a response begins to arrive [79]. Occasionally, this is also known as Server Response Time (SRT)

2.2 Search Engine Optimization

Web Vitals are connected to Search Engine Optimization (SEO) as performance is a part of page ranking [60]. Traditionally, SEO has meant optimizing websites for them to be easy for search engines to find, but as highlighted by [23], the meaning has changed to search experience optimization, putting focus to the user instead of search engines as focusing on user has tangible benefits on the visibility of a website. Technically, SEO can be split into on-page SEO, off-page SEO, and technical SEO [23]. Technical SEO includes performance, and search engines evaluate it in various ways in their search ranking as a part of other factors. In other words, if you improve the performance of your website, you will likely improve its technical SEO as well.

2.3 Transport Protocol

Starting from HTTP/2, the transport protocol of the web has been based on streaming, and the situation has evolved further with HTTP/3 as the protocol has moved to use UDP over TCP/IP underneath, allowing earlier limitations to be lifted [76]. As noted by [76], even with a streaming protocol, performance is still limited by a phenomenon called render-blocking, which occurs when files that were streamed to the client have to be reassembled before the rendering of the page can continue; therefore, highlighting the need to avoid render-blocking operations at the client. As a potential solution, [76] proposes reorganizing page content and structure to be streaming-friendly. The benefit of this approach can be seen in a reduced FCP as the browser can get to the content faster [76].

It can be argued that with the prevalence of streaming, earlier best practices, such as avoiding HTTP requests, may have to be re-evaluated [64]. As shown by the example of [76], gaining most of the benefit from streaming may require adaptation at the developer, or perhaps even framework, side to avoid the render-blocking problem.

2.4 Offloading Work

Although JavaScript runtimes have been traditionally single-threaded, introducing new primitives, such as Web Workers, has enabled developers to overcome this limitation [53]. Current browsers support Web Workers well, and global support is around 98.4% [12]. As highlighted by [53], client and server share computational tasks, and due to the ambiguity of what kind of logic to execute and where apart from security-related controls, [53] have dubbed this type of computation as gray computing. Gray computing comes with ethical challenges related to how much client resources should be consumed, and there are costs and benefits to be considered as gray computing techniques may enable benefits, for example, in website responsiveness [53].

Service Worker (SW) is a specific type of Web Worker that can perform operations in the background, enabling, for example, offline support and push notifications [17][2]. SW is closely related to Progressive Web Applications (PWAs) as it is a vital part of allowing web applications that operate as their mobile counterparts [69]. Given SW can intercept traffic heading to a website, it represents a new vector for attacks [17], highlighting that additions to the web platform may raise new concerns.

It is well understood that many websites depend on third-party resources, which comes with user tracking problems, for example, [32]. Using third-party resources can also lead to render-blocking and increased resource usage [11]. One way to solve the problem is to push third-party-related loading to Web Workers as illustrated by Partytown [11]. Partytown provides an excellent example of tooling that can help developers leverage the capabilities of modern web browsers with low effort.

2.5 Impact of Server Latency on User Experience

With the server requirement of web applications, you get the latency problem, which has haunted the web since its early days [75]. Server latency is also visible in Google's Web Vitals [83] as a factor for SEO optimizers to consider. The negative impact of high latencies on User eXperience (UX) and conversion are well understood [8,9]. To address the problem, Content Delivery Networks (CDNs) were introduced early on [49]. Since then, the approach has begotten the concept of edge computing [58] as a programmable generalization.

[2] More specifically, Service Worker can act as a proxy between the traffic between the client and the server enabling techniques such as prefetching and even testing [45].

2.6 Caching at Different Levels of Development Stack

There are several ways how to overcome latency. The primary solution provided by edge computing is to push the server close to the user, as then there is less work to do by definition. That said, edge computing has additional needs, such as needing edge databases compatible with the globally distributed approach [16]. Another approach to reducing latency is to decrease the amount of processing on the server side, for example, by caching [63]. Latency can also be alleviated by sending smaller data payloads to the client and deferring the sending of data that is not needed immediately.

Caching can be applied on different levels of a client-server arrangement, as seen in Fig. 1. For example, clients can have a cache of their own, and there can be a server cache [73]. To complicate things further, the server cache can be populated differently. For example, in Distributed Persistent Rendering (DPR) or Incremental Static Regeneration (ISR), the server renders a snapshot of a requested page to the cache served from the cache for the subsequent requests [73]. DPR and ISR have slightly different takes on stale content as ISR can return stale results while DPR cannot [73].

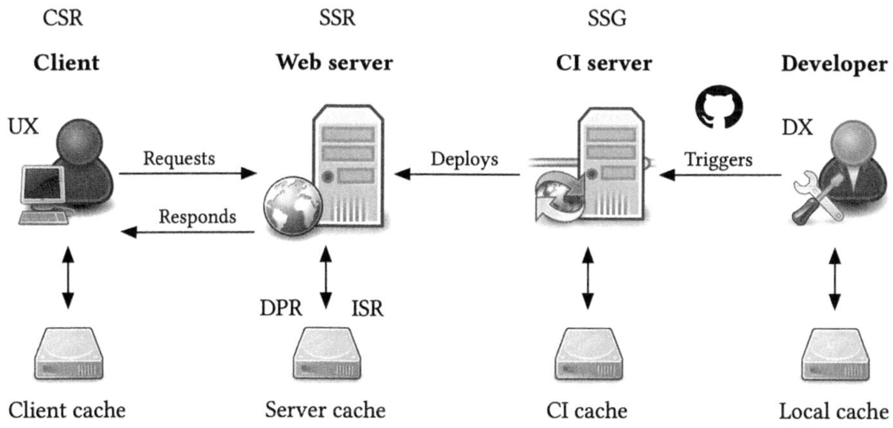

Fig. 1. In the workflow, a developer is working that triggers a build on a CI server, which deploys build artifacts to a web server that then serves a website to a client using a standard request/response flow. Caching is possible at each phase of the flow. In addition, a web server may generate a page per request using SSR or return it from a cache using DPR or ISR. During a CI deployment, SSG may be performed. Optionally, a page may be rendered on the client by leveraging CSR. [73]

To further complicate things, code that runs on a server has to be deployed somehow. A static file server can be used for Static Site Generation (SSG), or a static website could be deployed on CDN infrastructure [73]. Server-Side Rendering (SSR) or Client-Side Rendering (CSR) could be leveraged for more complex cases [73]. In the case of SSR, the web server would generate requests on the fly and then optionally cache them as described above. Alternatively, a service such as Memcached or Varnish could be used between the client and the server [63].

Developers exist at the other extreme as they develop their code, most likely locally, and then push it to a version control system, which triggers a Continuous Integration

(CI) server, which then deploys the built artifacts to a web server. The exciting implication of the flow is that a developer may experience the performance of a system differently locally than in a deployed environment due to differences in hardware and environment. To experience the impact of latency in the production environment, you would have to deploy it to equivalent infrastructure or simulate the conditions locally by, for example, throttling connection speed.

2.7 Overview of Aspects of Web Performance

Given that the web platform has evolved substantially during the past thirty years, there are many opportunities for web developers and web framework authors to improve their work quality, especially regarding performance. Table 2 recaps the directions discussed in the context of this article.

Table 2. The table discusses the main aspects of web application performance. As aspects, we highlight particularly server latency, transport protocol, offloading work, and caching as optimization directions.

Aspect	Description	Related examples
Server latency	Server latency is a well-understood and studied problem, given that it cannot be avoided in traditional web applications.	CDNs [49] and edge computing [58] provide ways to address server latency by moving assets and computation closer to the client. With Local-First Software [37], the issue can be alleviated by moving computation to the client and minimizing server roundtrips
Transport protocol	As the web has evolved, so have the transport protocols underneath it. The biggest change has been the shift to streaming approaches [76].	To leverage streaming fully, [76] experimented with reorganizing pages as streaming-friendly to avoid the problem of render-blocking
Offloading work	Web platform allows performing secondary work within Web Workers [53] and Service Worker [17] by moving it out of the main thread.	Partytown [11] is a good example of a small library pushing third-party related work to Web Workers [11]
Caching	As a general technique, caching can be applied at different levels of a web application stack to reduce the work required.	On the server, Memcached and Varnish are well-known examples [63]. On the client, the Service Worker can act as a caching proxy [17]

3 Review of Web Application Performance Techniques

Building on the main aspects behind web application performance, we next focus on techniques for speeding up applications.

3.1 Progressive Enhancement

Although it was not conceived as a performance-oriented technique but rather as a design to focus on the user, Progressive Enhancement (PE) provides a baseline for building site fundamentals first [26]. In PE, the idea is to start markup (i.e., HTML) first and use the correct semantics, build up styling (i.e., CSS), and only last develop interactivity (i.e., JavaScript) to follow the best practice of graceful degradation of the early web [26]. The crux of the approach is not to rely on the availability of CSS or JavaScript on the user side. New approaches, such as disappearing frameworks [70], build on this idea and often provide a fallback that allows the usage of an application without JavaScript enabled. We argue that although the focus here is not on performance, it is more of a side effect of focusing on the fundamentals first, as the markup is the most valuable deliverable provided to the client while the rest follow.

3.2 HTML-First Development

HTML First manifesto [29] follows this line of thought as it argues that faster and more maintainable web applications can be built by leveraging the default capabilities of modern web browsers, the extreme simplicity of HTML's attribute syntax, and understanding there is power in the default behavior of web's "view source" functionality that allows developers to examine the markup of deployed applications. The manifesto aligns with the idea that modern web browsers are sufficiently advanced to be leveraged even without frameworks or perhaps with frameworks that do less, as shown by the examples of Enhance [22] or Gustwind [72], frameworks that build on web standards while containing a few opinions of their own. The example of W3C's Open UI [52] Community Group shows developers' interest in improving the web platform so that more can be done directly without using third-party solutions for fundamental user interface operations.

3.3 Utility-First Styling

Utility-first styling is a related approach in the sense that it is pushing style declarations to the HTML as styling tokens (e.g., `mx-2`[3]) that refer to associated CSS generated by a compiler [38]. Utility-first styling helps developers avoid traditional challenges of CSS by addressing issues such as specificity, resetting styles, location dependence, multiple inheritance, and nesting, to mention several [38]. Due to moving style declarations to the markup, it can be argued that utility-first styling may bloat HTML to some extent. Conversely, the generated CSS is generally a tiny amount compared to conventional techniques according to our experience with libraries like Tailwind [66] or Twind [68].

3.4 HTML-First State Management

In traditional web applications, so-called Multi-Page Applications (MPAs), the application state was maintained on the server side [62]. Although the approach comes with its

[3] `mx-2` means that the browser should apply a margin of two units on the horizontal axis to the given element. Furthermore, the two units refer to a design scale that maps to the figure passed to the browser, offering a degree of abstraction and flexibility.

benefits in terms of simplicity and usable URLs[4] [1] for example, Single-Page Applications (SPAs) [62] emerged to allow creation of complex web applications. Moving to the SPA model brought a new problem in the form of client-side state management [1], and contemporary web application frameworks each address the issue in their own way.

Instead of solving state management through JavaScript, HTML-first state management approaches the problem from another angle by applying utility-first styling to the state. HTML-first state management solutions ask, what if the state was managed directly within HTML markup instead of separate JavaScript files? Generally, these types of solutions rely on a small runtime that makes state declarations associated with HTML alive, and they may also leverage techniques such as resumability [71]. HTML-first state management is consistent with progressive enhancement, and due to its library-like nature, it can often be integrated within other frameworks, for example, to provide interactivity to an otherwise static page at a low cost.

3.5 Local-First Software: Shifting Responsibilities Between the Client and the Server

As pointed out by [37], most current web applications are server-driven as a large part of the application state lives on the server. In local-first software, the client and server responsibilities are shifted, and the server gains a supporting role instead [37]. The key issues in distributed models like this are data ownership and the responsibility of processing. The interest in decentralization is visible in the advent of new social media platforms, such as Bluesky, that leverage new models of computation and state management [36].

3.6 Partial Loading

Partial loading is a technique where content is loaded later [46] or, for example, on demand. As a more specific example, islands architecture allows developers to define interactive sections of a page and associate loading strategies to them [74]. Through added control, developers can defer or even avoid loading functionality that may not be immediately needed. The concept of resumability [71] can be combined with a code-splitting [42] approach to leverage the idea of partial loading on a granular level that was impossible earlier.

3.7 Performance Tooling

While moving to a newer framework utilizing the latest techniques is an option, it may also be costly, especially for large codebases. Another orthogonal approach is to apply supporting tooling that makes the chosen framework faster. Examples of this direction are Million.js [7] and AUTRATAC/Waiter [77]. Million.js is an optimizing compiler for the popular virtual DOM, a technique utilized by libraries like React.js, and the compiler may lead to loading and rendering performance improvements up to orders of

[4] Uniform Resource Locators (URLs) give a way to identify unique resources available on the web through addresses [1].

magnitude while providing a compatibility layer for React.js [7]. In the approach pioneered by AUTRATAC and Waiter, the idea was to delay code execution by transpiling code to asynchronous code that can be downloaded later, resulting in reduced amount of render-blocking JavaScript as well as reduced amounts of downloaded code [77]. The benefits of AUTRATAC/Waiter were visible in lowered FCP, especially in poor connection speeds [77], and in general, the same applies for techniques oriented at reducing initial payloads as there is less work to do then. Partytown [11] is another notable example as it allows web developers to push third-party dependencies, such as analytics, to be loaded through Web Workers in the background to avoid blocking JavaScript main thread.

3.8 Dead Code Elimination

Another common approach to reducing the amount of JavaScript shipped to the client is Dead Code Elimination (DCE), and the target of this type of tooling is to analyze application code and drop unused parts from it [43]. More specifically, these types of tools are called minifiers [40], and to maintain the debugging benefits of unminified code, typically, source maps are shipped in addition [56]. The risk is that minifiers can break a website by accidentally eliminating code that is needed, and that is a problem AUTRATAC/Waiter addresses by approaching the issue from a different angle [77]. Tree-shaking is a DCE technique enabled by ES2015 modules, and the fact that it is possible to analyze which parts of them are used statically, but the limited adoption of the new standard may restrict the usefulness of this style of DCE [50].

3.9 Compression

One way to reduce the amount of data sent to the client is to compress it before sending and decompress at the recipient [24]. Many compression algorithms exist, where perhaps the most well-known options for web are gzip [24] and Brotli [2], which improves further compared to earlier gzip. According to [13], 97.52% of users have access to Brotli through their browsers, meaning that applying Brotli compression is an easy way for developers to speed up their sites.

New image formats, such as WebP, promise better compression ratios over earlier formats, such as JPEG or PNG [61]. As with Brotli, the browser support for WebP is on a good level as 97.05% of users have access to it [14]. The situation had significantly changed from 2015 when only roughly half of users had access to WebP, and its use cases were, therefore, more limited [61]. Due to the improved situation, WebP is a good option, especially for image-heavy applications. JPEG XL [3] is another contender still being adopted by browsers, as shown by the early adoption rate of 10% [15].

3.10 Overview of Performance Optimizations

Given that the web platform has evolved substantially during the past thirty years, there are many opportunities for web developers and web framework authors to improve their work quality by applying specific performance optimization techniques. Table 3 summarizes the directions discussed in this section.

Table 3. The table discusses web application performance optimization directions for contemporary web developers. It is good to note that the list is likely non-exhaustive as new techniques become available constantly.

Technique	Description	Related examples
Rendering approach	Web applications can be rendered either on the client or server side, and some of these techniques may complement each other [73].	Most well-understood examples of rendering techniques are CSR, SSR, DPR, ISR, and SSG [73]
HTML-first development	HTML-first development captures the sentiment related to using the web platform directly as much as possible. It builds on the earlier ideas of graceful degradation and progressive enhancement.	The best examples are HTML elements that can be used directly or techniques where CSS can be leveraged to provide interactivity without JavaScript
Utility-first styling	Utility-first styling rethinks CSS by moving styling primitives to HTML markup. It may improve performance as a side effect due to smaller and more efficient CSS payloads.	Tailwind is perhaps the most well-known example of a utility-first library
HTML-first state management	HTML-first state management solutions move state management directly to HTML markup and typically ship with a small JavaScript runtime, reducing the amount of code that must be shipped to the client.	Alpine.js [55] and HTMX [30] are good examples with different takes on the topic, as Alpine.js is mainly frontend-focused while HTMX comes with backend opinions
Local-first software	Local-first software re-evaluates how web applications should be built by shifting logic and data to the client and by treating the server in a supporting role [37].	Trellis [35], a Trello clone, was an early example of the local-first approach and is still being explored as a technical option
Partial loading	Partial loading allow developers to defer or even avoid work.	Islands architecture applies the idea and lets developers define islands of interactivity and associated loading strategies [74]. It is possible to leverage partial loading on top of resumability to a great degree [71]. Code-splitting [42] is yet another example of how to apply partial loading on code
Performance tooling	Performance-oriented tooling can provide benefits by integrating the tools into a project and then applying them.	Examples of performance-oriented tooling include AUTRATAC/Waiter [77], Million.js [7], and Partytown [11]
Dead Code Elimination	DCE can be used to eliminate code that the client does not need. Tree-shaking is a commonly used safe technique to achieve DCE.	Minifiers [43] and tree-shaking bundlers [50]
Compression	Given it is well understood that smaller payloads sent to the client result in better performance, compression techniques can be applied to the payloads.	Well-known compression techniques include [24] and Brotli [2]. In addition, code can be minified to reduce its size [40], and new image formats, such as WebP [61] or JPEG XL [3], can be leveraged

4 Impact of Web Frameworks on Performance

Developers often use a web framework for any application with a modest amount of complexity to save development time. The need for frameworks says something about the web platform itself. However, it can be argued that perhaps one day, frameworks won't be as needed as today as the platform covers enough functionality.

4.1 Why Web Frameworks Are Needed

Given that the web platform is powerful yet simultaneously complex, multiple web frameworks exist to capture the complexity and make it more approachable for developers. That is not to say there is no value in using the platform directly, and as the platform matures, there is likely less reason to use frameworks, at least for some use cases. Frameworks and their needs tend to inspire standards, and over time, the platform is catching up in terms of capabilities so that frameworks can do less or can do what they do even better.

4.2 Server-, Client-, and Full-Stack Frameworks

Generally, web frameworks can be split into server and client-oriented ones. Occasionally, a web framework covers both ends, but usually, there is a bias towards one end. So-called meta-frameworks go a step further by providing basic structure yet optionally requiring another framework to complement them. For example, Astro framework [6] has grown popular partially thanks to its meta-approach as it has enabled developers to bring familiar tools to a new environment with its unique benefits. Phoenix framework [54] built using Elixir [21] is a good example of a server-oriented web framework. Django [19] developed with Python and Ruby on Rails [27] are two more examples of a framework starting server-first. Astro framework has a heavier focus on the front end, but technically, it is possible to build full-stack applications using it. Next.js [47], a full-stack framework built around Meta's React [57], is another example that can cover both ends.

Svelte [28], Vue [85], and Angular [4] are examples of client-oriented frameworks. However, full-stack solutions are available for each, allowing the development of complex applications around them. As mentioned earlier, solutions like Astro allow the integration of these frameworks within another meta-framework.

The mentioned frameworks are only a few out of many, and you might ask why so many exist. Generally, frameworks capture several opinions on everyday tasks, such as routing, and exist to make these tasks more accessible for developers to perform. Occasionally, frameworks can help to pinpoint issues in platforms. For example, today's web frameworks can leverage standard features inspired by earlier frameworks and perhaps do less work independently.

4.3 Impact of Web Frameworks on Performance

Given frameworks capture opinions, they can also help establish a baseline for web performance. Historically, SPA frameworks have focused on Developer eXperience (DX), perhaps at the cost of UX [70]. With the advent of disappearing frameworks [70], it seems change is in the air, and techniques, such as resumability [71], enable developers to write code that follows the currently understood best practices and is performant out of the box without the need for further manual optimizations, such as code-splitting where code is separated into multiple chunks that can be loaded for example based on user interaction [42].

Conversely, a framework can work against developers by leading to technical costs that are not always apparent. For example, hydration, a common technique to turn pages interactive, comes with a loading cost that may not always be apparent to developers [71]. Therefore, adopting a framework does not automatically mean an application will be more performant; the opposite may be true.

4.4 Adoption of Performance Optimizations in Web Application Frameworks

To give a fuller view of the topic, Table 4 shows how web application frameworks have adopted performance optimization directions discussed in this article. Refer to Table 3 for brief explanations of the techniques.

5 Discussion

Optimizing web application performance is not an easy task as there are many different ways to approach the problem. Most likely, some Pareto principle [20] applies for many cases where a considerable performance improvement may be gained through a small amount of effort. Given the size of websites these days, addressing the size problem [31] may be an excellent place to start.

5.1 Addressing Size of Websites

Some aspects discussed in this article, particularly web frameworks, development techniques, tooling, and compression, touch on website size. The interesting question is, what is the contribution of each factor to the whole, and how much can be done about it by improving a specific section or by, for example, applying a particular technique?

Perhaps the bigger underlying question is why websites are so large in the first place and what the main reasons behind their growth are. Even with enhanced approaches, could the growth be at least slowed down, or is there something more fundamental going on driving the development?

5.2 Towards Lower Latency and Distributed Models

A central underlying theme we noticed was the shift in the server space. The problem of server latency is well understood, so edge computing has gained interest in the web space during the past few years as commercial offerings keep maturing. Simultaneously, there has been interest in offloading work to the client and developing distributed models for web applications, as shown by local-first software [37]. The more prominent underlying theme in this rethinking has to do with aspects of privacy and data ownership.

Table 4. The table combines web performance optimization techniques discussed in this article with web application frameworks to show their relationship. Notably, techniques such as applying different rendering techniques, partial loading, or applying performance tooling, DCE, and compression are common among frameworks.

Technique	Relationship with web application frameworks
Rendering approach	Each web application framework is forced to implement some rendering approach by definition. Pure client-side frameworks implement CSR, while full-stack and server-side frameworks can leverage SSR, DPR, ISR, and SSG
HTML-first development	HTML-first development approach has been adopted specifically by Enhance [22] and Gustwind [72]
Utility-first styling	Utility-first styling is a technique that may be used by a framework through external libraries and tooling. Typically styling approach a framework is up to the developer to decide
HTML-first state management	HTML-first state management are similar to styling in that, so far, frameworks have not adopted the idea. Instead, it can be implemented on top of them. A notable exception of deeper integration is Laravel Livewire, which includes first-class support for Alpine.js [41]
Local-first software	So far, no known example of a local-first framework exists as the approach is meant to be used on top. That is not to say there might not be value in a framework focusing solely on the approach as Astro did with islands architecture [74] earlier
Partial loading	Partial loading has been adopted by frameworks, such as Astro [74], Qwik through resumability [71], and numerous other frameworks where it is possible to use code-splitting. Notably, in the latter case, significant developer effort may be required to gain major benefits
Performance tooling	Performance tooling exist separately from web application frameworks; typically, it is not shipped with them
Dead Code Elimination	DCE is commonly used in frameworks that leverage a bundler in their build process. A good example is the integration of minification to Next.js in production mode [25]
Compression	Compression is typically exposed to the developer for full-stack or server frameworks. For example, in Next.js, compression is available through configuration [48]

5.3 Evolutionary Pressure of Web Frameworks

To develop complex web applications, developers often resort to web frameworks that capture opinions, which avoids effort and hopefully enables best practices. The problem is that best practices tend to change over time, and the question is whether mainstream frameworks keep up with the change. Another evolutionary pressure comes from the standardization side as standards keep capturing critical ideas to the web platform itself, meaning there is less work for frameworks to do. A good example is the introduction of the ES2015 version of JavaScript language, which is now available in web browsers without external tooling. Also, the introduction of standards like Web Workers has opened new venues for technical development and rethinking best practices as new solutions, and even paradigms, can be developed on top of them.

5.4 The Way Performance Is Measured Shapes Performance

Undoubtedly, what you measure is what you get, particularly when the measurements are used as metrics for decisions such as page rankings. Therefore, it has been interesting to notice that, for example, Google Web Vitals still keeps evolving, as shown by the introduction of INP [78] as a web vital in favor of FID. Since web frameworks follow performance metrics closely to remain competitive, improving how we measure performance is likely good for the web platform overall.

6 Conclusion

In this article, we sought to find answers to the question *What are the currently available ways for improving web application performance?* and found surprisingly many ways. At the core of it all is the aspect of measuring web performance, and it turns out Google's Core Web Vitals provide a good approximation of that. With the advent of edge computing, some focus has shifted to minimizing server latency. With the introduction of local-first software, there is interest in distributed models that move computation and data to the client, leaving servers more to a supporting role.

Web frameworks allow codifying best practices, but the concern is that although they save developer effort this way, they might also remain on the conservative side, at least when there are many developers to support, meaning it is up to the bleeding edge frameworks to innovate and try new techniques to gain market share. A more specific study into the way ideas between mainstream and bleeding edge frameworks transfer would be worthwhile as that would give more insight within the space to understand how frameworks adopt new standards and techniques that become available. Another direction of inquiry would be to consider how standardization can capture innovation happening at frameworks and how this affects the whole ecosystem.

In the scope of this article, we did not consider which factors make the most sense for individual developers to apply in their daily work. There are also questions related to which techniques web framework authors should adopt to disseminate current best practices to their userbases. It is a good question about which practices should be encoded into the frameworks and which should be applied by the users on top of them. Another related issue has to do with the web platform and its adoption of worthwhile techniques to raise the baseline for frameworks.

References

1. Akse, J.A.: Applying client-side state management on server-generated web pages. Master's thesis, University of Oslo (2016)
2. Alakuijala, J., Kliuchnikov, E., Szabadka, Z., Vandevenne, L.: Comparison of brotli, deflate, zopfli, lzma, lzham and bzip2 compression algorithms. In: Google Inc., pp. 1–6 (2015)
3. Alakuijala, J., et al.: Jpeg xl next-generation image compression architecture and coding tools. In: Applications of Digital Image Processing XLII, vol. 11137, pp. 112–124. SPIE (2019)
4. Angular—angular.io (2024). https://angular.io/. Accessed 18 Mar 2024
5. Aqeel, W., Chandrasekaran, B., Feldmann, A., Maggs, B.M.: On landing and internal web pages: the strange case of jekyll and hyde in web performance measurement. In: Proceedings of the ACM Internet Measurement Conference, pp. 680–695 (2020)
6. Astro (2024). https://astro.build/. Accessed 18 Mar 2024
7. Bai, A.: Million. js: a fast compiler-augmented virtual dom for the web. In: Proceedings of the 38th ACM/SIGAPP Symposium on Applied Computing, pp. 1813–1820 (2023)
8. Bai, X., Arapakis, I., Cambazoglu, B.B., Freire, A.: Understanding and leveraging the impact of response latency on user behaviour in web search. ACM Trans. Inf. Syst. (TOIS) **36**(2), 1–42 (2017)
9. Basalla, M., Schneider, J., Luksik, M., Jaakonmäki, R., Vom Brocke, J.: On latency of e-commerce platforms. J. Organ. Comput. Electron. Commer. **31**(1), 1–17 (2021)
10. Berners-Lee, T., Cailliau, R., Groff, J.F., Pollermann, B.: World-wide web: the information universe. Internet Res. (1992)
11. Introduction - Partytown—partytown.builder.io (2024). https://partytown.builder.io/. Accessed 13 Feb 2024
12. Can I use... Support tables for HTML5, CSS3, etc—caniuse.com. https://caniuse.com/?search=web%20workers. Accessed 13 Feb 2024
13. Can I use... Support tables for HTML5, CSS3, etc—caniuse.com (2024). https://caniuse.com/?search=brotli. Accessed 12 Feb 2024
14. Can I use... Support tables for HTML5, CSS3, etc—caniuse.com (2024). https://caniuse.com/?search=webp. Accessed 12 Feb 2024
15. Can I use... Support tables for HTML5, CSS3, etc—caniuse.com (2024). https://caniuse.com/?search=jpeg%20xl. Accessed 16 Feb 2024
16. Chen, X., et al.: Achieving low tail-latency and high scalability for serializable transactions in edge computing. In: Proceedings of the Sixteenth European Conference on Computer Systems, pp. 210–227 (2021)
17. Chinprutthiwong, P., Vardhan, R., Yang, G., Zhang, Y., Gu, G.: The service worker hiding in your browser: the next web attack target? In: Proceedings of the 24th International Symposium on Research in Attacks, Intrusions and Defenses, pp. 312–323 (2021)
18. Dedehayir, O., Steinert, M.: The hype cycle model: a review and future directions. Technol. Forecast. Soc. Chang. **108**, 28–41 (2016)
19. Django (2024). https://www.djangoproject.com/. Accessed 18 Mar 2024
20. Dunford, R., Su, Q., Tamang, E.: The pareto principle (2014)
21. The Elixir programming language (2024). https://elixir-lang.org/. Accessed 18 Mar 2024
22. Enhance: Enhance (2024). https://enhance.dev/. Accessed 18 Mar 2024
23. Eslamdoust, A.: An overview of search engine optimization techniques. Future Gener. Commun. Internet Things **1**(4), 52–57 (2022)
24. Ferragina, P., Manzini, G.: On compressing the textual web. In: Proceedings of the Third ACM International Conference on Web Search and Data Mining, pp. 391–400 (2010)

25. How to minify a next js app · Issue #3632 (2018). https://github.com/vercel/next.js/issues/3632. Accessed 18 Mar 2024
26. Gustafson, A.: Adaptive web design: crafting rich experiences with progressive enhancement. New Riders (2015)
27. Hansson, D.H.: Ruby on Rails (2024). https://rubyonrails.org/. Accessed 18 Mar 2024
28. Harris, R.: Svelte (2024). https://svelte.dev/. Accessed 18 Mar 2024
29. HTML First (2024). https://html-first.com/. Accessed 12 Feb 2024
30. htmx - high power tools for html (2024). https://htmx.org/. Accessed 18 Mar 2024
31. Page Weight | 2022 (2022). https://almanac.httparchive.org/en/2022/page-weight. Accessed 28 Aug 2024
32. Ikram, M., Masood, R., Tyson, G., Kaafar, M.A., Loizon, N., Ensafi, R.: The chain of implicit trust: an analysis of the web third-party resources loading. In: The World Wide Web Conference, pp. 2851–2857 (2019)
33. Johannes Bahrke, T.R.: Press corner (2023). https://ec.europa.eu/commission/presscorner/detail/en/ip_23_3718. Accessed 25 Jan 2024
34. Killelea, P.: Web Performance Tuning: speeding up the web. O'Reilly Media, Inc., Newton (2002)
35. Kleppmann, M.: GitHub - automerge/trellis (2024). https://github.com/automerge/trellis. Accessed 18 Mar 2024
36. Kleppmann, M., et al.: Bluesky and the at protocol: usable decentralized social media. arXiv preprint arXiv:2402.03239 (2024)
37. Kleppmann, M., Wiggins, A., Van Hardenberg, P., McGranaghan, M.: Local-first software: you own your data, in spite of the cloud. In: Proceedings of the 2019 ACM SIGPLAN International Symposium on New Ideas, New Paradigms, and Reflections on Programming and Software, pp. 154–178 (2019)
38. Klimm, M.C.: Design systems for micro frontends-an investigation into the development of framework-agnostic design systems using svelte and tailwind Css. Bachelor's thesis, Hochschulbibliothek der Technischen Hochschule Köln (2021)
39. Korpal, G., Scott, D.: Decentralization and web3 technologies. Authorea Preprints (2023)
40. Król, K.: Comparative analysis of selected online tools for javascript code minification: a case study of a map application. Geomat. Landmanag. Landsc. (2), 119–129 (2020)
41. Alpine | Laravel — livewire.laravel.com (2024). https://livewire.laravel.com/docs/alpine. Accessed 18 Mar 2024
42. Livshits, B., Kiciman, E.: Doloto: code splitting for network-bound web 2.0 applications. In: Proceedings of the 16th ACM SIGSOFT International Symposium on Foundations of Software Engineering, pp. 350–360 (2008)
43. Malavolta, I., et al.: Javascript dead code identification, elimination, and empirical assessment. IEEE Trans. Softw. Eng. (2023)
44. Morris, J.J., Alam, P.: Analysis of the dot-com bubble of the 1990s. SSRN 1152412 (2008)
45. Mock Service Worker (2024). https://mswjs.io/. Accessed 13 Feb 2024
46. Netravali, R., Goyal, A., Mickens, J., Balakrishnan, H.: Polaris: faster page loads using fine-grained dependency tracking. In: 13th USENIX Symposium on Networked Systems Design and Implementation (NSDI 16) (2016)
47. Next.js by Vercel (2024). https://nextjs.org/. Accessed 18 Mar 2024
48. next.config.js Options: compress (2024). https://nextjs.org/docs/app/api-reference/next-config-js/compress. Accessed 18 Mar 2024
49. Nygren, E., Sitaraman, R.K., Sun, J.: The akamai network: a platform for high-performance internet applications. ACM SIGOPS Oper. Syst. Rev. **44**(3), 2–19 (2010)
50. Obbink, N.G., Malavolta, I., Scoccia, G.L., Lago, P.: An extensible approach for taming the challenges of javascript dead code elimination. In: 2018 IEEE 25th International Conference on Software Analysis, Evolution and Reengineering (SANER), pp. 291–401. IEEE (2018)

51. Oluwatosin, H.S.: Client-server model. IOSR J. Comput. Eng. **16**(1), 67–71 (2014)
52. Open UI (2024). https://open-ui.org/. Accessed 18 Mar 2024
53. Pan, Y., White, J., Sun, Y., Gray, J.: Gray computing: an analysis of computing with background javascript tasks. In: 2015 IEEE/ACM 37th IEEE International Conference on Software Engineering, vol. 1, pp. 167–177. IEEE (2015)
54. Phoenix Framework (2024). https://www.phoenixframework.org/. Accessed 18 Mar 2024
55. Porzio, C.: Alpine.js (2024). https://alpinejs.dev/. Accessed 18 Mar 2024
56. Rack, J., Staicu, C.A.: Jack-in-the-box: an empirical study of javascript bundling on the web and its security implications. In: Proceedings of the 2023 ACM SIGSAC Conference on Computer and Communications Security, pp. 3198–3212 (2023)
57. React (2024). https://react.dev/. Accessed 18 Mar 2024
58. Satyanarayanan, M.: The emergence of edge computing. Computer **50**(1), 30–39 (2017)
59. Scoccia, G.L., Autili, M.: Web frameworks for desktop apps: an exploratory study. In: Proceedings of the 14th ACM/IEEE International Symposium on Empirical Software Engineering and Measurement (ESEM), pp. 1–6 (2020)
60. Sellamuthu, K., Ranjithkumar, S., Kavitha, K., Gowtham, S.: On page seo techniques for better ranking in search engines. In: 2022 8th International Conference on Smart Structures and Systems (ICSSS), pp. 01–06. IEEE (2022)
61. Si, Z., Shen, K.: Research on the WebP image format. In: Xu, M., Yang, L., Ouyang, Y., Ouyang, Y. (eds.) Advanced Graphic Communications, Packaging Technology and Materials. LNEE, vol. 369, pp. 271–277. Springer, Singapore (2016). https://doi.org/10.1007/978-981-10-0072-0_35
62. Solovei, V., Olshevska, O., Bortsova, Y.: The difference between developing single page application and traditional web application based on mechatronics robot laboratory onaft application. Autom. Technol. Bus. Processes **10**(1) (2018)
63. Song, X., Shi, J., Chen, H., Zang, B.: Revisiting software {Zero-Copy} for web-caching applications with twin memory allocation. In: 2012 USENIX Annual Technical Conference (USENIX ATC 12), pp. 355–360 (2012)
64. Souders, S.: High-performance web sites. Commun. ACM **51**(12), 36–41 (2008)
65. Internet and social media users in the world 2023 (2023). https://www.statista.com/statistics/617136/digital-population-worldwide/. Accessed 28 Aug 2023
66. Tailwind CSS (2024). https://tailwindcss.com/. Accessed 18 May 2024
67. Taivalsaari, A., Mikkonen, T.: The web as an application platform: the saga continues. In: 2011 37th EUROMICRO Conference on Software Engineering and Advanced Applications, pp. 170–174. IEEE (2011)
68. Tandel, S.: Twind.style (2024). https://twind.style/. Accessed 18 May 2024
69. Tandel, S., Jamadar, A.: Impact of progressive web apps on web app development. Int. J. Innov. Res. Sci. Eng. Technol. **7**(9), 9439–9444 (2018)
70. Vepsäläinen, J., Hellas, A., Vuorimaa, P.: The rise of disappearing frameworks in web development. In: International Conference on Web Engineering, pp. 319–326. Springer, Heidelberg (2023). https://doi.org/10.1007/978-3-031-34444-2_23
71. Vepsäläinen, J., Hevery, M., Vuorimaa, P.: Resumability-a new primitive for developing web applications. IEEE Access (2024)
72. Vepsäläinen, J.: Gustwind (2024). https://gustwind.js.org/. Accessed 18 Mar z2024
73. Vepsäläinen, J., Hellas, A., Vuorimaa, P.: Implications of edge computing for static site generation. In: WEBIST 2023: 19th International Conference on Web Information Systems and Technologies, pp. 223–231. Springer (2023)
74. Vepsäläinen, J., Hellas, A., Vuorimaa, P.: The state of disappearing frameworks in 2023. In: WEBIST 2023: 19th International Conference on Web Information Systems and Technologies. Springer, Heidelberg (2023)

75. Viles, C.L., French, J.C.: Availability and latency of world wide web information servers. Comput. Syst. **8**(1), 61–91 (1995)
76. Vogel, L., Springer, T.: How streaming can improve the world (wide web). In: Companion Proceedings of the ACM Web Conference 2023, pp. 140–143 (2023)
77. Vogel, L., Springer, T.: Waiter and autratac: don't throw it away, just delay! In: Garrigós, I., Murillo Rodríguez, J.M., Wimmer, M. (eds.) Web Engineering, pp. 278–292. Springer, Cham (2023)
78. Wagner, J.: Interaction to Next Paint (INP) (2023). https://web.dev/articles/inp. Accessed 13 Feb 2024
79. Wagner, J., Pollard, B.: Time to First Byte (TTFB) (2021). https://web.dev/articles/ttfb. Accessed 13 Feb 2024
80. Wagner, J., Viscomi, R.: Interaction to Next Paint becomes a Core Web Vital on March 12 (2024). https://web.dev/blog/inp-cwv-march-12. Accessed 12 Feb 2024
81. Walton, P.: First Contentful Paint (FCP) (2023). https://web.dev/articles/fcp. Accessed 13 Feb 2024
82. Walton, P.: Time to Interactive (TTI) (2023). https://web.dev/articles/tti. Accessed 28 Feb 2024
83. Walton, P.: Web Vitals (2024). https://web.dev/articles/vitals. Accessed 12 Feb 2024
84. Wu, W.: React Native vs Flutter, Cross-platforms mobile application frameworks. Master's thesis, Metropolia Ammattikorkeakoulu (2018)
85. You, E.: Vue.js—vuejs.org (2024). https://vuejs.org/. Accessed 18 Mar 2024

Interest and Challenges of Students Regarding Web and Mobile Technologies

Manuela Andreea Petrescu(✉), Adrian Sterca, and Ioan Bădărînză

Department of Computer Science, Babes-Bolyai University, Str. M. Kogalniceanu, Cluj-Napoca, Romania
{manuela.petrescu,adrian.sterca,ioan.badarinza}@ubbcluj.ro

Abstract. This study investigates the career aspirations and concerns of second-year computer science students interested in web and mobile development. A survey questionnaire was administered to a randomly selected group of students enrolled in a web development course. The findings reveal that financial aspects are a major concern for students, with job availability and salary expectations heavily influencing their career choices. This financial focus represents a change compared to previous years, possibly linked to recent economic trends. Although technical challenges such as keeping up with evolving technologies remain a concern, interestingly, the reported prevalence of psychological challenges have not changed. The study also explores the perceived value of a bachelor's degree in computer science. Students acknowledge the benefits of structured learning offered by universities, but believe that online resources are sufficient for entry-level positions. Overall, the findings suggest a growing emphasis on financial considerations and a potential decline in the perceived necessity of a traditional university education in the minds of aspiring web and mobile developers.

Keywords: Learning · Educational study · Web · Mobile · Students · Interest · Challenge · Financial

1 Introduction

Software development for the world wide web space and for mobile devices is crucial to modern business as it allows organizations to communicate with their clients via digital channels. The population of the world is currently 7.91 billion people [8], and with a 1% annual growth rate, it will reach 8 billion in 2024 and it is predicted to reach 9 billion by mid-2025. The same report states that 57.0%, or more than half, of the world's population lives in metropolitan areas. The technology also spread to the most isolated and difficult-to-reach places, including rural regions. Due to this, a growing number of people are using the Internet, including those living in developing countries. According to [8], in the first months of 2022, there were 4.95 billion Internet users worldwide, which is equivalent to 62.5% of the global population. Considering that the data shows 4.0% increase, or 192 million new users, in internet penetration over the last year, it is crucial to take note of the strong trend. A considerable rise in mobile phone usage has been observed worldwide. Some of the new users connect to

M. Marchiori and F. García Peñalvo (Eds.): WEBIST 2023, LNBIP 543, pp. 63–76, 2025.
https://doi.org/10.1007/978-3-031-89621-7_4

the Internet using specialized devices such as laptops, computers, tablets, and mobile phones. Currently, 67.1% people on the planet use mobile phones, making up more than two-thirds of the worldwide population. Due to the increasing affordability and performance of technology, there is a growing trend in the adoption of mobile phones. The total number of mobile users has increased by 1.8% in the last year; in fact, 95 million more people have signed up than during the same period in the previous year [8].

Due to their use of infrastructure and the large number of potential customers, web and mobile development are crucial elements of contemporary business, allowing organizations to engage with their clients through digital channels. With the majority of people utilizing smartphones and other digital devices in today's world, having a web and mobile presence is crucial for reaching a wider audience. The need to design and develop websites for the Internet is growing, according to market trends. There is a growing trend in the development of mobile applications. Promotion of goods, services, and brands online is greatly aided by the development of the web and mobile, helping companies reach a wider audience by becoming more visible online. With the help of cutting-edge technology, such as artificial intelligence (AI) and machine learning, web and mobile applications can provide consumers with a personalized experience that is tailored to their individual requirements and tastes. Consequently, businesses can use targeted advertising efforts to increase sales by focusing on particular clients. As a result, there are more loyal customers and higher levels of customer satisfaction, which eventually translate into higher sales and profits.

Web and mobile development are essential elements of a modern business, according to market trends, as more and more potential customers are located online, particularly from mobile devices. In 2020, mobile devices represented 55% of website traffic, surpassing the usage of other devices [16]. The amount of data transferred over the Internet increased from approximately 50 exabytes to more than 300 exabytes in 2022, according to the Statista global Internet traffic figures [16]. The use of various applications, which offer a variety of advantages, such as individualized customer experiences, insightful data, and a competitive edge, was the primary source of traffic. Software developers with specific skills will be in high demand as companies continue to invest in creating online web and mobile applications.

With this configuration in mind, our objectives were to ascertain (1) whether second-year computer science students at universities are interested in pursuing careers in web or mobile development, and if so, which technologies they would prefer to work with; (2) what obstacles they face; and (3) why do they decide to work in these fields.

The structure of this paper is as follows: Sect. 2 examines recent studies on related topics. The methodology of the experiment, including the list of participants and methods for collecting and interpreting the data, is thoroughly described in Sect. 3. Section 4 describes the data collection and analysis of the research questions. Sections 5, 6, and 7 present our findings after analyzing the collected responses to our research questions. In Sect. 5, the primary obstacles students face when pursuing a profession in web or mobile development are discussed. The interests of students in such a career are examined in Sect. 6, and their opinions concerning the significance of a bachelor's degree in this field are discussed in Sect. 7. Section 8 presents some ideas about our findings while interpreting the results of the surveys. Possible threats to validity are discussed in Sect. 9. Finally, Sect. 10 presents the concluding remarks.

2 Literature Review

There is a great need for qualified developers in the web and mobile application development business, which is expanding daily. During their academic years, teaching students how to create web applications is a crucial component. Professors have made several attempts to organize the curriculum so that students are aware of the range of Web development environments and technologies. For example, the authors of [7] describe a cooperative and centralized method of teaching Web development. They attempted to create a centralized web development learning process after observing that students struggle to comprehend and troubleshoot the environments utilized for online applications. Following this modification, they observed that students' comprehension of the online settings improved. To better prepare students for the next step of their career as junior software developers in a company, several studies propose requiring team projects as part of university courses [4,6].

However, the rapidly evolving nature of technology remains one of the main obstacles for computer science students learning to develop a web application [11]. Recent research indicates that graduates are well prepared for the labor market despite the difficulties they face when learning new technologies. After being employed, they still have to go through an initial training procedure, although this is primarily to know the corporate culture and how its staff operate [9].

Prior research has demonstrated that while recruiting new personnel, the IT business is looking for a specific set of competencies. The ability to solve problems is the most crucial of these abilities [13]. The other crucial ones are cooperation and communication [1,3].

3 Methodology and Setup

We conducted an online survey to obtain answers to the research questions we posed in the introduction of this paper. Second-year computer science students enrolled in the Web Programming course made up the set of participants. The primary goal of the web programming course is to educate students on how to use front-end and back-end technologies to create responsive web applications (for desktop layouts or mobile). Its curriculum is long and covers more than a dozen languages.

The course covers HTML and XML languages for the web (such as XHTML and XSLT), CSS, Javascript, Javascript libraries like jQuery, and some Javascript frameworks like Angular on the front end. On the back end, it offers a foundational understanding of how to create an application back end using PHP, Java, and .Net. To have some experience with all of these technologies and completely determine their interests and career path, students must complete a great amount of homework. The course also covers popular web application architectures, such as the REST API architecture, and prominent design patterns, such as the Model-View-Controller. The course also covers web application security, which is a very difficult topic these days. Thus, there are many technologies covered in the Web Programming course. Given how quickly web technologies are evolving, this course is also among the most dynamic.

The survey research method used is described in the ACM Sigsoft Empirical Standards for Software Engineering Research [14]. To give students the opportunity to express themselves and gain a better understanding of their interests, we also asked open questions. To verify certain aspects of their declared interests and measure them, we also used closed-ended questions. We carefully considered the sequence in which the survey's questions were asked.

The students were requested to complete the survey at their own discretion, during or after the laboratory, and we sent them the URL to do so. Approximately 47% of the responses were completed after the lab was over. We made note of the survey's voluntary nature and anonymity when we provided the URL and explained why and how we will use the collected data.

3.1 Participants

Second-year college students, from two lines of study, enrolled in the Web Programming course made up the intended participant group for this survey; the students were randomized to work in laboratories with two of the study's authors. The two processes that made up the allocation procedure were as follows: in the first phase, students were placed into groups according to a list ordered alphabetically by their last name. In the second step, the groups were randomly assigned to the authors of this study by third parties (when the schedule was created).

There were 200 students in total who were informed about the survey, received the survey link, and were asked to participate. With 109 responses, we thought that was a fair amount that would allow us to get valid results: results that accurately represent their opinions, thoughts, and interests.

4 Data Collection

We used quantitative methodologies for this study, more precisely questionnaire surveys as specified by community standards [14]. We chose to use Google Forms to collect data because it is friendly and easy to use and most of the students have already interacted with it in other surveys. The students were aware of the anonymity provided by it, so they could freely express their opinions, an essential element to obtain relevant data. The survey remained open for a two-week period.

The questions were written in English and Romanian, so the students could choose to respond in the language they were most comfortable with. The responses written in Romanian were translated into English with automatic translation applications and the result was verified by one of the authors to check the meaning and context of the language. As the questions were optional to improve the response rate, not all students responded to all the questions. So when we analyzed the responses to a question, we mentioned the percentage of students who did not answer that particular question.

For text interpretation, we applied thematic analysis, as described by [2]. Previous computer science studies have used this study survey method, which involved thematic analysis of acquired data for interpretation [10, 15]. The following actions and steps were taken during the data analysis process according to the standard:

- collect data
- perform a brief analysis of the open answers. The purpose is to reallocate them to other survey questions (if they were better fitted to other questions)
- determine the key items and group them into classes
- an author verifies the classification of the key items, makes observations
- the observations are discussed, and analyzed by all authors, and the outcomes are used for the final results.

There were questions for which we received no response, and for some questions, the open answers comprised multiple keywords. Due to this aspect, we had answers that contained more than one key item, so the total number of key items is larger than the number of answers. The results are shown as a proportion between the number of key items and the total number of responses to provide a clear picture of the prevalence of important key items. As a result, the total percentages of key items will be greater than 100%. Table 1 displays the survey questions, with the first two questions being used to categorize the group of participants.

Table 1. Survey Questions.

C1. What is your line of study?
C2. How do you define yourself (men/women/other)
Q1. Would you like to pursue a career in web/mobile development? (Yes/No)
Q2. Please, mention the reasons for having a career in web/mobile development.
Q3. Please, mention the reasons for NOT having a career in web/mobile development.
Q4. Which are in your opinion the main challenges for a web/mobile developer?
Q6. Do you think you need a university degree to work as a web/mobile developer?
Q7. Specify your reasons for the previous question.

5 Students' Challenges in a Career in Web/Mobile Development

We analyzed the student's challenges and grouped them by specificity. By far, the most mentioned challenges are related to technologies: keeping up-to-date, complicated technologies, fixing bugs, or application design. A second important group is related to time management, perceived stress, work-life balance, motivation, and self-learning. Time constraints and stress have an impact on students; some of them even mention challenges related to emotional health: emotional issues, lack of social interactions, impostor syndrome, or burnout. Soft skills such as customer communication challenges, customer satisfaction, or team integration challenges suggest that students already have some work experience in software companies. The same idea is enforced by the fact

Table 2. Challenges reported by students for working in WEB/mobile.

Tech related	keep up to date 35.78%, complicated technologies 6.42%, fixing bugs 8.26%, application design 7.34%, to have new/efficient ideas 8.25%
Personal	time management 10.09%, work-life balance 4.59%, stress 4.59%, challenges motivation 3.67%, self learning 2.75%
Work challenges	customer communication 8,26%, customer satisfaction 4.59%, team integration, motivation 2.75%, cooperation skill 1.83%
Financial	finding a job 8,26%, too much competition 6.42%, AI competition 2.75%
Psychological	emotional issues, lack of social interaction, impostor syndrome, burnout (appeared in 8.26% of the answers)

that many answers reflect an interest in financial aspects: too much competition, hard to find a job, or AI competition. The percentages of the main challenges are detailed in Table 2.

By far, the challenge most mentioned in the student responses was the difficulty of keeping up with new technologies, having a small increase of 3% compared to the percentages obtained in the previous year [12]. Correlated with this aspect, *complicated technologies* are mentioned by 6.42% of students, and *fixing bugs and errors* by 8.26% of students, an increase of approximately 2% from the previous year [12]. Student responses sometimes mentioned one challenge, sometimes more than one challenge, according to their previous experience: *"to keep up with new emerging technologies"*, *"To satisfy clients and stay at the forefront of tech"*, *"Being up to date with technologies and mental health"*.

Personal challenges such as time constraints doubled compared to the last year: 10.09 in 2024 compared to 4.92 [12], having an impact on work-life balance, burnout, and perceived stress: *"Time and task management, cooperation skills"*, *"the stress"*, *"Handling the pressures of the job and balancing it with a healthy personal lifestyle and family life"*, *"From what I see around me, lack of motivation, having too much to stress about and feeling overwhelmed"*. Students in computer science seem to complain about the job characteristics that require soft skills and involve communication with the client, understanding their needs and ideas, raising up to their expectations, and also integration and communication in the development team: *"deal with customers that are not clear about their demands"*, *"Clarifying client requirements and proper design of the software"*, *"Satisfy the customer"*, *"I think the main challenge of a software developer is to meet the expectations of the client"* and team work related: *"teamwork, many do not know how to socialize and work in a team"*, *"In addition to technical challenges, I think a lot of developers lack social skills and do not have the ability to communicate in a concise way"*.

Financial aspects related to financial security, such as having a job, market competition, and AI competition are also present in the student's answers. They mentioned the difficulty of finding a job, not only related to web or mobile development, but also, in general, as a programmer in the IT industry, the tough competition due to an influx of computer science students and AI competition: *"Finding a job, adapting to working conditions that are periodically affected by industry movements", "having to constantly improve and juggle with positions/opportunities/salaries", "The 'flood' of juniors caused by the hype that "IT pays well", "the wider spread of AI in the field"*.

Due to these challenges and probably other not mentioned reasons, psychological issues appear in their answers: *"Impostor syndrome, burnout, the lack of social interactions", "Impostor syndrome and seeing others doing good and getting discouraged"*. Too much information, unstructured, and lack of clear documentation also pose a problem to some students: *"Tutorial hell, the large size of information that takes a lot of time to fully process and understand, which to me seems really energy draining and easy to give up on"*. As expected, we got empty answers, about 5.50% to this question, some students acknowledging the fact that they cannot provide a relevant answer: *"I don't know, I have not worked in the field"*.

As conclusion, compared to the previous year, the main challenges remain the same, even if the percentages are sightly different. The main challenge continues to be keeping up with the rapidly evolving technologies, personal challenges related to time management, stress remains in the top, and developing soft skills is crucial for students (even if it is for teamwork, communication, or client comprehension). The reported psychological related challenges, such as burnout, emotional issues, and lack of social interactions did not evolve in percentages over these years, being around 8.20%. However, the students in the current participant set are more aware of the changes in the market dynamics and are more preoccupied about job opportunities and competition.

6 Students' Interest in a Career in Web/Mobile Development

In order to determine the students' interest in pursuing a career in web or mobile development, we examined the responses to three questions: the first question asked about their declared intention, the second question asked about their reasons to have a career in web/mobile development, and the last question was related to the reasons for not having a career path in web/mobile development.

The percentages of students who stated that they would like to pursue a career path in web/mobile development are similar to the previous year: 59.63% compared to 60.7% [12]. A group of 34.86% students were undecided and 5.50% of them stated that they would not like such a career. The responses reflect an overall increase of the interest in having a career in web/mobile development.

Students' Reasons for Choosing Web or Mobile Development. After we analyzed the answers, we grouped the reasons into four categories: reasons related to financial stability, to domain specificity (interesting, creative domain), to their interest for a specific part (back end, user interface - UI, front end), and to personal reasons: easy to understand, flexible, fun. The three most mentioned categories are visualized in Fig. 1.

The most common motivation to select a profession in online and mobile development is financial in nature. The expanding field and the growing industry are the main motivating factor, followed by the likelihood of finding well-paying employment. Students appear to have a good understanding of market trends, as evidenced by their

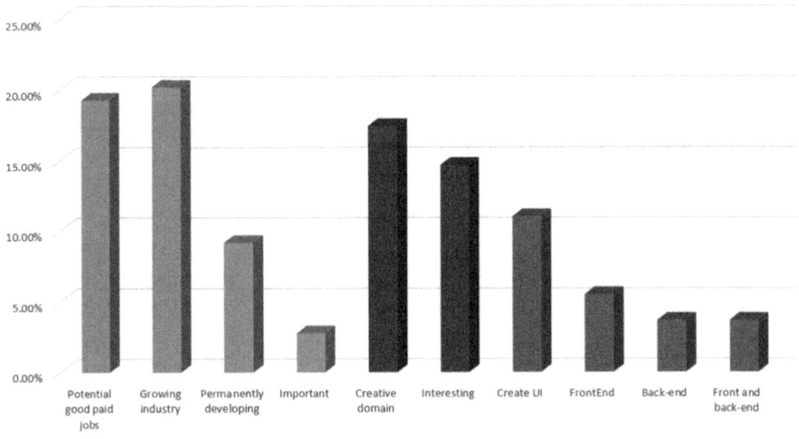

Fig. 1. Reported reasons for having a career in web/mobile development, grouped by categories.

responses, suggesting a positive correlation between market demand and their intention to work in a particular profession. Some answers are more detailed than others, but reflect the same concern related to job stability and job demand on the market; the answers point out the same topic: *"The constantly evolving technology, the wide market, and the large selection of potential jobs"*, *"any company or legal entity needs a website"*, *"The high demand on the market"*, *"It is a very accessible field. There are always pretty well paid job offers"*. The shortest answer received for this question was *"$"*.

Creative domain and *Interesting* scored next for their generic reasons, 17.43% respectively 14.68%, *"it offers creativity"*, *"seems like an interesting domain"*, *"Money and passion"*. Other reasons were related to their personal opinions and personal skills sets, some like to *"Create user interfaces"*, others prefer to work on the back end, other students considered it easy to understand, flexible and fun: *"Easy to understand and love design"*, *"I like to work on web sites and implement the back-end part"*. Approximately 5.50% of the students answered that they prefer to work in other technologies.

Students' Reasons for NOT Choosing Web or Mobile Development. Compared to the previous question, many students 17,36% decided not to answer this question, either leaving the field empty or selecting a character such as "-" or "*". Other rather large group 9.17%, specified that they see no reasons for working in web/mobile related field: *"None"*, *"There are no reasons"*, others were more elusive and mentioned *"depends on conditions"*.

Other relevant group, around 7.34% students, mentioned that they are interested in something else: AI, devOps, or simply: *"I also like cyber security"*, *"I would also be interested in other areas of software development (backend)"*. Rapidly changing technologies were not only present in responses related to challenges encountered in development, but also as main reasons for not having a web/mobile related career, around 7,34% mentioning it: *"It may become annoying since the web frameworks are changing pretty fast"*. Other responses are related to the market competition 5.50%: *"There is a lot of competition for web developer positions"* or to the AI competition 2.75%: *"AI will take over web programming"*.

Other reasons, summarizing 14,78%, were personal reasons; likes and dislikes, considering the web/mobile not interesting, boring, repetitive, or leading to burnout: *"I don't like front-end", "It's not the most interesting topic", "For me, this career is a little bit monotonous"*.

In conclusion, the most common motivation for selecting a profession in web or mobile development is financial in nature. The expanding field is the main motivating factor, followed by the likelihood of finding well-paying employment. The students appear to have a good understanding of market trends, as evidenced by their responses, suggesting a positive correlation between market demand and their intention to work in a particular profession. Most of the students provided few reasons for not having a career in web development, the reasons were mainly related to domain challenges: rapidly changing technologies and high domain concurrency.

7 Is a Bachelor Degree Needed to Work in Web/Mobile Development?

To determine students' opinions related to the usefulness and necessity of having a bachelor's degree to work as a web/mobile developer, we asked them to state their opinion and provide the reasons for it. We used the following questions, we did not limit the number of reasons, therefore, there were answers containing one reasons and answers containing up to four reasons:

- ⋆ Do you think you need a bachelor degree to work as a developer in web or mobile?
- ⋆ Please detail the reasons for your response to the previous question.

The largest majority of students 78.89% considered that having a bachelor's degree is not necessary to work as a web/mobile developer. When asked about reasoning, the provided reasons can be grouped into two main categories: reasons that support the assumption that a diploma is not needed and reasons in favor of having one. The main reasons and their percentages are detailed in Fig. 2. A small group of students, 3.67% did not give any reason or answered the second question.

A large group of students, 38.53% considered that online courses and other free documentation from the Internet suffice to learn for a career as a web/mobile developer. Effort, interest, and knowledge count more than a computer science diploma for 15.60% of them, as 12.84% state that they or their colleagues from work have a web/mobile developer career without having a bachelor's degree in computer science. Some students, 19.27% clearly state that no diploma is needed. Students have access to free online information and courses to complete or start their learning: *"All the information we need is on the Internet. It is an application problem, not an information problem", "I just don't think you need a diploma to understand react, js, html, etc.", "You can probably get a sufficient understanding of this stuff with just online courses"*. However, nothing can replace the learning effort, so what matters in the end is the knowledge obtained and the capacity to apply it: *"if you allocate the necessary time plus quality resources, the final result can be even better than that of a person who only obtained a*

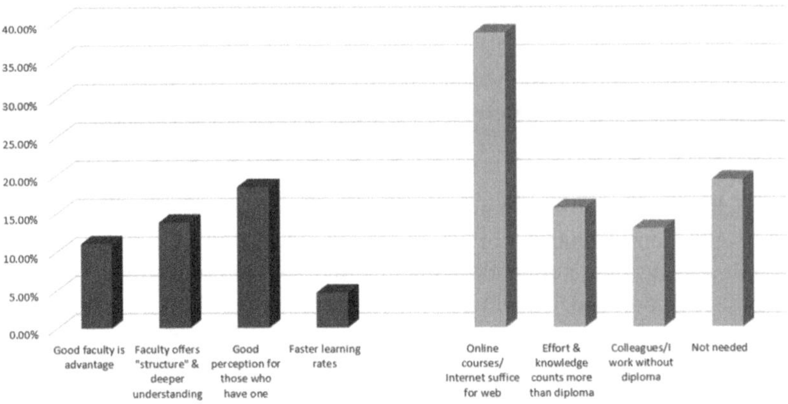

Fig. 2. Reported reasons for having a computer science diploma when working as a web/mobile developer.

diploma". Their arguments are enforced by having acquaintances of being themselves in the position to work in the computer science industry without having a computer science bachelor degree; some answers are more succinct: *"Because I work as a full stack developer without a diploma"*, others are more insightful: *"There are numerous examples in my circle of acquaintances who work in Web Development but do not have a diploma. There are indeed benefits to having a college degree, but I don't think it is a necessity"*.

Even if the largest majority considers that a degree is not needed for a job in web/mobile development, students are aware of the benefits of having one. Most of them recognize two major benefits: the first is related to knowledge acquisition, as universities provide structured information and a faster learning pace compared to self-learning. The second benefit is related to the job market, where a bachelor degree can help someone land on an interview and could prove a minimum set of knowledge. Students acknowledge that following an university curriculum helps them learn in depth, acquire, and understand more than an online course could provide. This aspect is mentioned not only related to web development, but also to programming in general: *"You don't need a degree to learn the technologies needed to code. But it is important to note that studying at a university does offer a proper structure, which is significantly more difficult to achieve on your own, but certainly not impossible"*, *"A web development course provides you with basic knowledge, but the faculty helps you form more solid foundations"*, *"you need to have a deeper understanding of how things work, and you cannot get that from just a Java Spring or React course"*. The financial aspects are present in the student's responses, a bachelor degree represents a method to help landing on a job: *"CV boost as it puts you in front"*, *"the employer can know that you possess a minimal set of knowledge and skills"*, *"the employers are looking for people with higher education"*.

In conclusion, students do not consider necessary to have a computer science bachelor degree to work as a developer in web/mobile, their beliefs are enforced by the fact that they have acquaintances, work colleagues, or even themselves that do not have a diploma and work as software developers in computer science industry. However, a large percentage recognize the fact that having a diploma and attending an university program offers benefits for in-depth learning, and it also helps in the recruitment process.

8 Results Discussions

Financial Aspects. Financial aspects related to programming, job concurrency, entry level, and payment represent a preoccupation that is prevalent in students' answers in all the questions. It does not matter whether we ask about challenges, reasons for having or not a career in web/mobile development, there are around 10% answers that mention financial aspects. Financial reasons appear as a benefit for having a bachelor degree, as it can help someone pass the first phase of the interview (CV checks) and to be taken into consideration for the later phases. So, the changes in the computer science job-related market dynamics impact student's preferences and preoccupations. The latest trends [5] and the last year layoffs in the USA seem to have repercussions on student perception and financial risk tolerance. In conclusion, compared to the previous year, students are more preoccupied with financial aspects: finding a job, payment, and market concurrency.

Challenges and Reasons Dynamics. The percentages reported by the students for the challenges were not noticeable different compared to those reported in the previous year [12]. The primary challenges are still the same as they were the year before. Keeping up with rapidly changing technologies remains the greatest challenge, but students must also focus on developing soft skills (related to teamwork, communication, or client comprehension) and dealing with personal issues such as stress and time management. Over the years, the percentage of psychological challenges reported, such as exhaustion, emotional problems, and lack of social interactions, did not change; it remained at 8.20%. The only notable changes in terms of challenges were related to the financial aspect, as detailed in the previous paragraph. Compared to the previous year [12], one reason was mentioned more frequently: the expanding web domain factor, which implies the likelihood of finding a job. Other reasons that ranked high in both studies were related to domain specificity: an interesting and creative domain, or their personal preferences: some students like the UI, others stated that would prefer back-end of full-stack. However, in both studies, the students were relatively interested in working as a web/mobile developer.

The Relation Between Web Career and Bachelor's Degree According to responses received, a bachelor's degree in computer science is not required to work as a web or mobile software developer. The reasons behind their statements is the fact that based on their life experience, they have models (colleagues) that work without a diploma. Also, it seems that even if going to university helps faster learning and acquiring more structured knowledge, they are aware of the free online information available that can help improve their knowledge. Do online courses pose a threat to replace university courses as main learning methods? In students' opinions, the online courses and information are enough to prepare for entry jobs in programming, web and mobile development.

9 Threats to Validity

We were aware of the possible threats that could influence the results, so we followed the survey research guidelines in community standards, as given in [14] as we developed our study. We analyzed the guidelines, identified potential validity risks, and tried to mitigate them. We analyzed construct validity, internal validity, and external validity.

Construct Validity. We looked for coherence, relevance, and pertinence in the survey questions. Three steps were involved in developing the set of questions: first, the authors suggested a series of questions. To improve clarity, each author reviewed and discussed the questions. In the second phase, we selected a sample of questions that were deemed most pertinent to our investigation. In the last stage, we checked the questions and determined which ones were necessary and which were not. After these steps were completed, we prepared and sent the survey to the participants; we collected the information anonymously.

For **internal validity**, we considered the target population, participant selection, dropout rates, subjective methodology of the authors, and study ethics. We sent the survey to every second-year computer science students registered in the Web Programming course who was randomly assigned to a specific teacher; by doing so, we addressed concerns regarding the target demographic and participant set. Because of this, there was no participant selection process; everyone was simply included. The participation of the students was optional and anonymous; they had no direct benefits. Teachers were not involved in the allocation process; the target participant set was randomly assigned. Additionally, we reduced the number of online questions in the survey so that participants could complete it quickly and to minimize dropout rates. We acknowledged the possibility of bias in the students' favorable reactions, so we took measures to reduce the risks. We told the students that all answers are anonymous, so there will be no benefits or repercussions for anything they might answer.

We were aware that our method may be subjective, but we reduced the chance of subjectivity by following the community standards protocols for free-text interpretation. Taking care of the ethical aspects, we made it clear to the students that participation in the survey was voluntary. Additionally, we informed them of the reasons for data collection and how we plan to use the information received.

External Validity. We examine the possibility of expanding and applying the study findings to IT students. Because the target participant set was formed by students from three lines of study, we could conclude that the set of participants was quite diverse and the results could be representative for IT students.

10 Conclusion and Future Work

The purpose of this study was to determine the main obstacles faced by second-year computer science students who want to work in web or mobile development as well as their interest in these fields. Also, we wanted to find out their perception related to the importance of having a bachelor's degree in computer science. The research method used was a survey questionnaire. When we developed and executed the study, we followed the scientific requirements for this type of research as stated in [14], and took measures to mitigate the potential threats to validity. We interpreted the free text obtained as responses to open questions using thematic analysis. The set of participants was randomly selected from second-year students enrolled in Web Programming course.

We found that students are interested in a job as a web or mobile developer. The main motivating factors for this interest in the profession were market expansion, as more and more people are using online applications and "trendiness". Some responses specifically addressed the financial implications and the expectation among students that they will find well-paying employment in this field. The challenges can be grouped in classes: technical challenges (such as rapidly hanging technologies), personal challenges (such as time management), work related challenges (client communication), financial and psychological (burnout, emotional issues). Compared to the previous year, the only notable difference is related to the financial aspect, which is now more mentioned. The main motivating factor is the field's expansion, followed by the possibility of getting a well-paying job. Domain related challenges were the primary cause of not having a career: rapidly evolving technologies as well as high domain concurrency.

Based on their life experience, as students know and/or have colleagues who work without a computer science bachelor degree, students consider that a degree is not required to have a job as a web/mobile developer. However, going to university has benefits: faster learning and acquiring more structured knowledge. Students believe that the resources and courses offered online are enough to prepare them for entry-level positions in web/mobile programming.

Future research could explore the long-term career outcomes of web and mobile developers who rely primarily on online learning resources compared to those with a bachelor's degree. Furthermore, investigating the specific online learning platforms and resources most valued by students would provide valuable insights. Furthermore, examining the impact of economic fluctuations on student perceptions of financial security within the field of web and mobile development could offer valuable data for career guidance programs. Overall, a deeper understanding of the evolving educational landscape and its influence on career aspirations in this dynamic field is crucial.

Disclosure of Interests. The authors have no competing interests.

References

1. Begel, A., Simon, B.: Novice software developers, all over again. In: Proceedings of the Fourth International Workshop on Computing Education Research, pp. 3–14 (2008)
2. Braun, V., Clarke, V., Hayfield, N., Terry, G.: Thematic Analysis, pp. 843–860. Springer, Singapore (2019). https://doi.org/10.1007/978-981-10-5251-4_103
3. Clokie, T.L., Fourie, E.: Graduate employability and communication competence: are undergraduates taught relevant skills? Bus. Prof. Commun. Q. **79**(4), 442–463 (2016)

4. Heberle, A., Neumann, R., Stengel, I., Regier, S.: Teaching agile principles and software engineering concepts through real-life projects. In: Proceedings of the 2018 IEEE Global Engineering Education Conference (EDUCON), pp. 1723–1728 (2018)
5. Hughes, N.C.: Tech layoffs predictions 2024: when will the job cuts end? Technical report, Techopedia (2022). https://www.techopedia.com/tech-layoffs-predictions
6. Iacob, C., Faily, S.: Exploring the gap between the student expectations and the reality of teamwork in undergraduate software engineering group projects. J. Syst. Softw. **157**, 1–18 (2019)
7. Jevremovic, A., Ristic, N., Veinovic, M., Shimic, G., Stanisic, N.: Wide: centralized and collaborative approach to teaching web development. J. Internet Technol. **19**(4), 1003–1014 (2018)
8. Kemp, S.: Digital 2022 global overview report. Technical report, DATAREPORRER (2022). https://datareportal.com/reports/digital-2022-global-overview-report
9. Lundberg, G.M., Krogstie, B.R., Krogstie, J.: From employable to fully operational: the need for training of computer science graduates (2021)
10. Motogna, S., Suciu, D., Molnar, A.J.: Investigating student insight in software engineering team projects. In: Proceedings of the 16th International Conference on Evaluation of Novel Approaches to Software Engineering - ENASE, pp. 362–371. INSTICC, SciTePress (2021). https://doi.org/10.5220/0010478803620371
11. Park, T.H., Wiedenbeck, S.: Learning web development: challenges at an earlier stage of computing education. In: Proceedings of the Seventh International Workshop on Computing Education Research, pp. 125–132 (2011)
12. Petrescu, M., Adrian, S., Badarinza, I.: Students' interests related to web and mobile technologies study, pp. 242–249 (2023). https://doi.org/10.5220/0012174900003584
13. Radermacher, A., Walia, G., Knudson, D.: Investigating the skill gap between graduating students and industry expectations. In: Companion Proceedings of the 36th International Conference on Software Engineering, pp. 291–300 (2014)
14. Ralph, P.: ACM Sigsoft Empirical Standards for Software Engineering Research, version 0.2.0 (2021). https://github.com/acmsigsoft/EmpiricalStandards
15. Redmond, K., Evans, S., Sahami, M.: A large-scale quantitative study of women in computer science at Stanford University. In: Proceeding of the 44th ACM Technical Symposium on Computer Science Education, pp. 439–444 (2013)
16. Statista: Data volume of global consumer internet traffic from 2017 to 2022, by subsegment. Technical report, Statista (2022). https://www.statista.com/statistics

SWI (Soft Web Intelligence) Powered by User-Defined Fuzzy Operators and Aggregators

Paolo Fosci and Giuseppe Psaila(✉)

DIGIP, University of Bergamo, Viale Marconi 5, 24044 Dalmine, BG, Italy
{paolo.fosci,giuseppe.psaila}@unibg.it

Abstract. We proposed the notion of Soft Web Intelligence in our previous works: in short, it interprets the general notion of Web Intelligence in a modern key, by explicitly considering the current technological panorama. Specifically, *JSON* data sets are acquired from the Internet, stored within *JSON* document stores and then processed and queried by means of soft computing and soft querying methods.

In this paper, we present the joint adoption of fuzzy operators and fuzzy aggregators to perform data-analysis tasks on aggregated data. Indeed, when data are processed through fuzzy sets to perform soft querying, data items with membership degrees to fuzzy sets could be aggregated, to obtain a novel membership to another fuzzy set belonging to a different universe; in this context, User-defined Fuzzy aggregators become critical for actually performing Soft Web Intelligence based on soft computing.

Keywords: Soft web intelligence · User-defined fuzzy operators and fuzzy aggregators · Novel constructs in J-CO-QL+ · Soft querying on JSON data sets

1 Introduction

After two decades, the concept of Web Intelligence is still a hot topic of research in computer science. The paper [23] proposed Web Intelligence for the first time: the idea is to move from the well-known approaches that fall under the hat of Business Intelligence (or BI), so as to adapt them for gathering and analyzing possibly semi-structured data sets from the Word-Wide Web, so as to discover useful knowledge from Web sites and Web sources. In [23], the authors argued that Artificial Intelligence (or AI) is the key technology to get rid of the complexity of the Web, in terms of variety and complexity of formats which data must be extracted from.

Among AI techniques, "soft computing" and "fuzzy logic" have played a significant role, due to their capabilities of easily dealing with imprecision and vagueness both

This study was funded by the European Union - *NextGenerationEU, in the framework of the GRINS – Growing Resilient, INclusive and Sustainable project (GRINS PE00000018 - CUP F83C22001720001)*. The views and opinions expressed are solely those of the authors and do not necessarily reflect those of the European Union, nor can the European Union be held responsible for them.

M. Marchiori and F. García Peñalvo (Eds.): WEBIST 2023, LNBIP 543, pp. 77–100, 2025.
https://doi.org/10.1007/978-3-031-89621-7_5

within data and within queries over data. We demonstrated these capabilities in our recent research works [5,7,12], whose goal is to develop a software framework, named *J-CO* Framework, which integrated advanced capabilities of managing and integrating large collections of *JSON* documents with sophisticated capabilities of soft querying *JSON* documents through fuzzy sets.

As far as Web Intelligence is concerned, in our recent papers [7,9,10] we introduced and explored the notion of *Soft Web Intelligence*, i.e., a precise and focused interpretation of the general notion of Web Intelligence, in which the *J-CO* Framework plays a key role, so as to exploit its capabilities of soft querying o actually extract useful information from Web source that provides *JSON* data sets.

In *Soft Web Intelligence*, user-defined fuzzy operators and fuzzy aggregators are the key linguistic tools to perform soft queries on *JSON* data sets acquired from Web sources. The main contribution of this paper is to unify previous contributions to the exploitation of either fuzzy operators or of fuzzy aggregators into a homogeneous framework; this way, we show the potential of applying complex soft processes executed by the *J-CO* Framework towards performing tasks of *soft Web Intelligence*.

The paper is organized as follows. Section 2 presents the background of our work as far as related work is concerned. Section 3 introduces the vision of *Soft Web Intelligence*. Section 4 presents the main features of the *J-CO* Framework (in Sect. 4.1) and how fuzzy sets are supported in the *J-CO-QL*$^+$ language (in Sect. 4.2). Section 5 introduces the novel construct that we recently added to *J-CO-QL*$^+$ to declare user-defined fuzzy aggregators, together with the semantic model. Section 6 addresses a practical case study by exploiting user-defined fuzzy aggregators. Finally, Sect. 7 draws conclusions and future work.

2 Related Work

In this section, we present the background from which this work has originated.

The concept of Web Intelligence was introduced in [23]. It can be conceived as the evolution of "Business Intelligence" [17] towards the World Wide Web; "Artificial Intelligence" (AI for short) was supposed to be the key tool for performing Web Intelligence, but what is Artificial Intelligence?

In the current common perception, AI means "Neural Networks" and "Deep Learning", due to the success they have recently obtained. Nevertheless, techniques such for "Data Mining" belongs to AI too and are very popular, so they have been successfully applied to Web Intelligence [14,29].

Nonetheless, in the literature, many papers try to give a more specific yet wider interpretation of Web Intelligence, such as "Computational Web Intelligence" [27], i.e., the adoption of "Computational Intelligence" in Web Intelligence, as well as "Brain Informatics" [28], i.e., fostering Web Intelligence through techniques that come out from the study on the human brain.

Fuzzy Logic and Soft Computing belong to AI too: indeed, their capabilities of approximate reasoning based on "linguistic predicates" provides a significant contribution towards AI.

Remember that a Fuzzy Set A in a universe U is a mapping for each $x \in U$, $A : x \rightarrow [0,1]$, also denoted as $\mu_A : x \rightarrow [0,1]$. The co-domain is the set of "membership degrees" (or simply "memberships", for brevity in the following): each item x belongs to A with a degree; when the degree is $0 < \mu_A(x) < 1$, x belongs to A only partially; obviously, $\mu_A(x) = 1$ denotes full membership of x to A, while $\mu_A(x) = 0$ means that x does not belong at all to A.

If fuzzy sets represent either approximate or imprecise properties that are easily synthesized by "linguistic terms" (e.g., "young people", "rich people", and so on), then selection predicates become "linguistic predicates", corresponding to fuzzy sets. Therefore, the membership degree of an item to a fuzzy set denotes how the item satisfies the represented linguistic predicate.

The reader can guess that fuzzy logic can be exploited for Web Intelligence. In this regard, Zadeh, the creator of Fuzzy-Set Theory [24], in [25,26] proposed the concept of "WebIQ" or "WIQ", i.e., a formal framework in which Fuzzy-Set Theory provides the formal tools to query Web sources in an extensive and imprecise way.

Nevertheless, when performing Data Intelligence activities [1] data must be aggregated; if we use linguistic predicates and fuzzy logic, "fuzzy aggregations" become essential. We are focused on groups (or categories) $G_j = \{x_{j,1}, x_{j,2}, \dots\}$) of items $x_{j,i}$ that belong to the same group G_j because they share some common properties or are samples of the same category of items. Each $x_{j,i}$ singularly may belong to a fuzzy set A, thus it is provided with a membership $\mu_A(x_{j,i})$. Consequently, the fuzzy set $\overline{A}$ of groups G_j can be seen as a partition of A: the membership degree of a group G_j to $\overline{A}$ should be derived by somehow aggregating memberships $\mu_A(x_{j,i})$, group by group. Alternatively, if $x_{j,i}$ has not a membership, its values might have to be aggregated to obtain the final membership of the G_j group.

Popular fuzzy aggregators of this type are "Weighted aggregation" (see [3]) and "Ordered Weighted Aggregation" (OWA) [16,22].

As a final remark for this section, we want the reader to be aware that very few papers are available in the literature that talk about fuzzy logic and soft computing in Web Intelligence. We found the paper [15]: it exploits soft computing in a group decision-making system to express preferences, but Web Intelligence activities were not supported by soft computing. We also found the paper [18]: it uses `FUZZYALGOL`, a fuzzy procedural programming language [21], for soft querying Web sources.

3 Soft Web Intelligence

In [9], we conceived the notion of *Soft Web Intelligence* as an interpretation of the general notion of Web Intelligence on the basis of the current technological panorama. Indeed, the current technological panorama has been characterized by the advent of *JSON* as *de-facto* standard format for information interchange over the Internet, in place of XML. One important consequence of this phenomenon has been the development of a specific category of *NoSQL* databases (i.e., databases that are not relational and do not use SQL as query language) known as "*JSON* document stores", because they directly manage collections of *JSON* documents; *MongoDB* appears to be the most popular one.

Hereafter, we present our vision of *Soft Web Intelligence*, by relying on Definition 1 (which comes from our previous work [10]).

Definition 1. *"Soft Web Intelligence" is the continuous acquisition, integration and querying of JSON data sets coming from or representing Web sources, by exploiting*

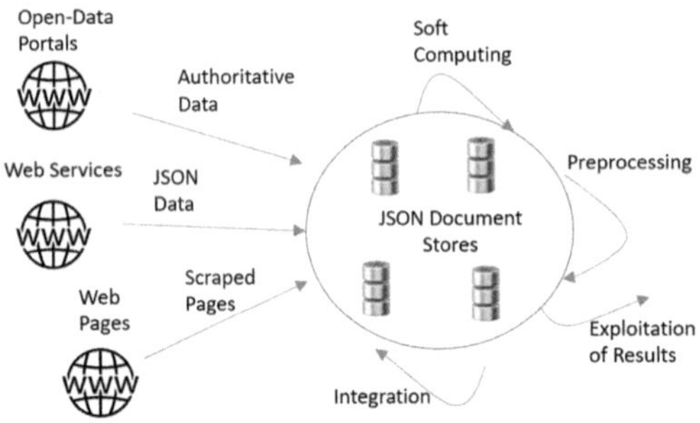

Fig. 1. Vision of Web Intelligence.

Soft Computing and Soft Querying, to use them for decision-making and knowledge discovery.

As the reader can see, Definition 1 instantiates the general concept of *Web Intelligence* (provided more than 20 years ago [23]) by considering the current technological panorama.

Through Fig. 1, we illustrate how we figure out *Soft Web Intelligence.*

- **Web Sources.** Web technology has significantly evolved in the last two decades. Consequently, many different types of Web sources are now available, apart from classical Web sites. Here, we focus on Web sources that provide *JSON* data sets.
  - Web Services plays a key role in the modern Web panorama: they are expected to provide either complete data sets or single pieces of data. If the Web service works only in this latter mode and many piece of information must be acquired, this means that the full data set of interested must be built by singularly collecting all data items. Typically, social media expose Web services that can be exploited to interact with the system; nowadays, *JSON* is the data format on which most of Web services rely.
  - Common channels that public administrations exploit to publish data sets (often referred to as "Authoritative Data Sets") concerning the administered territory are Open-Data portals. Among all formats, *JSON* is becoming more and more popular in this context too.
  - The content of Web pages (i.e., HTML pages) could be seen as a giant repository of precious information. Techniques for "Web Scraping" could acquire the content of HTML pages and represent it as *JSON* data sets.
- ***JSON* Document Stores.** Data acquired from Web sources as *JSON* data sets should be stored within a pool of databases managed by (potentially different) *JSON* document stores. Currently, *MongoDB* is (in our perception) the most popular *JSON* store. Nevertheless, other *JSON* stores are available and widely diffused: for example, *CouchDB* has been chosen as the DBMS for the *HyperLedger Fabric* permissioned BlockChain platform [2]).
- **Processing Activities.** Processing is the key factor for the success of *Soft Web Intelligence.*

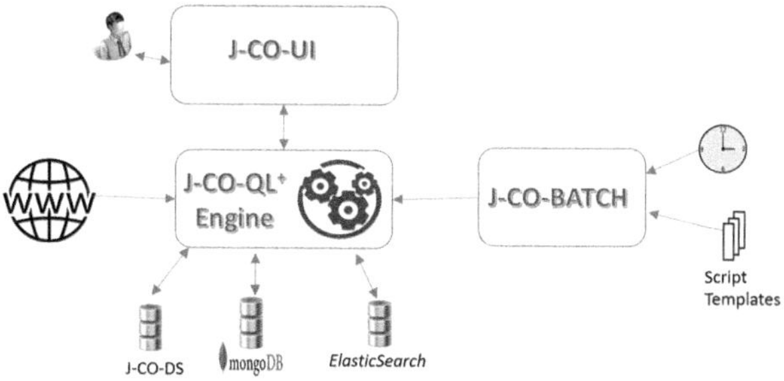

Fig. 2. The *J-CO* Framework.

- The first processing activity to perform is "Preprocessing": indeed, it is necessary to remove noise from data and make their formats homogeneous.
- The second activity to perform on data is "Integration": indeed, the different data sets (possibly stored in different databases) must be integrated, so as to unite those pieces of information that came from different sources.
- The final activity to perform on data sets is "Soft Computing": during this activity, techniques based on soft computing and soft querying are used to extract knowledge from integrated data.

Notice that the tool that provides soft computing capabilities strongly determines the level of complexity that the analysis tasks can reach. This is why we are introducing fuzzy aggregators within the framework we used to develop the concept of *Soft Web Intelligence*, as it will be illustrated in the next sections.

4 Fuzzy Sets in the J-CO Framework

In this section, we present the *J-CO* Framework and how its language (named *J-CO-QL*$^+$) provides users with constructs to define fuzzy operators, to perform complex soft queries on *JSON* data sets. Then, in Sect. 5 we will extensively present the novel concept of "fuzzy aggregator".

4.1 The J-CO Framework

The *J-CO* Framework is a pool of software tools whose goal is to provide analysts with a powerful support for gathering, integrating and querying possibly-large collections of *JSON* data sets. The core of the framework is its query language, named *J-CO-QL*$^+$.

The current organization of the framework, which is the result of [9], is depicted in Fig. 2. We explain it hereafter.

- *J-CO-QL*$^+$ *Engine*. This component actually executes *J-CO-QL*$^+$ scripts (i.e., queries). It is able to retrieve data from document databases (for example, managed by *MongoDB*) and saves results into them; it also can send HTTP requests to Web sources to get *JSON* data sets directly from them.

Listing 1. *J-CO-QL*$^+$: Fuzzy Operator `casePopulationFO`.

```
1. CREATE FUZZY OPERATOR casesPopulationFO
     PARAMETERS    cases TYPE NUMERIC,
                   pop   TYPE NUMERIC
     PRECONDITION pop > 0 AND cases >= 0
     EVALUATE      100000 * cases / pop    // cases per 100k inhabitants
     POLYLINE      [ (  0, 0.0), (100, 0.1),
                     (300, 0.9), (900, 1.0) ];
```

- *J-CO-DS*. This component is a simple document store specifically introduced in [19] to store large or very large single *JSON* documents (such as many *GeoJSON* documents that cannot be stored within other *JSON* stores, such as *MongoDB*, which has a 16Mb limit per single document).
- *J-CO-BATCH* (introduced in [9]) is an off-line executor of *J-CO-QL*$^+$ scripts; in particular, it is possible to schedule the repeated execution of scripts. Another feature is the concept of "template": *J-CO-QL*$^+$ scripts can be parameterized to reuse them with different settings/configurations.
- *J-CO-UI* is the user interface, by means of which analysts can write *J-CO-QL*$^+$ scripts, submit them to the *J-CO-QL*$^+$ *Engine* and inspect results.

4.2 Fuzzy Sets and Operators

The *J-CO-QL*$^+$ language manipulates "collections" of *JSON* documents.

For each single document, it is possible to evaluate its memberships to many fuzzy sets in an independent way. These degrees are represented within the same document, by adding a special root-level field named `~fuzzysets`. It is a nested document that behaves as a key/value map: a field within it denotes the membership to a fuzzy set, in such a way the field name is the name of the eponym fuzzy set, while the value (in the range $[0, 1]$) is the membership degree.

In order to deal with fuzzy sets, *J-CO-QL*$^+$ provides constructs to derive membership degrees of *JSON* documents from their fields. Indeed, the membership degree of a document to a fuzzy set certainly depend on properties described by its field; through "user-defined fuzzy operators", it is possible to derive membership degrees of documents to fuzzy sets.

Consider Listing 1: it reports the definition of a user-defined fuzzy operator in the *J-CO-QL*$^+$ language. The goal of the operator is to provide a membership degree moving from two numerical parameters. Hereafter, we present it in details.

- The name of the fuzzy operator is `casesPopulationFO`. Its goal is to derive a membership degree as follows: given the number of cases of a disease and the number of inhabitants of a country, it computes the number of cases per 100,000 inhabitants and uses this value to obtain the membership degree.
- The clause `PARAMETERS` defines the formal parameters of the fuzzy operator. Specifically, two parameters are defined, i.e., the number of cases (the parameter `cases`) and the population (the parameter `pop`), both as integer values.

- The clause `PRECONDITION` specifies a Boolean condition that must be satisfied before evaluating the operator: if the condition is not satisfied, the operator is not evaluated and an error is raised. Specifically, the precondition says that the parameter `pop` must be greater than zero, while the parameter `cases` must be no less than zero.
- The clause `EVALUATE` computes the value that is later used to obtain the membership degree. It is a mathematical expression that exploits the parameters. Specifically, the number of cases for every 100,000 inhabitants is computed.
- The clause `POLYLINE` specifies the actual membership function.
  J-CO-QL$^+$ provides users with the possibility to define polylines as membership function to express complex semantics. The polyline is defined by a sequence of points, in the form (x_i, y_i): x_i is any real number, while y_i is a real number in the range $[0, 1]$; given a point with index $i > 1$, $x_i >= x_{i-1}$.
  If we denote the value provided by the `EVALUATE` clause with x, it is matched on the x-axis of the polyline, to obtain the corresponding value on the y-axis, as the corresponding membership degree. Note that if $x < x_1$ (respectively, $x > x_n$, with n the greatest i index), then y_0 (respectively, y_n), will be returned as membership degree.
  Figure 3a depicts the polyline defined for the fuzzy operator under definition. The "S-shape" of the function allows for reducing the relevance (in terms of membership degree) for a number of cases that is less than 100 per 100,000 inhabitants; at the same time the shape quickly enhances the relevance when the cases grow up from 100 to 500 per 100,000 inhabitants; after 500, the membership slowly grows up from 0.9, reaching 1 for 900 cases per 100,000 inhabitants.

The statements of *J-CO-QL*$^+$ that manages *JSON* documents, provide specific clauses whose goal is to evaluate membership degrees of documents to fuzzy sets, so as to specify "soft queries" (in Sect. 6.1, we will show how to use fuzzy operators in practice).

In our previous research works, we explored the adoption of fuzzy sets for soft querying in many contexts. For example, in [7] we considered soft querying for spatial data sets, on the basis of complex geometries and spatial relationships between them. We also addressed the definition of a high-level domain-specific language named *GeoSoft* [4, 12], whose execution relies on fuzzy capabilities of *J-CO-QL*$^+$ to process spatial data. Furthermore, we demonstrated that soft querying is effective to integrate geotagged data sets based on registry data and coordinates [7].

5 User-Defined Fuzzy Aggregators

In data analysis activities, it may be necessary to aggregate data items.

If a resulting aggregated item a belongs to a fuzzy set f, it is necessary to evaluate the membership degree of a to f. In [11], we started addressing the aggregation of membership degrees of elementary items, while in [12] we extended our vision to any kind of fields in elementary items. In both cases, the resulting aggregate item a must be provided with a membership degree. In this section, we report the both the linguistic and formal framework that we introduced in *J-CO-QL*$^+$ for user-defined fuzzy aggregators.

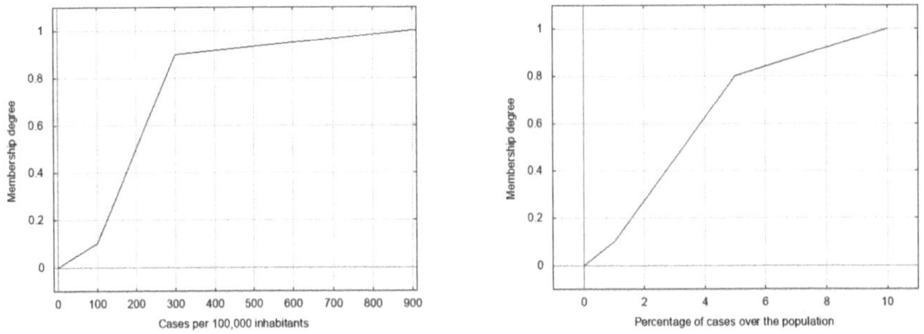

(a) Membership function for the fuzzy operator `casesPopulationFO`.

(b) Membership function for the fuzzy aggregator `integrateCasesFA`.

Fig. 3. Membership functions defined for aggregators in *J-CO-QL*$^+$.

Listing 2. *J-CO-QL*$^+$: Fuzzy Aggregator `integrateCasesFA`.

```
2. CREATE FUZZY AGGREGATOR integrateCasesFA
     PARAMETERS    cases TYPE  ARRAY,
                   pop   TYPE  NUMERIC
     FOR ALL       cs    IN    cases
       AGGREGATE   cs    AS    av
     EVALUATE      100 * av / pop                // % of infected people
     POLYLINE      [ (0, 0.0), ( 1, 0.1),
                     (5, 0.8), (10, 1.0) ];
```

5.1 Basic Structure

In *J-CO-QL*$^+$ fuzzy aggregators are defined through the statement `CREATE FUZZY AGGREGATOR`. A first example can be found in Listing 2, while other examples are in Listing 3 and Listing 4. In comparison with the statement `CREATE FUZZY OPERATOR` (see Listing 1), the reader can see that the external structure is the same, i.e.:

- the clause `PARAMETERS` defines the formal parameters (for fuzzy aggregators, at least one parameter must be an array);
- the clause `PRECONDITION` is evaluated on parameters and must be satisfied in order to evaluate the aggregators;
- the clause `EVALUATE` evaluates the final numerical value that is matched against the membership function defined by the clause `POLYLINE` (if this clause is missing, the default polyline `[(0,0),(1,1)]` is considered), so as to return the membership degree.

The difference in comparison with the statement `CREATE FUZZY OPERATOR` is the presence, between the clause `PRECONDITION` and the clause `EVALUATE`, of a pool of specific clauses whose goal is to derived aggregated values to be combined within the clause `EVALUATE`.

We are ready to introduce the fuzzy aggregator `integrateCasesFA` defined in Listing 2. Its goal is to receive an array of integer values to aggregate: the array is named cases, since each item denotes the number of cases of a disease; the second parameter is named pop (shortcut for "population"), because the aggregator must provide a membership degree based on the aggregated cases in relation with the population. In the following, we present the novel clauses.

- The clause FOR ALL scans all values in an array parameter, so as to perform aggregations. Specifically, all items cs in the array cases are explored.
- The sub-clause AGGREGATE evaluates, for each item cs in the array cases, an expression, whose value is aggregated into the "variable" specified after the keyword AS.
  Specifically, the expression to evaluate refers only to the item cs; this means that all the values in the array cases are aggregated (summed) into the final value av.
- The clause EVALUATE exploits the aggregated value av and the parameter pop. Specifically, it computes the percentage of cases over the number of inhabitants. The resulting value is matched against the polyline functions (depicted in Fig. 3b), so as to obtain the wished membership degree. The shape of the polyline function is defined in such a way it enhances degrees when the number of cases is greater than 1% of inhabitants.

5.2 Adding Local Derived Values

The fuzzy aggregator integrateCasesFA is a simple case of aggregator; in contrast, widely-used aggregators, such as the "Ordered Weighted Aggregator" (OWA, see [22]) are more complex to define and novel features must be provided by the language. Specifically, an OWA aggregation is a weighted aggregation, where weights are obtained through a monotone function; furthermore, since the greatest weights are used for the greatest values, the array of values must be suitably sorted. Thus, clauses to compute weights and sort arrays must be provided by the language.

To this end, consider the fuzzy aggregator named owaFA that is reported in Listing 3. The aggregator works on an array memberships of membership degrees, so as to aggregate them into a unique membership degree. To multiply items with the greatest values by the greatest weights, the array must be, at first, sorted in ascending order. As a monotone function, a parabolic function is used: $f(x) = x^2$. Weights are computed as follows: given the size s of the array memberships, denoting with $1 \leq i \leq s$ the position of each item in the array, the weight w_i of the i-th item is $w_i = f(i/s) - f((i-1)/s)$ Denoting with μ_i the i-th value in the array memberships, the final aggregated membership degree $\overline{\mu}$ is $\overline{\mu} = \Sigma_{u=1}^{s}(w_i \times \mu_i)$ (assuming that $f(1) - f(0) = 1$).

Hereafter, we present the sub-clauses exploited by the aggregator owaFA.

- The clause SORT generates a novel array, named sMemberships, sorting each numeric item ms in the array memberships in ascending order.
- The clause LOCALLY evaluates, for each item sms in the array sMemberships, a derived value that is used in the sub-clause AGGREGATE. The locally-derived value is named w, since it is the weight for each item in the sorted array sMemberships. Notice that special symbol POS, that denotes the actual position of the processed item in the array.

Listing 3. *J-CO-QL*$^+$: Fuzzy Aggregator owaFA.

```
3. CREATE FUZZY AGGREGATOR owaFA
     PARAMETERS   memberships  TYPE ARRAY
     SORT         ms           IN   memberships
       BY           ms         TYPE NUMERIC ASC  AS sMemberships
     FOR ALL      sms          IN   sMemberships
       LOCALLY      (POS^2 - (POS-1)^2) / (COUNT(sMemberships)^2) AS w
       AGGREGATE    sms * w    AS av
     EVALUATE     av;
```

Listing 4. *J-CO-QL*$^+$: Fuzzy Aggregator weighetedMembershipsFA.

```
4. CREATE FUZZY AGGREGATOR weightedMembershipsFA
     PARAMETERS      memberships   TYPE   ARRAY,
                     w             TYPE   ARRAY
     PRECONDITION    COUNT (memberships) = COUNT (w)
     FOR ALL         m             IN     memberships
       LOCALLY       m * w[POS]    AS     wm
       AGGREGATE     wm            AS     am,
                     w [POS]       AS     aw
     EVALUATE        am / aw;
```

- finally, the clause AGGREGATE, for each item in the array multiplies its value by its weight and sums the results into the av value. Since we are obtaining an aggregated membership degree, the clause POLYLINE is not specified, so as to return the av value (that is a membership degree) as it is.

5.3 Multiple Aggregated Values

In the fuzzy aggregators so far presented, only one aggregated value is computed. However, more complex (and quite common) aggregators are based on more than one aggregated value. A typical example, is a "weighted aggregation", i.e., the array of items is accompanied by an array of weights. For example, if we have an array of memberships μ such as $s = |\mu|$ (size of the array μ) and a corresponding array of weights w, such that $|w| = s$. The final weighted aggregation of memberships $\overline{\mu}$ is $\overline{\mu} = (\Sigma_{i=1}^{s} \mu[i] \times w[i])/(\Sigma_{i_1}^{s} w[i])$. Clearly, both the numerator and the denominator are aggregated values.

Listing 4 reports the fuzzy aggregator named weightedMembershipsFA, which performs the above-presented aggregations. Hereafter, we describe it in details.

- The aggregator receives two parameters: memberships is the array of membership degrees to aggregate; w is the array of weights.
  Consequently, the clause PRECONDITION must verify that both array have the same size.
- Within the clause FOR ALL, the sub-clause LOCALLY computes the products of the single membership value with the corresponding weight (i.e., w[POS]).
  Then, the sub-clause AGGREGATE computes two aggregated values: am is the numerator, while aw is the denominator.
- Finally, the clause EVALUATE actually computes the final aggregated membership degree, by computing the fraction am / aw. Since this value is by definition in the range $[0, 1]$, no polyline function is defined and it is returned as it is.

5.4 Semantic Model

After we presented the statement provided by *J-CO-QL*$^+$ to define fuzzy aggregators, we report the semantic model behind, as presented in [10].

- A fuzzy aggregator is defined by means of a tuple

$$\langle LV, AV, evalExpr, Points, Params, Precond \rangle$$

  where LV is the (possibly empty) set of "local values", AV is the (non empty) set of "aggregated values", $evalExpr$ is the expression to evaluate with aggregated values, $Points$ is the array of points that constitute the polyline, $params$ is the list of parameters of the aggregator. *Precond* is the precondition: if it does not hold on *Params*, the aggregator is not evaluated.
- Each local value $lv_j \in LV$ is obtained as

$$lv_j = le_j(e, POS, Params)$$

  where le_j is the expression used to evaluate the local value; le_j can be seen as a function of an item e and of its position POS in the array, as well as of the list of parameters.
- An aggregated value $av_k \in AV$ is evaluated as

$$av_k = \Sigma_{e \in V} ae_k(e, pos(e), LV, Params)$$

  where ae_k is the expression used to evaluate the local value to aggregate; notice that ae_k depends on the item e, its position in the array, the set of local values and the list of parameters. The sum is performed for each item e in the V vector, where $V \in Params$.
- The final membership μ is obtained as

$$\mu = Polyline(Points, evalExpr(AV, Params))$$

  where the expression $evalExpr$ depends on the set AV of aggregated values and on the list of parameters.
  The resulting value passes through the function$Polyline$, which receives the list $Points$ of points specified in the clause `POLYLINE` (if missing, we assume that $Points = [(0,0), (1,1)]$).

The reader can see that, based on this semantic model, in *J-CO-QL*$^+$ it is possible to define a broad range of fuzzy aggregators. As far as we know, this is a novelty in the panorama of (soft) query languages.

In Sect. 6, we will show how to exploit fuzzy aggregators for soft querying data in *J-CO-QL*$^+$.

6 Case Study and Query

In this section, we show how to use fuzzy operators and fuzzy aggregators for soft querying a data set acquired from a Web portal.

6.1 Case Study

In this paper, we propose a case study that shows a possible application whose context were already considered in [8], i.e., acquiring and analyzing data concerned with the recent COVID-19 pandemic. Before going on, **it is worth clarifying** that the goal of the paper is only to show how it would be possible for non-programmers to perform the analysis task we considered with limited effort; we do not want at all to make any medical-political considerations in any way (which is outside the scope of this paper). Since the data set we used is publicly available on the Web, it is useful to show the final results.

Since the beginning of the COVID-19 pandemic, European countries have collected weekly data about the number of cases and the number of deaths due to COVID-19. The data set is publicly available on the web site of the European Centre for Disease Prevention and Control (ECDC)[1] and it is still updated[2] Clearly, this data set could provide interesting hints, if properly analyzed.

Consequently, we identified the following problem to address.

Problem 1. *Given the weekly cases of COVID-19 infections, observe the period from October 2020 (week* 40*) to March 2021 (week* 13*), so as to identify those countries that showed the highest amount of cases (per 100,000 inhabitants) and also possibly showed a significant amount of weekly peaks of cases.*

Problem 1 can be thought as a soft query.

Query 1. *Consider the universe of weekly cases records per each single country. In this universe, we consider the fuzzy set named* `ManyCasesRecords`*, which contains weekly cases records that are numerous (the greater the number of cases, the greater the membership degree). Consider now the universe of countries, on which two fuzzy sets are defined, named* `ManyCasesCountries` *and* `ManyPeaksCountries`*. The membership degree of a country to the fuzzy set* `ManyCasesCountries` *ranks countries on the bases of the aggregated recorded cases in the observed period (the greater the number of cases, the greater the membership degree). The membership degree of a country to the fuzzy set named* `ManyPeaksCountries` *ranks countries on the basis of the number of weekly peaks of cases (the greater the number of peaks, the greater the membership degree); the membership degree of a country to the fuzzy set* `ManyPeaksCountries` *is obtained by aggregating the membership degrees of weekly records to the fuzzy set* `ManyCasesRecords` *for that country, in such a way high membership degrees get the highest weights (OWA). We are interested in computing a third fuzzy set on the universe of countries, whose name is* `WantedCountries`*, in which the membership degrees ranks countries on the basis of a weighted aggregation of membership degrees to the fuzzy set* `ManyCasesCountries` *(with weight* 2*) and to the fuzzy set* `ManyPeaksCountries` *(with weight* 1*). Specifically, we are interested in countries that, in the observed period, have a membership degree to the fuzzy set* `WantedCountries` *that is no less than* 0.80.

[1] ECDC web site: https://www.ecdc.europa.eu/en/covid-19 accessed on 12/04/2024.

[2] Data set: https://opendata.ecdc.europa.eu/covid19/nationalcasedeath/json/ accessed on 12/04/2024.

Listing 5. *J-CO-QL*$^+$: Java Functions `getYear` and `getWeek`.

```
5. CREATE JAVA FUNCTION getYear
     PARAMETERS yearWeek      TYPE String
     CLASS      GetYearClass
     CLASS BODY {
       // yearWeek : is supposed to be in the form "YYYY-WW"
       public static int getYear(String yearWeek)  {
         try { return Integer.parseInt(yearWeek.substring(0,4)); }
         catch (Exception e) { return 0; }
       }
     } END BODY;

6. CREATE JAVA FUNCTION getWeek
     PARAMETERS yearWeek      TYPE String
     CLASS      GetWeekClass
     CLASS BODY {
       // yearWeek : is supposed to be in the form "YYYY-WW"
       public static int getWeek(String yearWeek)  {
         try { return Integer.parseInt(yearWeek.substring(5,7)); }
         catch (Exception e) { return 0; }
       }
     } END BODY;
```

Clearly, in order to evaluate the memberships of countries to the fuzzy sets, it is necessary to aggregate records of weekly cases.

6.2 User-Defined Functions

The *J-CO-QL*$^+$ script that implements Query 1 has to process string fields that denote the week for which cases are recorded in the form `"YYYY-WW"`, where `YYYY` denotes the year (four digits) and `WW` denotes the number of week (two digits) in the year. A couple of functions to manipulate these strings and extract year and week number are necessary.

As extensively presented in [12], *J-CO-QL*$^+$ provides constructs to define either Java functions or JavaScript functions, i.e., user-defined functions that are implemented either in the Java language or in the JavaScript language. Listing 5 reports the definition of the functions named `getYear` and `getWeek`, implemented in the Java language The interested reader can refer to [12] for details: here, we simply note that the two functions receive one formal parameter named `yearWeek` and return, respectively, the number corresponding to the year and the number corresponding to the week.

User that are more familiar with the JavaScript language can use this language; Listing 6 reports the JavaScript implementation of the same user-defined functions presented in Listing 5. Simply notice that the execution of JavaScript functions is significantly slower than the execution of Java functions.

Listing 6. *J-CO-QL*$^+$: Alternative Javascript Functions `getYear` and `getWeek`.

```
5 alt. CREATE JAVASCRIPT FUNCTION getYear
         PARAMETERS yearWeek      TYPE String
         BODY {
           // yearWeek: is supposed to be in the form "YYYY-WW"
           if (yearWeek == null)
             return 0;
           var y = parseInt (yearWeek.substring (0,4), 10);
           return y;
         } END BODY;

6 alt. CREATE JAVASCRIPT FUNCTION getWeek
         PARAMETERS yearWeek      TYPE String
         BODY {
           // yearWeek: is supposed to be in the form "YYYY-WW"
           if (yearWeek == null)
             return 0;
           var w = parseInt (yearWeek.substring (5,7), 10);
           return w;
         } END BODY;
```

6.3 Query

Before we present the *J-CO-QL*$^+$ query for the case study, it is worth introducing shortly the data and execution models of the language (readers can find details in [6,20]).

- A *JSON* document is the basic computational unit. *JSON* documents are grouped within "collections": an instruction takes one or two collections as input and generates a new collection.
- A query (i.e., script) in *J-CO-QL*$^+$ is a sequence of instructions. They constitute a "piped flow", in such a way an instruction receives a "temporary collection" (generated by the previous instruction) as input and possibly generates a novel temporary collection as output. The "temporary" adjective denotes that the collection is not saved in any database, but is a temporary result of the process. Instructions can also acquire data either from *JSON* databases or from Web sources.

Listings 7 and 8 actually perform Query 1 (we divide the script in two parts for editorial issues); notice that, the script in Listing 7 starts from the line 7, because the definitions of the fuzzy operator, the three fuzzy aggregators and the two Java functions reported in previous listings are part of the query itself. Hereafter, we present how the script works.

Acquiring Data. On Line 7, the instruction `GET COLLECTION FROM WEB` retrieves the starting collection from the ECDC web site.

The collection is actually composed by one single *JSON* document and Fig. 4 shows an excerpt of it. Notice that the document, apart from meta-information fields such as `url` and `timestamp`, holds an array field (named `data`) that contains sub-documents, such that each contained-document reports a weekly record for a single country. We shortly present some fields, leaving out uninteresting fields for this case study: the field `country` reports the European country to which the weekly record is referred to; the field `population` reports the population of the country; the field `year_week` reports the referring week of the of the record; finally, the field `indicator` can assume only two values, `"cases"` and `"deaths"`, in order to specify that the value reported in the field `weekly_count` is referred, respectively, to the number of COVID-19 cases or to the number of COVID-19 deaths. Furthermore, notice that sub-documents in the array field `data` may not have always the same structure, as shown by the two sub-documents structures highlighted by yellow boxes in Fig. 4. For the sake of completeness, the array field `data` contains 12, 648 weekly records sub-documents.

Listing 7. *J-CO-QL*$^{+}$: Script 1.

```
7.   GET COLLECTION FROM WEB
      "https://opendata.ecdc.europa.eu/covid19/nationalcasedeath/json";

8.   EXPAND
       UNPACK   WITH ARRAY .data
       ARRAY    .data TO  .unpackedData;

9.   FILTER
       CASE WHERE WITH .unpackedData.item.weekly_count
             AND .unpackedData.item.indicator = "cases"
         GENERATE
           BUILD {
             .country      : .unpackedData.item.country,
             .population   : .unpackedData.item.population,
             .weeklyCases  : .unpackedData.item.weekly_count,
             .year         : getYear(.unpackedData.item.year_week),
             .week         : getWeek(.unpackedData.item.year_week)
           };

10.  FILTER
       CASE WHERE (.year = 2020 AND .week >= 40)
               OR (.year = 2021 AND .week <= 13)
         GENERATE
           CHECK FOR
             FUZZY SET ManyCasesRecords
               USING casesPopulationFO(.weeklyCases, .population);

11.  GROUP
       PARTITION  KNOWN FUZZY SETS ManyCasesRecords
             AND WITH   .country, .population,
                        .weeklyCases, .week, .year
         BY .country, .population INTO .countryData
         DROP GROUPING FIELDS;
```

Preparing Data. The role of the instruction `EXPAND` on Line 8 of Listing 7 is to unnest all the *JSON* sub-documents in the field `data` of the lonely document in the temporary collection depicted in Fig. 4; Fig. 5 depicts one unnested document, specifically the sub-document surrounded by the red-dashed box in Fig. 4, related to the COVID-19 cases in Luxembourg during the week 51 of 2020. From Fig. 5, notice how the instruction `EXPAND` works: for each document within the array field `data`, a novel document is generated into the output temporary collection; in this document (see Fig. 5), the novel field `unpackedData`) is further structured as a document with two fields, where

```
{
  "url"    : "https://opendata.ecdc.europa.eu/covid19/nationalcasedeath/json/",
  "timestamp" : "2024-04-12T17:30:47.110",
  "data"      : [
    {
      "continent"            : "Europe",
      "country"              : "Austria",
      "country_code"         : "AUT",
      "indicator"            : "cases",
      "note"                 : "",
      "population"           : 8978929,
      "source"               : "TESSy COVID-19",
      "year_week"            : "2020-01"
    },
    { … },                                  // other sub document
    {
      "continent"            : "Europe",
      "country"              : "Luxembourg",
      "country_code"         : "LUX",
      "cumulative_count"     : 37094,
      "indicator"            : "cases",
      "note"                 : "",
      "population"           : 645397,
      "rate_14_day"          : 908.588,
      "source"               : "TESSy COVID-19",
      "weekly_count"         : 2623,
      "year_week"            : "2020-51"
    },
    { … },                                  // other sub document
    {
      "continent"            : "Europe",
      "country"              : "Sweden",
      "country_code"         : "SWE",
      "cumulative_count"     : 25568,
      "indicator"            : "deaths",
      "note"                 : "",
      "population"           : 10452326,
      "rate_14_day"          : 15.4033,
      "source"               : "TESSy COVID-19",
      "weekly_count"         : 1,
      "year_week"            : "2023-47"
    }
  ]
}
```

Fig. 4. Excerpt of document collection retrieved by instruction 7.

the field `item` contains the unnested document, while the field `position` denotes the position occupied by the unnested document in the original array. Thus, after the execution of the instruction `EXPAND`, the generated temporary collection contains 12, 648 documents reporting both weekly records about COVID-19 cases and weekly records about COVID-19 deaths.

The role of the instruction `FILTER` on Line 9 of Listing 7 is to select only documents that describe weekly records related to COVID-19 cases from the current temporary collection. The instruction also restructures documents in order to be easily managed afterwards.

Hereafter, we explain it in details. For each document in the current temporary collection, the clause `CASE WHERE` selects only those documents related to

```
{
  "url" : "https://opendata.ecdc.europa.eu/covid19/nationalcasedeath/json/",
  "timestamp"    : "2024-04-12T17:30:47.110",
  "unpackedData" : {
     "position"      : 8211,
     "item"          : {
       "continent"             : "Europe",
       "country"               : "Luxembourg",
       "country_code"          : "LUX",
       "cumulative_count"      : 37094,
       "indicator"             : "cases",
       "note"                  : "",
       "population"            : 645397,
       "rate_14_day"           : 908.588,
       "source"                : "TESSy COVID-19",
       "weekly_count"          : 2623,
       "year_week"             : "2020-51"
     }
  }
}
```


Fig. 5. Example of document in collection provided by instruction 8.

weekly records of COVID-19 cases that have been actually recorded (the field `expandedData.item.weekly_count` must be present), while the other documents are discarded. The remainder of the instruction works on the selected documents: for each of them, the following section `GENERATE` generates a novel document, by means of the block `BUILD`, where only the fields named `country`, `population` and `weekly_count` (renamed as `weeklyCases`) are reported at root-level. Moreover, two novel fields, named `year` and `week`, are appended by exploiting, respectively, the Java functions `getYear` and `getWeek`, previously defined in Listing 5.

Figure 6 shows how the document depicted in Fig. 5 is selected and transformed by the `FILTER` instruction. After the execution of the `FILTER` instruction on Line 9, the current temporary collection contains 6, 016 documents.


```
{
  "country"      : "Luxembourg",
  "population"   : 645397,
  "week"         : 51,
  "weeklyCases"  : 2623,
  "year"         : 2020
}
```


Fig. 6. Example of document in collection provided by instruction 9.

The role of the following `FILTER` instruction on Line 10 of Listing 5 is to select, from the temporary collection, only those weekly records related to COVID-19 cases in the period of interest.

Hereafter, we explain the instruction in details. The clause `CASE WHERE` selects only those documents that describe cases recorded in the period from October 2020 (week 40) to March 2021 (week 13). For each selected document, the following clause GENERATE leaves the document unchanged. Furthermore, it "fuzzifies" the document through the clause CHECK FOR is used: the branch FUZZY SET evaluates the membership degree of the document to the fuzzy set ManyCasesRecords; the USING sub-clause provides the soft condition, that is based on the fuzzy operator casesPopulationFO, previously defined in Listing 1; the operator is called by passing (as actual parameters) the values of the fields weeklyCases and population, so as to compute a membership degree. The fuzzification of the document implies that the special field ~fuzzysets is added at root-level: this field contains a field named ManyCasesRecords, whose value represents the membership degree to the fuzzy set.

Figure 7 shows how the document depicted in Fig. 6 is selected and fuzzified by the FILTER instruction on Line 10: notice the presence of the newly added special field ~fuzzysets, highlighted in the red box, holding inside the field ManyCasesRecords with the membership degree of the eponym fuzzy set.

After the execution of the instruction FILTER on Line 10, the current temporary collection contains 837 documents.

```
{
  "country"       : "Luxembourg",
  "population"    : 645397,
  "week"          : 51,
  "weeklyCases"   : 2623,
  "year"          : 2020,
  "~fuzzysets"    : {
    "ManyCasesRecords"    : 0.917736066341084
  }
}
```

Fig. 7. Example of document in collection provided by instruction 10.

```
{
  "country"       : "Luxembourg",
  "population"    : 645397,
  "countryData"   : [
    { ... },                              // other sub document
    ...,
    {
      "week"          : 51,
      "weeklyCases"   : 2623,
      "year"          : 2020,
      "~fuzzysets"    : {
        "ManyCasesRecords"    : 0.917736066341084
      }
    },
    ...,
    { ... }                               // other sub document
  ]
}
```

Fig. 8. Example of document in collection provided by instruction 11.

The role of the GROUP instruction on Line 11 of Listing 5 is to group together all the weekly records in the temporary collection related to the same country.

Hereafter, we explain it in details. The clause PARTITION selects all those documents that belong to the fuzzy set ManyCasesRecords and have the fields country, population, weeklyCases, year and week. The clause BY creates, for all documents with the same value in the fields country and population, one single novel document with the fields country and population at root-level; furthermore, as specified by the clause INTO, an array field named countryData, which contains all source documents grouped together (as specified by the option DROP GROUPING FIELDS, from these documents the grouping fields country and population are removed).

Figure 8 shows an example of documents generated by the instruction GROUP. Notice how the document depicted in Fig. 7 is now included within the array field countryData (highlighted with a yellow box). After the execution of the GROUP instruction, the current temporary collection contains 31 documents, one for each country of the European Union plus a few others of particular interest like Norway.

Listing 8. *J-CO-QL*$^{+}$: Script 2.

```
12.  FILTER
       CASE WHERE WITH .countryData
         GENERATE
           CHECK FOR
             FUZZY SET ManyCasesCountries
               USING integrateCasesFA (EXTRACT_ARRAY(.weeklyCases
                              FROM ARRAY .countryData), .population),
             FUZZY SET ManyPeaksCountries
               USING owaFA (MEMBERSHIP_ARRAY (ManyCasesRecords
                              FROM ARRAY .countryData)),
             FUZZY SET WantedCountries
               USING weightedMembershipsFA (MEMBERSHIP_ARRAY(
                   [ManyCasesCountries, ManyPeaksCountries]), [2, 1]);

13.  FILTER
       CASE WHERE KNOWN FUZZY SETS WantedCountries
         GENERATE
           ALPHACUT 0.80 ON WantedCountries
           BUILD {
             .country     : .country,
             .population  : .population,
             .ranking     : MEMBERSHIP_TO (WantedCountries)
           }
           DEFUZZIFY;

14.  USE DB Webist2024
       ON SERVER MongoDB 'http://127.0.0.1:27017';

15.  SAVE AS ecdcCovidCasesAnalysis@Webist2024;
```

Soft Querying with Fuzzy Aggregators. The instruction FILTER on Line 12 of Listing 8 actually performs the soft query. Hereafter, we explain it in details.

The clause CASE WHERE selects those documents that have the field countryData. The block GENERATE leaves the documents unchanged, apart from the fact that it evaluates their belonging to fuzzy sets. This time the clause CHECK FOR contains three different branches, each one devoted to compute the membership degree to a specific fuzzy set:

- The first branch FUZZY SET evaluates the membership to the fuzzy set ManyCasesCountries. To do this, the soft condition specified after the USING keyword exploits the fuzzy aggregator integrateCasesFA, defined in Listing 2. To call the fuzzy aggregator, an array of numbers must be provided as first actual parameter: since the array field countryData contains nested documents, the built-in function EXTRACT_ARRAY creates a novel array of numbers by projecting the array field countryData on the inner (numerical) field weeklyCases. The second actual parameter for the fuzzy aggregator is the field population.
  The evaluation of this first fuzzy set implies adding, at root level, the special field ~fuzzysets; within it, the field ManyCasesCountries reports the membership degree.
- The second branch FUZZY SET evaluates the membership of the current document to the fuzzy set ManyPeaksCountries, through the fuzzy aggregator owaFA defined in Listing 3.
  Apart from the fact that the aggregator owaFA performs an OWA aggregation (instead of a cumulative aggregation), the branch behaves similarly to the previous one: this time, the aggregator is called by passing an "array of memberships" obtained by the built-in function MEMBERSHIP_ARRAY, which projects the array countryData on the fuzzy set ManyCasesRecords contained in the inner special field ~fuzzysets.
  After the evaluation of this second fuzzy set, the field ManyPeaksCountries is added to the special field ~fuzzysets at root-level.
- The third branch FUZZY SET evaluates the degree of the current document to the fuzzy set WantedCountries. The goal is to combine the memberships to the fuzzy sets ManyCasesCountries and ManyPeaksCountries in such a way the former contributes with weight 2, while the latter contributes with weight 1.
  To perform this task, the fuzzy aggregator weightedMembershipsFA (reported in Listing 4) is exploited. This time, the "array of memberships", needed as first actual parameter, is retrieved by exploiting a different behaviour of the built-in function MEMBERSHIP_ARRAY: the function creates a novel array by taking the previously-calculated memberships to the listed fuzzy sets (i.e., ManyCasesCountries and ManyPeaksCountries), whose values are taken from the special field ~fuzzysets. The second actual parameter is a constant array that reports the weights to apply (i.e., 2 and 1, respectively); this way, requirements in Query 1 are met.
  The third and final field is added to the special field ~fuzzysets, with the membership to the fuzzy set WantedCountries.

Figure 9 shows the document presented in Fig. 8 after the execution of the FILTER instruction on Line 12. Notice the special field ~fuzzysets, with the membership degrees to the fuzzy sets ManyCasesCountries ManyPeaksCountries and WantedCountries that are highlighted by a blue box.

After the execution of the instruction FILTER on Line 12, the temporary collection still contains 31 documents. Table 1 reports, for each document/country, the fields country and population, as well as the membership degrees to the three fuzzy sets ManyCasesCountries ManyPeaksCountries and WantedCountries.

```
{
  "country"         : "Luxembourg",
  "population"      : 645397,
  "countryData"     : [
    { … },                                    // other sub document
    …,
    {
      "week"              : 51,
      "weeklyCases"       : 2623,
      "year"              : 2020,
      "~fuzzysets"        : {
        "ManyCasesRecords"      : 0.917736066341084
      }
    },
    …,
    { … }                                     // other sub document
  ],
  "~fuzzysets"  : {
    "ManyCasesCountries"    : 0.900150146097813,
    "ManyPeaksCountries"    : 0.753436097315063,
    "WantedCountries"       : 0.802340780242646
  }
}
```

Fig. 9. Example of document in collection provided by instruction 12.

At this point, we want to repeat the disclaimer we made at the beginning of Sect. 6.1: any medical/political consideration is out of the scope of this paper.

The instruction `FILTER` on Line 13 of Line 12 of Listing 8 concludes the soft query.

The clause `CASE WHERE` selects only those documents that belong to the fuzzy set `WantedCountries`.

In the section `GENERATE`, the clause `ALPHACUT` discards those documents whose membership to the fuzzy set `WantedCountries` is less than 0.80. This way, only documents that describe countries that actually had a high number of COVID-19 cases, together with many peaks of weekly COVID-19 cases during the monitored period (or very close to this situation) are selected.

The final block `BUILD` restructures the output documents and the option `DEFUZZIFY` removes the special filed `~fuzzysets`, so as documents become again classical crisp *JSON* documents.

Figure 10 shows how the document shown in Fig. 9 is selected and restructured by the instruction `FILTER` on Line 13 of Listing 8.

After the execution of the instruction `FILTER` on Line 13, the temporary collection now contains 7 documents, i.e., those corresponding to the rows above the thick line in Table 1.

Saving the Results. The instruction `USE DB` on Line 14 of Listing 8 specifies the database to connect with. Specifically, notice that the database `Webist2024` is managed by *MongoDB*.

Table 1. Tabular representation of the temporary collection generated by Line 12 of Listing 8.

country	population	ManyCasesCountries	ManyPeaksCountries	WantedCountries
Czechia	10,516,707	1.00	0.95	0.97
Slovakia	5,434,712	1.00	0.92	0.95
Slovenia	2,107,180	1.00	0.90	0.93
Sweden	10,452,326	0.88	0.82	0.84
Estonia	1,331,796	0.90	0.78	0.82
Luxembourg	645,397	0.90	0.75	0.80
Netherlands	17,590,672	0.87	0.77	0.80
Portugal	10,352,042	0.88	0.74	0.79
Lithuania	2,805,998	0.90	0.72	0.78
Hungary	9,689,010	0.87	0.73	0.78
Croatia	3,862,305	0.86	0.71	0.76
Liechtenstein	39,308	0.86	0.69	0.75
Poland	37,654,247	0.85	0.67	0.73
Latvia	1,875,757	0.82	0.68	0.72
Italy	59,030,133	0.83	0.66	0.71
Belgium	11,617,623	0.87	0.63	0.71
Cyprus	904,705	0.81	0.65	0.71
France	67,871,925	0.84	0.64	0.71
EU/EEA (total)	452,576,117	0.82	0.62	0.69
Austria	8,978,929	0.83	0.61	0.68
Spain	47,432,893	0.82	0.61	0.68
Bulgaria	6,838,937	0.78	0.61	0.67
Malta	520,971	0.80	0.57	0.65
Romania	19,042,455	0.71	0.52	0.58
Ireland	5,060,004	0.63	0.37	0.46
Denmark	5,873,420	0.45	0.30	0.35
Germany	83,237,124	0.47	0.28	0.35
Greece	10,459,782	0.37	0.24	0.28
Norway	5,425,270	0.20	0.08	0.12
Finland	5,548,241	0.15	0.06	0.09
Iceland	376,248	0.10	0.08	0.08

```
{
  "country"      : "Luxembourg",
  "population"   : 645397,
  "ranking"      : 0.802340780242646
}
```

Fig. 10. Example of document saved in the collection `ecdcCovidCasesAnalysis` by instruction 15.

Finally, the resulting collection, which contains documents describing countries satisfying the soft query formalized by Query 1, is finally saved by the instruction `SAVE AS` on Line 15, into the database `Webist2024` with the name `ecdcCovidCasesAnalysis`.

7 Conclusions

In this paper, we present the potential application of *Soft Web Intelligence* for acquiring and analyzing *JSON* data sets from Web sources, by exploiting soft-querying capabilities provided by fuzzy operators and fuzzy aggregators within the *J-CO-QL*$^+$ language. We showed that users can define their own operators and aggregators, to deal with possibly complex situations in which ranking capabilities are necessary to deal with imprecise contexts.

The paper resumes the vision of *Soft Web Intelligence*; then, it briefly describes the *J-CO* Framework and how fuzzy sets are basically supported by *J-CO-QL*$^+$. The concept of fuzzy aggregator is specifically presented (it is the novel contribution of our current research) in a dedicated section. Finally, a practical case study that exploits a real data set about records of COVID-19 cases is presented, to show the practical application of the concepts and illustrate the potential use of *J-CO-QL*$^+$ for non programmers in *Soft Web Intelligence* tasks.

As a future work, we will investigate the unification of fuzzy operators and fuzzy aggregator in one single concept, so as to provide a cleaner semantic model. We will also investigate the integration of aggregation within the novel concept, recently introduced in *J-CO-QL*$^+$ [13], of "multi-grade models for fuzzy sets", i.e., models of fuzzy sets in which multiple degrees characterize the belonging of an item to a fuzzy set.

The framework is available on a Github page[3].

References

1. Alahakoon, D., Yu, X.: Smart electricity meter data intelligence for future energy systems: a survey. IEEE Trans. Ind. Inf. **12**(1), 425–436 (2015)
2. Garcia Bringas, P., Pastor, I., Psaila, G.: Can blockchain technology provide information systems with trusted database? The case of hyperledger fabric. In: Cuzzocrea, A., Greco, S., Larsen, H.L., Saccà, D., Andreasen, T., Christiansen, H. (eds.) FQAS 2019. LNCS (LNAI), vol. 11529, pp. 265–277. Springer, Cham (2019). https://doi.org/10.1007/978-3-030-27629-4_25
3. Dombi, J., Jónás, T.: Weighted aggregation systems and an expectation level-based weighting and scoring procedure. Eur. J. Oper. Res. **299**(2), 580–588 (2022)
4. Fosci, P., Marrara, S., Psaila, G.: Geosoft: a language for soft querying features within geojson information layers. In: International Conference on Web Information Systems and Technologies, pp. 196–219. Springer, Cham (2020)
5. Fosci, P., Psaila, G.: J-co, a framework for fuzzy querying collections of json documents. In: Flexible Query Answering Systems: 14th International Conference, FQAS 2021, Bratislava, Slovakia, 19–24 September 2021, Proceedings 14, pp. 142–153. Springer, Cham (2021)
6. Fosci, P., Psaila, G.: Towards flexible retrieval, integration and analysis of json data sets through fuzzy sets: a case study. Information **12**(7), 258 (2021)
7. Fosci, P., Psaila, G.: Soft integration of geo-tagged data sets in J-CO-QL+. ISPRS Int. J. Geo Inf. **11**(9), 484 (2022)
8. Fosci, P., Psaila, G.: Soft web intelligence with the j-co framework. In: International Conference on Web Information Systems and Technologies, pp. 142–165. Springer, Cham (2022)

[3] Github repository of the *J-CO* Framework: https://github.com/JcoProjectTeam/JcoProjectPage, accessed on 12/04/2024.

9. Fosci, P., Psaila, G.: Towards soft web intelligence by collecting and processing json data sets from web sources. In: Proceedings of the 18th International Conference on Web Information Systems and Technologies (2022)
10. Fosci, P., Psaila, G.: Enhancing soft web intelligence with user-defined fuzzy aggregators. In: WEBIST, vol. 1, pp. 258–267. Science and Technology Publications, Lda (2023)
11. Fosci, P., Psaila, G.: Fuzzy aggregators in practice: meta-model and implementation. In: International Conference on Soft Computing Models in Industrial and Environmental Applications, pp. 56–68. Springer, Cham (2023)
12. Fosci, P., Psaila, G.: Soft querying powered by user-defined functions in J-CO-QL+. Neurocomputing **546**, 126311 (2023)
13. Fosci, P., Psaila, G.: A unified view of multi-grade fuzzy-set models in J-CO-QL+. Neurocomputing **565**, 126968 (2024)
14. Han, J., Chang, K.C.: Data mining for web intelligence. Computer **35**(11), 64–70 (2002)
15. Kacprzyk, J., Zadrożny, S.: Soft computing and web intelligence for supporting consensus reaching. Soft. Comput. **14**(8), 833–846 (2010)
16. Li, H., Yen, V.C.: Fuzzy Sets and Fuzzy Decision-Making. CRC Press (1995)
17. Negash, S., Gray, P.: Business intelligence. In: Handbook on Decision Support Systems 2, pp. 175–193. Springer, Cham (2008)
18. Poli, V.S.R.: Fuzzy data mining and web intelligence. In: International Conference on Fuzzy Theory and Its Applications (iFUZZY), pp. 74–79. IEEE (2015)
19. Psaila, G., Fosci, P.: Toward an anayist-oriented polystore framework for processing json geo-data. In: International Conference on Applied Computing 2018, Budapest; Hungary, 21–23 October 2018, pp. 213–222. IADIS (2018)
20. Psaila, G., Fosci, P.: J-co: a platform-independent framework for managing geo-referenced json data sets. Electronics **10**(5), 621 (2021)
21. Reddy, P.: Fuzzyalgol: fuzzy algorithmic language for designing fuzzy algorithms. J. Comput. Sci. Eng. **2**(2), 21–24 (2010)
22. Yager, R.R.: On ordered weighted averaging aggregation operators in multicriteria decision making. IEEE Trans. Syst. Man Cybern. **18**(1), 183–190 (1988)
23. Yao, Y., Zhong, N., Liu, J., Ohsuga, S.: Web intelligence (WI) research challenges and trends in the new information age. In: Asia-Pacific Conference on Web Intelligence, pp. 1–17. Springer, Cham (2001)
24. Zadeh, L.A.: Fuzzy sets. Inf. Control **8**(3), 338–353 (1965)
25. Zadeh, L.A.: A note on web intelligence, world knowledge and fuzzy logic. Data Knowl. Eng. **50**(3), 291–304 (2004)
26. Zadeh, L.A.: Web intelligence, world knowledge and fuzzy logic–the concept of web IQ (WIQ). In: International Conference on Knowledge-Based and Intelligent Information and Engineering Systems, pp. 1–5. Springer, Cham (2004)
27. Zhang, Y.Q., Lin, T.Y.: Computational web intelligence (CWI): synergy of computational intelligence and web technology. In: World Congress on Computational Intelligence, vol. 2, pp. 1104–1107. IEEE (2002)
28. Zhong, N., et al.: Web intelligence meets brain informatics. In: International Workshop on Web Intelligence Meets Brain Informatics, pp. 1–31. Springer, Cham (2006)
29. Zuccala, A., Thelwall, M., Oppenheim, C., Dhiensa, R.: Web intelligence analyses of digital libraries: a case study of the national electronic library for health (NeLH). J. Documentation (2007)

Matching Pre-processed Database Records Using Natural Language Queries on Advertisements

Yiu-Kai Ng(✉)

Computer Science Department, Brigham Young University, Provo, UT 84602, USA
ng@compsci.byu.edu

Abstract. Many commercial websites, such as Target.com, which aspire to increase client's transactions and thus profits, offer users easy-to-use pull-down menus and/or keyword searching tools to create advertisement (ads for short) queries that can be posted at their sites to extract the desired products. These websites, however, cannot handle natural language queries, which are formulated for specific information needs and can only be processed properly by natural language query processing (NLQP) systems. To solve the problem encountered by these commercial websites, we have developed a novel NLQP system, denoted *AdProc*, which retrieves database (DB for short) records that match information specified in ads queries on multiple ads domains. *AdProc* relies on an underlying database, which contains pre-processed (ads) records that provides the source of answers to users' queries. *AdProc* automates the process of populating a DB using online ads and answering user queries on multiple ads domains. If there is no existing DB record that satisfies all the constraints specified in a user query, partially-matched DB records are extracted and ranked to meet the user's information need, instead of retrieving no results. With the developed partial-match approach, *AdProc* allows users to examine partially-matched DB records, in addition to exact-matched records. The additional DB records are appealing to the end users, assuming that there is no exact-matched results or there are only a few retrieved results. Besides verifying the novelty of our exact-match approach, we have also included in the empirical study an evaluation to verify the merit of *AdProc* in retrieving partially-matched DB records. The performance evaluation shows that exact-matched records, as well as partially-matched records, are perceived as preferable by ordinary users based on real users' assessments and the proposed natural language querying system outperforms a number of baseline ranking systems.

Keywords: Query processing · Databases · Classification · Data extraction

This work is based on an earlier work, "Easy to Find: A Natural Language Query Processing System on Advertisements Using an Automatically Populated Database", appeared in Proceedings of the 19th International Conference on Web Information Systems and Technologies (WEBIST), 2023.

M. Marchiori and F. García Peñalvo (Eds.): WEBIST 2023, LNBIP 543, pp. 101–132, 2025.
https://doi.org/10.1007/978-3-031-89621-7_6

1 Introduction

The Web is a perfect forum for information exchange and access, since most websites allow users to freely extract archived and newly-created documents anytime and anywhere. The number of (un-/semi-)structured web pages has made the Web a huge repository of information of various kinds, which include advertisements (ads for short). While users can search for information using keyword-based and phrase-based queries, or even advanced searches that include simple Boolean operators on websites, these websites cannot process complex queries, which can only be handled properly by natural language query processing (NLQP) systems. In this paper, we introduce a closed-domain NLQP system, denoted *AdProc*, which retrieves exactly- and partially-matched answers to (Boolean, incomplete, or ambiguous) queries on ads. *AdProc* can answer queries on multiple ads domains using (i) an enhanced Naïve Bayes classifier to determine to which *ads domain* a user's query belongs and (ii) an underlying database (DB for short) with populated ads records, which serves as the source of answers to ads queries. Manually populating online data to a DB is not feasible, since it is a labor-intensive, time-consuming, and impractical process. *AdProc* applies the (i) Joint Beta-Binomial classifier to *classify* ads according to their domains, (ii) Support Vector Machine (SVM) to *label* non-stop keywords in ads based on their types, (iii) decision trees to *extract* previously-labeled keywords in an ad that are valid DB attribute values to generate a DB record, and (iv) feature set of attribute values in answering ads queries.

AdProc utilizes the *efficiency* of relational DB systems in managing data (records) and avoids the burden of *analyzing* the syntax and semantics of a user's natural language query Q such that interpreting the semantics of the keywords in Q employs at most the simple context-switching analysis, which facilitates the process of *transforming* the information needs expressed in Q into a SQL query to be evaluated against the underlying data(base records). Furthermore, *AdProc* combines the tasks of *classifying* and *extracting* data from ads into a single, automated process for populating a DB and retrieving relevant data.

AdProc applies information retrieval techniques and machine learning approaches that are simple and widely-adapted in its design. Compared with using popular deep learning models, the latter need massive data for training, require high computational power for processing data, and are lack of transparency, which are excessive, undesirable, intricate, and vague in designing *AdProc*.

The effectiveness and merits of *AdProc* have been experimentally verified in terms of achieving high accuracy of interpretation for complex queries. Furthermore, it has been validated that *AdProc* retrieves relevant results to the user with the accuracy ratio in the upper 90 percentile.

The remaining of this paper is organized as follows. In Sect. 2, we present works related to *AdProc*. In Sect. 3, we introduce the process of populating data from online ads in multiple domains to a relational DB and determining to which domain a query belongs. In Sect. 4, we detail the design of retrieving exact-matched DB records and ranking partially-matched DB records in answering a natural language query Q. In Sect. 5, we include the empirical study conducted to verify the effectiveness and efficiency of *AdProc* in populating ads and answering ads queries. In Sect. 6, we give a conclusion and discuss how to extend the current works.

The contributions of our research work include

(i) Developing an ads NLQP system that automates the process of *classifying*, *extracting*, and *populating* data from online ads.
(ii) Relying on simple probabilistic models to determine to which *domain* an ad query belongs.
(iii) Handling the retrieval of exact- and partially-matched DB records on multiple closed ads domains as answers to a user query.

2 Related Work

A number of methods have been proposed in the past in solving the classification problem. Dayanik et al. [9] perform classification using a Bayesian logistic regression framework that combines domain knowledge and supervised learning, whereas Xue et al. [34] extend the Probabilistic Latent Semantic Analysis algorithm by integrating labeled training data and unlabeled test data. As opposed to [9,34], the classifier of *AdProc* adopts a Beta-Binomial model, which considers the varying probability of word occurrence among documents in the same class for text classification. While *AdProc* can classify documents in various domains, the Relaxed Online SVM in [31] has only been evaluated for spam email classification, which involves only highly separable data.

Nguyen et al. [26] introduce a machine learning approach for labeling attributes from web form interfaces. Similar to our proposed tagging method, the labeling approach in [26] matches a form element (i.e., a keyword in an ad in our case) with its corresponding textual description (i.e., a type in our case). *AdProc*, however, avoids learning structural patterns required by the labeling approach in [26] for performing the keyword-tagging task.

AdProc relies on Support Vector Machine (SVM) for labeling keywords, which differs from the Hidden Markov Model (HMM) approach in [13] that labels components of web interfaces, such as text labels or text boxes. HMMs have been successfully adopted for extracting data from text documents which are later populated into DBs [17]. Since training a HMM with data from multiple domains is ineffective, Liu et al. [17] first group analogous data and train separate HMMs, which however is a complex process.

Rajput and Haider [29] apply ontologies, various information extraction techniques, and Bayesian Networks for extracting and predicting missing information from unstructured, ungrammatical, and incoherent data sources, such as online ads extracted from Craigslist.org. Although effective, the proposed model has not been validated for text classification using more than one ads domain.

In developing a closed-domain NLQP system for the construction business, Zhang et al. [36] rely on a thesaurus, which consists of a set of domain-specific concepts and the relations among them organized into a hierarchy, whereas the NLQP system in [10] re-ranks publications retrieved by the PubMed search engine for a given query Q so that the publication abstracts that contain medical terms relevant to the ones in Q are positioned higher in the answer set. Wang and Luo [32] present a Chinese NLQP system that answers queries in the telecom product domain by (i) applying semantic role tagging and (ii) using Hownet, an online knowledge based which describes the inter-conceptual and inter-attribute relations of Chinese and English words. On the contrary, *AdProc* depends only on a few predefined attribute types to retrieve potential answers to a user query, which is simple and effective.

Ko et al. [14] introduce a graphical model which relies on (i) the probability of an answer being correct and (ii) the joint probabilities of answers to establish the correlation among them and determine an answer list. The probabilistic approach in [28] relies on existing search engines to retrieve a set of relevant documents S to a query Q and extracts candidate answers to Q from S and ranks each answer based on a proximity and signature score. The strategies adopted in [14, 28] and the learning-based ranking strategy applied in [2] require user relevance feedback, which imposes additional workload.

A similar work on natural-language query-processing system was presented by Y. Ng in [25]. The author of [25] proposes an exact-matching approach that extracts from DB any product matches exactly the constraints specified in a user query. This behavior is sufficient if there are many exactly-matched DB records. However, in the case that no exact matches or very few are found, it may be frustrating to the user. Our natural-language query processing system, however, supplements exact-matched query results with partial matches. Relevant products are searched for, first exact matches and then partial matches, until some threshold number of products is achieved. This approach significantly enhances the diversity and quantity of the results retrieved by the system proposed by Y. Ng [25].

3 An Ads NLQP System

In this section, we discuss ads data types (in Sect. 3.1) and describe the three consecutive tasks of *AdProc*: (i) *classifying* online ads into their respective domains (in Sect. 3.2), (ii) *tagging* keywords in online ads according to their types (in Sect. 3.3), and (iii) *extracting* the previously-tagged non-Types IV keywords in ads and populate them as attribute values in the corresponding DB records (in Sect. 3.4) to fully automate the process of extracting online ads data to generate the underlying DB records.

3.1 Data Types Used in *AdProc*

In populating online ads and answering queries in multiple domains, *AdProc* relies on the enhanced Naïve Bayes classification model (see Sect. 3.2 for details), which determines the ad domain to which a user's query or an ad belongs.

AdProc considers various attribute types (from Types I to IV) proposed in this paper. The domain of a *Type I* attribute is (alphanumerical) character strings, and each of these attributes in a DB record is assigned a *single value*, which is not a numerical value (as Type III attributes are). A sample Type I attribute in the jewelry ads domain is "Category" and "necklace" is one of its possible values. Values of *Type II* attributes are (alphanumerical) character strings that describe the *properties* of the product in an ad. "Description" is a Type II attribute in the house-for-sale ads domain, and "central air" is one of its possible values. *Type III* attributes are of type numeric, such as "Salary" in the job ads domain and "70,000" is one of its possible values. Attributes that are neither Type I, II, nor III are treated as Type IV.

- *Type I* attribute values in an ad A yield the *unique identifier* of A that are required values to be included in A (its DB record, respectively). Type I attributes are *primary-indexed* fields of the relational schema which defines the corresponding ads domain. Sample Type I attributes in the Cars-for-Sale ads domain are "Maker" and "Model", and "Toyota" and "Camry" are one of their respective values.
- *Type II* attribute values describe the *properties* of an ad A, which are not required values in A. Type II attributes are *secondary-indexed* fields in the corresponding relational schema. "Color" is a Type II attribute in the Car-for-Sale ads domain, and "Blue" is one of its domain values.
- *Type III* attribute values specify the *quantitative values* in an ad. A sample *Type III* attribute value is "usd", which identifies the unit of "Price" (a *Type III* attribute) in the Cars-for-Sale ads domain.
- *Type IV* attribute values are non-essential, cosmetic keywords in ads, such as "large", which are not extracted as attribute values from any online ads.

3.2 Classification of Advertisements

In classifying online ads, we adopt a Naïve Bayes classifier, which is a popular text classification approach, since it is simple, easy to implement, robust, highly scalable, and domain independent. Even though Naïve Bayes classifiers assume mutual independence of attributes, they achieve high accuracy in text classification [20].

Joint Beta-Binomial Sampling Model. The well-known Naïve Bayes' Theorem, which considers the conditional distribution of the class variable, computes the probability of assigning the natural class c to a document d as

$$P(c|d) = \frac{P(d|c)P(c)}{P(d)} \tag{1}$$

where $P(d)$ is the probability of a given document d, $P(c)$ is the probability of a particular natural class c, which is computed as the number of training instances assigned to class c over the total number of training instances, and $P(d \mid c)$ is the probability of d given that class c occurred[1].

To estimate $P(d \mid c)$ in the Bayes' Theorem, we use the *Joint Beta-Binomial Sampling Model* introduced in [1], denoted JBBSM, which considers the "burstiness" of a word, i.e., a word is more likely to occur again in d if it has already appeared once in d. JBBSM computes $P(d \mid c)$ as a sequence of probabilities of the form $P(d_j \mid c)$, which is the probability of the jth $(1 \leq j \leq n)$ keyword in d, i.e., d_j, given a particular class c[2]:

$$P_{bb}(d_j \mid \alpha_j, \beta_j) = \binom{n}{d_j} \frac{B(d_j + \alpha_j, n - d_j + \beta_j)}{B(\alpha_j, \beta_j)} \tag{2}$$

where n is the length of d, which is the total number of keyword counts in d, B is the Beta function of JBBSM, and α_j and β_j are the parameters[3] that estimate the presence and absence of d_j in a particular class c, respectively.

The following equations estimate the parameter values of α_j and β_j in the Beta-Binomial model as defined in [1]:

$$\frac{\alpha_j}{\alpha_j + \beta_j} = \frac{\sum_{i=1}^{k} \hat{\theta_{i,j}}}{k} \tag{3}$$

$$\frac{\alpha_j \beta_j}{(\alpha_j + \beta_j)^2 (\alpha_j + \beta_j + 1)} = \frac{\sum_{i=1}^{k} (\hat{\theta_{i,j}} - E[\theta_j])^2}{k} \tag{4}$$

where k is the size of the document collection used for training the classification model, $\theta_{i,j}$ is the expected distribution of the jth keyword in the ith $(1 \leq i \leq k)$ document in the collection, $\hat{\theta_{i,j}}$ is the maximum likelihood estimate of the parameter $\theta_{i,j}$, θ_j is the probability of selecting the jth word in d, $E[\theta_j]$ $(= \frac{1}{k} \times \sum_{i=1}^{k} \theta_{i,j})$ is the expected value for the distribution over θ_j, and $(\hat{\theta_{i,j}} - E[\theta_j])$ yields the error in estimating the parameter $\theta_{i,j}$.

Using JBBSM, the probability $P(d \mid c)$ is

$$P(d \mid \alpha, \beta) = \prod_{j} P_{bb}(d_j \mid \alpha_j, \beta j) \tag{5}$$

where α and β are parameters of the Beta-Binomial distribution[4] of d, and $P_{bb}(d_j \mid \alpha_j, \beta_j)$ is as defined in Eq. 2.

In choosing the ads domain to which an ad d should be assigned, we compute the conditional probability $P(c \mid d)$ for each predefined ad domain. We assign to d the domain that yields the highest $P(c \mid d)$ among the set of predefined ads domains, C, i.e., $Class(d) = argmax_{\ c \in C}\ P(c \mid d)$.

[1] From now on, unless stated otherwise, d and c in $P(c|d)$ denote an online ad and ads domain, respectively.

[2] Documents in JBBSM are represented as vectors of count-valued random variables, and thus in Eq. 2, d_j denotes the frequency count of the jth keyword in d.

[3] α_j and β_j are the parameters associated with a particular word j, which are computed using Eqs. 3 and 4 for each of the natural classes.

[4] α and β (α_j and β_j, respectively) are the parameters of the distribution of d (keyword j in d, respectively).

Feature Selection. As claimed by Yang and Petersen [35], one of the main problems in document classification is the high dimensionality of the feature space, i.e., the large number of unique keywords in documents, which affects the performance of classifier algorithms in terms of computational time. In solving this problem, we apply different feature selection strategies (given below), which select a subset of keywords to represent ads in a set of predefined ads domains, without affecting the accuracy of the chosen classifier. Furthermore, as claimed by Chouaib et al. [6], feature selection strategies are often applied to reduce irrelevant/misleading features.

During the feature selection process, we first remove *stopwords*, i.e., words with little meanings, such as articles, prepositions, punctuations, and conjunctions, which often do not represent the content of an ad, as well as numerical values. Hereafter, we apply the *gain* approach in [35], which defines a criterion for feature selection often adopted in the field of machine learning. Information gain, as defined below, determines the "goodness" of a keyword globally for classification, which yields keywords that are the most representative of ads in various predefined domains.

$$G(t) = -\sum_{i=1}^{m} P(c_i) log P(c_i) + P(t) \sum_{i=1}^{m} P(c_i|t) log P(c_i|t) + P(\bar{t}) \sum_{i=1}^{m} P(c_i|\bar{t})\ log\ P(c_i|\bar{t}) \quad (6)$$

where m is the number of distinct ads domains, $P(c_i)$ is the probability of domain c_i, $P(t)$ ($P(\bar{t})$, respectively) is the probability of occurrence (absence, respectively) of keyword t, and $P(c_i \mid t)$ ($P(c_i \mid \bar{t})$, respectively) is the probability of domain c_i given that t is present (absent, respectively).

After computing the gain of all the distinct keywords in a collection of ads used for training purpose, we select the top n keywords that have the *highest gain* for representing ads in the set of predefined ads domains. In defining the appropriate n, we conducted an empirical study using a total of 80,000 randomly selected online ads belonged to the eight domains introduced in Sect. 5.1. We considered alternative values of n, such that $n \in \{50, 100, 200, 500, 1000, 2000\}$, and set $n = 1000$. By using 1000 keywords on the 80,000 ads we achieve the *highest* classification accuracy and still maintain the classification processing time on ads within a minute.

3.3 Keyword Tagging Based on Types

To tag each non-stop, non-numerical[5] keyword in online ads according to their types, we rely on the Support Vector Machine (SVM) approach, since SVM is a robust methodology which has been shown to yield state of the art performance on classification [31]. SVM constructs hyper-planes as decision surfaces in a higher dimension space so that data becomes linearly separable and maintains a maximum margin of separation between positive and negative examples, i.e., binary training instances.

In describing SVMs, we use the notation in [31], which state that given a set S which contains N labeled input vectors, i.e., training instances, of the

[5] Numerical keywords are tagged as Type III by default.

form { (x_1, y_1), ..., (x_N, y_N) }, x_i $(1 \leq i \leq N)$ is an input vector of features describing a given example i, i.e., a keyword in an ad in our case, and $y_i \in$ {-1, 1} is the corresponding (binary) class label associated with x_i. Furthermore, $\varphi(x_i)$ is the kernel mapping that generates the corresponding vector in feature space, and $k(x_i, x_j) = \varphi(x_i) . \varphi(x_j)$ is the kernel function that computes the dot/scalar product between $\varphi(x_i)$ and $\varphi(x_j)$ to determine their distance in the feature space. In determining the support vectors of the (soft-margin) SVM, the following optimization problem[6] must first be solved:

$$max_\alpha\{ \sum_{i=1}^{N} \alpha_i - \sum_{i,j=1}^{N} \alpha_i\alpha_j y_i y_j k(x_i, x_j) \} \tag{7}$$

where $0 \leq \alpha_i \leq C$ and $\sum_{i,j}^{N} \alpha_i y_i = 0$, and C is a regularization parameter controlling the penalty for misclassification, α_i (α_j, respectively) is the Lagrange multiplier associated with x_i (x_j, respectively), and y_i (y_j, respectively) is the corresponding class of x_i (x_j, respectively). By solving Eq. 7, we can obtain the coefficients α_i $(1 \leq i \leq N)$, which determine the support vectors of the SVM. In implementing SVM, we empirically established C to be 1.0.

After training the SVM, the input vectors x_i with which the corresponding $\alpha_i > 0$ are the support vectors. Having selected the support vectors, we proceed to use the SVM to determine the proper tag of a keyword k in an ad. In our implementation of SVM we represent k as a vector v which consists of "1" and "0" values assigned to each of the features (introduced in Sect. 3.3) and the tag that should be assigned to k is determined by Eq. 8.

$$f(v) = sign(\sum_{i=1}^{N'} \alpha_i y_i . k(x_i, v) + b) \tag{8}$$

where b is the bias term determined during the training of the SVM and N' is the number of support vectors, i.e., input vectors for which $\alpha_i > 0$.

RBF Kernel. In implementing SVM for tagging ads keywords according to their types, we adopted Radial Basis Function (RBF) in Eq. 9 as the kernel function for the SVM, since RBF is one of the most typical kernels.

$$K(x_i, x_j) = exp(-\frac{||x_i - x_j||^2}{\sigma^2}) \tag{9}$$

where $||x_i - x_j||$ is the Euclidean distance computed between vectors x_i and x_j[7], and σ is the parameter that determines the area of influence of the corresponding

[6] As stated in [33], in solving constraint optimization problems it is a common practice to adopt the Lagrange multiplier method, since this method reduces a problem complexity with a minimal impact on optimality.

[7] x_i and x_j are two input vectors. In our case, each input vector represents the heuristics, i.e., features, of a keyword in an ad.

support vector. In our implementation of SVM, we have empirically determined the value of the parameter σ to be 500, which yields a smoother decision surface, since an RBF with σ set to be 500 allows a support vector to have a larger area of influence.

Multiclass-SVM. In tagging keywords based on their types, we cannot apply directly a binary SVM, since the adopted SVM must handle more than two types of attribute values. We consider one-against-all and one-against-one for solving the multi-class problem using binary SVM. Given j (≥ 1) different predefined types, the one-against-all approach constructs j binary SVM classifiers, each of which separates one type from the rest. Hereafter, the jth SVM is trained using the training instances in which the ones belonged to the jth type are given positive labels and the remaining instances negative ones [18]. The one-against-one approach, on the other hand, constructs $\frac{j(j-1)}{2}$ binary classifiers, one for each possible class pair. In classifying a new instance, a vote is added to the class selected by each of the $\frac{j(j-1)}{2}$ classifiers and the class with the most votes is the one chosen as the class of the new instance.

As claimed by [18], the most common strategy, which is the one we adopt, in implementing SVM is one-against-all. The classification of a new instance v, i.e., a feature-vector associated with a keyword in an ad, using our multiclass SVM is computed by selecting the class (among all the possible predefined classes, i.e., attribute values types) for which the corresponding $f(v)$ (in Eq. 8) is the highest.

Feature Representation. To train our multi-class SVM, each training instance, i.e., input vector, is a feature-vector associated with a non-stop, non-numerical keyword k in an ad, in which a "1" is assigned to each feature f (introduced below) if f applies to k and is assigned a value "0", otherwise.

- **Is-Plural.** This feature is set to '1' if a keyword is in a *plural* form and is '0', otherwise. Type I attribute values tend to be expressed in the singular form.
- **Is-Capitalized.** We have observed that, in general, the first character in keywords that represent Type I attribute values is *capitalized*, and this feature is assigned the value of '1', and '0', otherwise.
- **Is-Style.** This feature is set to '1' if the keyword being evaluated is either *bolded* or *italicized* in an ad and is '0', otherwise. The most important attribute values in an ad tend to be either bolded or italicized and are Type I attribute values.
- **In-Title.** Since the most descriptive attribute values of an ad d appear in the *title* or *first sentence* of d, this feature is assigned a value '1' if the keyword being analyzed is in the title or in the first sentence of d and is given a '0', otherwise. In-Title values are Type I attribute values.
- **Is-Adjective.** In implementing this feature we rely on the part-of-speech (POS) tagger[8], which assigns POS, such as noun or verb, to (key)words (in

[8] http://nlp.stanford.edu/software/tagger.shtml.

ads). A feature is assigned a '1' if the word is given an *"adjective"* POS tag and '0', otherwise. Most of Type II attribute values are adjectives which describe essential information of an ad.

- **Is-Measurement.** This feature takes a value of '1' if the keyword being evaluated represents a *measurement*, e.g., $, usd, miles, square feet, or inches, and'0', otherwise. The set of measurement terms was extracted from Hobby[9], a website that lists units of measurements for different categories, such as length, area, power, or speed. This feature is a clear indicator of Type III attribute values.
- **Is-Alphanumeric.** This feature is assigned a '1' if the keyword being evaluated is alphanumerical and is assigned a '0', otherwise. This feature assists in identifying Type II attribute values.
- **Is-Location.** This feature relies on a set of predefined locations to decide whether the keyword being evaluated is a *location*, in which case the feature is set to '1' and '0', otherwise. The list of locations, e.g., US cities, was extracted from Wikipedia[10]. This feature assists in identifying Type III attribute values.
- **Is-Acronym.** This feature is set to '1' if the word being evaluated represents an *acronym* and is '0', otherwise. In determining whether a word is an acronym, we adapt the approach proposed by Chieu and Ng [5] to look for sequences of capitalized words in an ad d that match (potential) acronyms found in d. If a sequence of capitalized letters in an ad matches the word k being evaluated, then we treat k as an acronym.

3.4 Extracting Data for Query Answering

Having identified the domain of an ad d (in Sect. 3.2) and assigned the corresponding type to each (non-stop) keyword k in d (in Sect. 3.3), *AdProc* proceeds to extract data in d and populate the underlying DB[11] employed by *AdProc* in answering queries on ads. In this regard, extracting information from unstructured data sources is a *classification* process, since k is either assigned as a value to its corresponding attribute in the DB record of d or a "not-valid" label, indicating that k will not be populated to the DB.

We apply the C4.5 decision tree algorithm [23] to construct a decision tree for extracting data from online ads. Decision trees are widely-used and employ a simple classification technique for inductive inference which utilizes the decision process as a set of if-then rules. The algorithm applies the divide-and-conquer strategy and recursively partitions the training instances into subsets according to a splitting criterion (test separation), which is predefined prior to constructing the tree. We construct a decision tree for each previously-defined ad domain (schema).

9 www.hobbyprojects.com/dictionary_of_units.html.
10 en.wikipedia.org/wiki/List_of_cities,_towns,_and_villages_in_the_United_States.
11 The DB schema for each ad domain is defined prior to invoking *AdProc* to automate the data extraction process.

Prior to extracting and populating data in online ads to the underlying DB, we first define the set of training instances S for constructing the corresponding decision tree. A training instance in S, which is associated with a keyword k in an online ad d, is a sextuple of the form $<f_1, f_2, f_3, f_4, f_5, A>$, where f_i ($1 \leq i \leq 5$) is one of the possible values which can be assigned to i, where $i \in$ *Feature-Set* with features defined below, and A is one of the DB attributes in the corresponding schema for which k is a valid value of A or A is the label "not-valid" (when k is not a valid attribute value). Each feature i in *Feature-Set* is defined below.

1. **Keyword-Type** is the attribute type of k.
2. **Previous-Keyword-Type** is the attribute type of the keyword immediately preceding k in d.
3. **Post-Keyword-Type** is the attribute type of the keyword immediately following k in d.
4. **Previous-Keyword-Attribute** is the DB attribute of the keyword immediately preceding k in d.
5. **Closest-Type-IV** is the Type IV keyword in d that is closest to k.

The possible values of *Keyword-Type*, *Previous-Keyword-Type*, and *Post-Keyword-Type* are Type I - IV, whereas the possible values of *Previous-Keyword-Attribute* include the set of attributes in the DB schema corresponding to the domain of d and the label "not-valid".

The features in *Feature-Set* are defined for capturing essential information to accurately identify each keyword k as an attribute value, which are based on the context in which k appears, i.e., based on other keywords that appear before and after k, in an online ad. Moreover, the *Closest-Type-IV* feature identifies keywords commonly associated with numerical values for data extraction. For example, given the phrase "25 acres", we rely on the keyword "acres", i.e., the closest Type IV keyword to "25", in assigning the numerical value "25" to its corresponding DB attribute.

AdProc constructs a decision tree for each ads domain of interest to perform the data extraction task. A major task in constructing a decision tree is to establish the criterion used for identifying the feature (in *Feature-Set*) that is the most effective in splitting training instances into different groups, i.e., various DB attributes in our case. A goodness measure used for indicating the classification power of a split is the *information gain*, which we adapt and as defined below.

$$Information\ Gain(S, F) = Entropy(S) - \sum_{f \in Values(F)} \frac{|S_f|}{S} Entropy(S_f) \quad (10)$$

where F is a feature, $Values(F)$ is the set of all possible values of F, S_f is the subset of training instances in S in which the value of F is f, $|S_f|$ ($|S|$, respectively) is the number of training instances in S_f (S, respectively), and

Entropy(S) is defined in Eq. 11.

$$Entropy(S) = \sum_{i=1}^{|A|} -p_i \; log_2 \; p_i \tag{11}$$

where p_i is the number of instances in S that include attribute i, and $|A|$ is the number of attributes in the corresponding schema plus one, the "not-valid" label.

4 Generating SQL Queries

Constraints specified by users in natural language queries must be translated into database query languages for execution. For the implementation of *AdProc*, we selected SQL as the query language of translation output. There are three common components to a *AdProc* SQL query: the SELECT-FROM, WHERE, and ORDER BY clauses.

4.1 The SELECT-FROM Clause

During the classification process, *AdProc* determines the category of the given query Q, which dictates the corresponding *table* name in the underlying database scheme. To retrieve all the relevant information to the user, *AdProc* extracts all the columns of the table for each ad that satisfies the constraints specified in Q using the wildcard (*), i.e., SELECT *.

4.2 The WHERE Clause

The WHERE clause is a logical combination of each of the parsed constraints stated in a user query. *AdProc* links each constraint to the corresponding attribute/column in the database table in the WHERE clause. Since in a natural language query Q the user often does not identify the correct table column, *AdProc* make inferences of the correct attribute based on the information provided in Q. *AdProc* performs the matching based on the "Type" constraints.

- For Type I and Type II constraints, the corresponding table column can be determined by a local search of the value, called *value matching*, using a *trie* data structure. For example, if "Toyota" and "Camry" are specified as constraints, *AdProc* looks up the corresponding trie (that is periodically updated), which is created according to the column values of the DB tables, to determine whether the values appear and their corresponding column names, i.e., "Make" and "Model", respectively. Value matching has the potential to return multiple rows for a single value, since a constraint value may appear in different rows in the corresponding table. For example, the word "new" may refer to the *condition* of the automobile, or it may refer to a part of a *location* (such as "New York" or "New Jersey"). If the ambiguity was not able to be resolved earlier in tokenization, *AdProc* allows either alternative for a match by using the "Boolean Or" operator. Such ambiguity results in a WHERE clause "`condition LIKE ‘‘new’’ OR region LIKE ‘‘new’’`".

– For Type III constraints, they cannot be identified by using their corresponding attributes based on value matching due to the infinite nature of numbers. A similar approach to the value matching, however, can be performed by using the unit of the Type III constraint instead of the value alone. For this strategy to be effective, two conditions must be satisfied: (i) each Type III attribute in the DB table includes a list of related units (which is anticipated) and (ii) there is an attached unit for each Type III constraint in a query (which is often the case).

The WHERE clause of the SQL query is composed of the translated constraints of Types I, II, and/or III. The values of Type I and Type II constraints are compared with the column values in SQL using the `LIKE` operator, which requires that the given substring appears in all matches. For Type III comparison constraints, if a comparison operator is not given, the equality operation (=) is assumed. For *range* Type III comparisons, the `BETWEEN` operator is used.

4.3 The ORDER BY Clause

The "ORDER BY" clause is created for *implicit sorting order* of Type III comparisons. These implicit constraints come from sorting each result by its distance from the optimal value of the comparison. Equality comparisons have an optimal value of the number to be matched on, and they are sorted by `ABS(x-y)`, where x is a Type III attribute value and y is an expected value. $<$ or $\leq$ operations have an optimal value of 0, whereas $>$ or $\geq$ operations have an optimal value of infinity, i.e., ∞. *Ranges* have no optimal value, and therefore do not create an implicit sorting constraint. Consider the constraints specified in Q, "*Toyota blue Camry or white Corolla 2015 less than $15000 in NY*". In the WHERE clause, the constraint "Toyota" is determined to be car `Make`, "blue" and "white" are `colors`, whereas "Camry" and "Corolla" are car `Models`. "2015" is assumed to be a year, since it is within the range of `Years`. The constraint "less than $15K" matches with `Cost` because of the dollar sign. Lastly, "NY", an abbreviation of "New York", is matched in the `Cars` table as a `Location` value. Hereafter, the ORDER BY clause is constructed in which "2015" is the optimal value for `Years` and `Price` is minimized. The constructed SQL statement for Q is

```
SELECT * FROM Cars
WHERE (make LIKE "Toyota"
          AND ((color LIKE "blue" AND model LIKE "Camry")
            OR (color LIKE "white" AND model LIKE "Corolla"))
          AND year = 2015 AND price < 15000 AND location LIKE "New York")
ORDER BY ABS(year - 2015), price;
```

4.4 Partial Matching

The DB query is generated through query translation (as shown in previous subsections) and any exact product matches to the query constraints are extracted from the DB. This behavior is sufficient if there are many exact matches, especially when the user query is vague. However, in the case that no exact matches or very few are found, it may be confusing or frustrating for the user. For this reason, *AdProc* supplements no or very few results with *partially-matched* records. Relevant product are searched for, first exact matches and then partial matches, until some threshold number of products is achieved. This threshold number of records to be extracted is set to be 15 based on the user's attention span determined by [12] to not overwhelm the user.

Similarity Measures Based on Replacements. Partial matches are conducted by iteratively altering search constraints one by one and searching for products not previously retrieved. These alterations occur in rounds, numbered by the count of search constraints specified in the original query Q. The 1st round considers exact matches on all the constraints N specified in Q. The 2nd round is a union of all combinations of subqueries with 1 altered constraint, i.e., N-1, whereas the 3rd round evaluates all unique subqueries with 2 altered constraints, i.e., N-2. This pattern continues for all needed rounds.

Example 1. *Consider the five constraints specified in* SQ_g*:*

1. *make LIKE "honda"*
2. *(color LIKE "red" AND model LIKE "accord") OR (color LIKE "silver" AND model LIKE "civic")*
3. *year = 2008*
4. *price < 6000*
5. *location LIKE "New York"*

The 1st *round would use a query of* $(1 \wedge 2 \wedge 3 \wedge 4 \wedge 5)$*, whereas for the* 2nd *round* $(\approx 1 \wedge 2 \wedge 3 \wedge 4 \wedge 5) \vee (1 \wedge \approx 2 \wedge 3 \wedge 4 \wedge 5) \vee (1 \wedge 2 \wedge \approx 3 \wedge 4 \wedge 5) \vee (1 \wedge 2 \wedge 3 \wedge \approx 4 \wedge 5) \vee (1 \wedge 2 \wedge 3 \wedge 4 \wedge \approx 5)$ *would be used, where "*$\approx x$*" indicates that the* x^{th} *constraint was altered, and so on.* □

Constraints can be altered in two different ways: a *similarity replacement* or a *removal*. A *similarity replacement* captures more information about the original constraint, and is therefore done first. If no similarity replacement can be found for the value or if the similarity replacement does not yield any results, the constraint is *removed*.

Similarity Measures Based on Attribute Values. In designing the N-1 partial-matching strategy for enriching the quality and quantity of retrieved answers to an ads query Q, we have defined the *degree of similarity* of an attribute value V in each record and the corresponding value T in Q that is excluded from the N-1 matching conditions for *ranking* partially-matched results.

Type I Values. To measure the degree of similarity between a partially-matched Type I attribute value in an DB record and the corresponding one expressed in Q belonged to an ad domain D, we consult the *TI-matrix* for D. To construct the

TI-matrix for *D*, which includes the similarity values between any two distinct Type I attribute values in *D*, *AdProc* relies on query logs obtained from local ads search engines. Each query log includes a number of query sessions, each of which captures a period of sustained user activities on the corresponding ads search engine. Each log session differs in length and includes (i) a *user ID*, (ii) the *query text*, (iii) the *date* and *time* of each search, and (iv) optionally *clicked documents*. A *user ID*, which is an anonymous identifier of the user who performed the search, determines the boundary of each session (as each user ID is unique and associated with one session). The *query text* are the keywords in a user query, and multiple queries can be created under the same session. The *date* and *time* of search can be used to determine the relative importance of the query results page based on the *time* the user spent on that page, and *clicked documents* are retrieved documents that the user has clicked on and are represented by their titles and ranked by the corresponding ads search engine.

The *similarity* value of any two distinct *Type* I attribute values *A* and *B*, denoted TI_Sim, in the *TI-matrix* is computed in Eq. 12 based on the following features that can be determined using a query log:

(1) Number of times *A* is modified to *B* in the query log or vice versa, denoted $Mod(A, B)$.
(2) Average time between submissions of *A* and *B* in the same session, denoted $Time(A, B)$.
(3) Average time spent on an ad containing *B* when *A* is searched or vice versa, denoted $Ad\_Time(A, B)$.
(4) Ranking of an ad (as determined by the ads search engine that provides the query log) containing *B* when *A* is searched for, or vice versa, which is averaged over the entire query log, denoted $Rank(A, B)$. The *higher* *B* is ranked, the *more likely* *B* is similar to *A* as determined by the corresponding search engine that ranks *B*.
(5) Number of times a document containing *B* is clicked when *A* is searched or vice versa, denoted $Click(A, B)$.

To normalize each of our feature values, we divide the resulting value of a feature *F* by the maximum possible value of *F* derived from our query log so that each factor value is in the range of [0..1].

$$TI\_Sim(A,B) = Mod(A,B) + Time(A,B) + Ad\_Time(A,B) + Rank(A,B) + Click(A,B) \tag{12}$$

Example 2. *Consider the query Q "Find an Accord car". In addition to retrieving all the Accords, if there are any in the car-ad table, AdProc applies the partial-matching strategy to retrieve similar Models that yield the alternative choices of Accord. Table 1 shows a portion of the TI-matrix for the car-ad domain, which includes the TI_Sim values of Accord and a few other Models.* □

Type II Values. In establishing the *similarity* values among the *properties* of various products in different ads of the same domain, we use the *word-similarity*

Table 1. Some of the TI_Sim values (in the TI-*matrix* for the car-ad domain) of Accord and its similar models in the same category

Model	Similar Model	TI_Sim	Model	Similar Model	TI_Sim
Accord	Malibu	1.73	Accord	Focus	0.91
Accord	Camry	1.43	Accord	Corolla	0.83
Accord	Fusion	1.11	Accord	Mazda	0.79

matrix, denoted WS-*matrix*, introduced in [15]. WS-*matrix* is a 54,625 × 54,625 symmetric matrix, which contains the similarity values of pairs of non-stop, stemmed words, i.e., words reduced to their grammatical root. The similarity value between any two non-stop, stemmed words w_i and w_j in WS-*matrix* is computed by using the (i) *frequency* of *co-occurrence* and (ii) relative *distance* of w_i and w_j in a document. WS-*matrix* was constructed using the documents in the Wikipedia document collection[12] that consists of 930,000 documents written by more than 89,000 authors on various topics, which are diverse in content and writing styles. The WS_matrix has been adopted successfully in solving various information retrieval problems [27]. The similarity between an attribute value T, such as 'white', in a query and an attribute value V, such as 'blue', in a data record, denoted $Feat\_Sim(T, V)$, is a similarity value in WS-*matrix*.

Type III Values. In establishing the *similarity* among *numerical attribute values* in different ads domains, we apply Eq. 13.

$$Num\_Sim(T, V) = 1 - \frac{|T - V|}{Attribute\_Value\_Range} \tag{13}$$

where $Attribute\_Value\_Range$ is the normalization factor of $Num\_\ Sim$, which is different for each numerical-value attribute in the DB, and the complement is applied so that the *closer* T and V are, the *higher* their Num_Sim value is.

The $Attribute\_Value\_Range$ of each Type III attribute A in an ads domain is determined by using the statistical data extracted from ebay.com. We obtain from ebay.com the 10 highest (lowest, respectively) values for A and subtract the averaged *minimum* values of A from the averaged *maximum* values of A to establish the attribute value range for A,

Example 3. *Consider query Q "Find all \$10,000 cars". A car ad that includes a price of \$11,000 is closer to the price specified in Q than another car ad which includes a price tag of \$7,500, since* $Num\_Sim(\$10{,}000, \$7{,}500) = 1 - 2{,}500/10{,}000 = 0.75 \leq Num\_Sim(\$10{,}000, \$11{,}000) = 1 - 1{,}000/10{,}000 = 0.90$*, assuming that 10,000 is the price range for cars determined by the 10 maximum (minimum, respectively) values extracted from ebay.com.* □

[12] en.wikipedia.org/wiki/Wikipedia: Databasedownload.

The *ranking similarity* value, $Rank\_Sim$, between a user's query Q and a partially-matched record r in which T and V occur, respectively is computed as follows:

$$Rank\_Sim(r, Q) = (N - 1) + \begin{cases} TI\_Sim(T, V) & \text{if } T \text{ and } V \text{ are Type I attribute values} \\ Feat\_Sim(T, V) & \text{if } T \text{ and } V \text{ are Type II attribute values} \\ Num\_Sim(T, V) & \text{if } T \text{ and } V \text{ are Type III attribute values} \end{cases} \quad (14)$$

where N is the number of attributes specified in Q. Num_Sim is in the range of [0..1], whereas TI_Sim and $Feat\_Sim$ are in different numerical scales. Hence, TI_Sim and $Feat\_Sim$ are normalized by the maximum possible value in each of their corresponding matrices. The value N-1 is added to account for the *exact* matches between the corresponding attribute values in Q and r, each of which is assigned the value of 1.

Example 4. *Table 2 shows a portion of the partially-matched answers to the question Q "Find Honda Accord blue less than 15,000 dollars" that AdProc retrieves and ranks using Eq. 14. Table 2 also includes the partially-matched attribute value, which is highlighted, in each record, in addition to the similarity-measure strategy adopted for computing the ranking.* □

Table 2. Top-5 ranked, partially-matched answers to query Q, "Find Honda Accord blue less than 15,000 dollars".

Ranking	Make	Model	Price	Features	$Rank\_Sim$
1	**Chevy**	Malibu	5899	blue, anti-lock brake, power steering	3.73
2	Toyota	**Camry**	8561	4 cylinder, automatic, blue, ...	3.43
3	**Ford**	Focus	6795	blue, cd player, radio, power door locks, ...	2.91
4	Honda	Accord	**16536**	4 cylinder, blue, 2 wheel drive, cassette player, ...	2.63
5	Honda	Accord	6600	GPS system, **gold**, auto-off headlights, ...	2.31

5 Experimental Results

In this section, we assess the performance of *AdProc* by evaluating its major tasks in populating ads and answering ads queries. We (i) first introduce the dataset and metrics used for the performance evaluation (in Sects. 5.1 and 5.2, respectively), and (ii) analyze its accuracy in classifying (in Sect. 5.3), tagging (in Sect. 5.4), extracting ads data (in Sect. 5.5), and evaluating its populating process (in Sect. 5.6), besides retrieving answers to ads queries (in Sect. 5.7).

5.1 The Dataset

To the best of our knowledge, there is no existing dataset for evaluating classification, labeling, and/or data extraction of online ads. Thus, we have created our own dataset, denoted *EData*, for assessing the performance of *AdProc* based on it.

EData consists of 80,000 uniformly distributed online ads, with 10,000 ads in each of the eight different ads domains, which were randomly extracted from Craigslist.org, a popular source of unstructured ads in various domains, and Coupons.com. The eight domains are cars(-for-sale), food (coupons), furniture, houses(-for-sale), jewelry, C(omputer) S(cience) jobs, motorcycles(-for-sale), and music(al instruments). These chosen ads domains vary in terms of their (i) diversity, which include ads in jobs, food, housing, transportation, and entertainment that offer our daily needs, (ii) ad size, from arbitrary long (such as houses) ads to relatively short (such as jewelry) ads, and (iii) word distribution, i.e., different word usage associated with different types of ads. Moreover, to test the versatility of *AdProc*, domains that are closely related, e.g., cars and motorcycles, and diverse, e.g., food and jobs, in content and nature were chosen.

To obtain a representative set of queries for verifying the *accuracy* of *AdProc* from (i) classifying ads queries to their corresponding domains to (ii) retrieving answers in multiple ads domains, we collected 1,750 queries on the eight ads domains from *Facebook users* between January 9, 2023 and May 11, 2023. These Facebook users were recruited by the (friends and students of the) author for the empirical studies.

5.2 Evaluation Measures

To evaluate the effectiveness of *AdProc* in classifying ads, tagging keywords, extracting ads data, and assigning users' queries to their corresponding domains, we rely on the accuracy ratio defined below.

$$Accuracy = \frac{Correctly\_classified\_instances}{|Dataset|} \tag{15}$$

where $|Dataset|$ is the size of a given dataset, i.e., the total number of instances considered for evaluation, and $Correctly\_classified\_instances$ is the number of instances correctly classified/labeled/extracted by the corresponding evaluated method.

To measure the accuracy of *AdProc* in *retrieving* correct answers, we compare the answers generated by *AdProc* on each one of the 1,750 Facebook queries with the ones on the same answer set provided by a group of 350 independent Facebook appraisers[13] which serve as the gold-standard, using the Spearman Rank Correlation Coefficient (SRCC) [4]. SRCC returns -1 or 1, where '1' indicates that the two given results to be compared are identical and '-1' implies that the results are not related.

$$SRCC = \frac{1 - \frac{6}{n^3 - n} \times (\sum d_i^2 + \frac{1}{12} \sum (f_k^3 - f_k))}{\sqrt{1 - \frac{\sum f_k^3 - f_k}{n^3 - n}}} \tag{16}$$

[13] To gather appraisers' feedback, we relied on the Facebook users and students at our university, who completed a survey between May 12, 2023 and February 22, 2024.

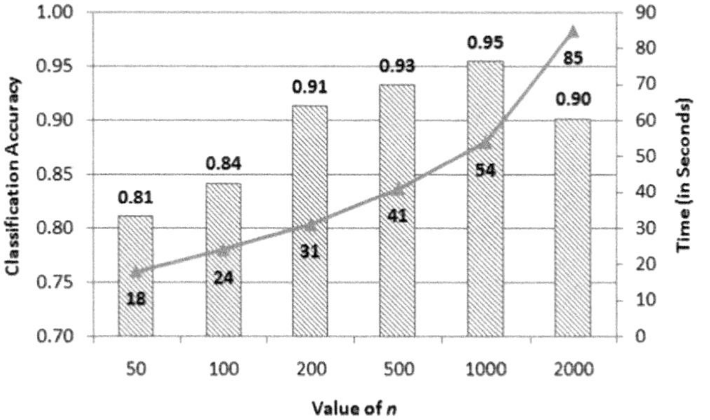

Fig. 1. Accuracy ratio and classification time for the 80,000 ads used for determining the ideal value of n, the size of the reduced vocabulary for ads classification [25].

where d_i is the difference between the two results for the same query i, n is the total number of queries, f_k is the number of ties in the kth (≥ 1) group of ties created by the appraisers.

To measure the performance of *AdProc* in correctly transforming the information needs expressed in a user's query into a correct SQL statement, we use *precision* as the evaluation metric.

$$Precision = \frac{\#\text{Correct Matches Retrieved by } AdProc}{\#\text{Records Retrieved by } SQL} \quad (17)$$

where a *correct match* is a retrieved answer that satisfies all the search criteria specified in a query.

5.3 Classification of Ads

Prior to performing the classification task, feature selection introduced in Sect. 3.2 is applied to reduce the size of the vocabulary, i.e., the number of distinct keywords in $EData$, so that the top n (≥ 1) non-stopwords and alphanumerical keywords are chosen for representing ads in the corresponding domains. Figure 1 depicts the accuracy ratio and classification time for determining the ideal value n ($= 1{,}000$).

Figure 2 shows that the accuracy ratio of classifying $EData$ ads in each domain, and the average achieved by *AdProc* are in the ninety percentile. Ads in the jobs, food coupons, and houses domains achieve higher classification accuracy, whereas cars and motorcycles achieve lower accuracy ratios. We observed that domains in which (i) ads descriptions are *short*, such as jewelry ads, and (ii) their word usage is *similar* to the one in others, such as cars and motorcycles ads, tend to yield *lower* accuracy.

To analyze the effectiveness of the proposed classifier in assigning ads to their corresponding domains, we used a *confusion matrix*, denoted CF. CF

(= $[I_{i,j}]$) displays the total number of instances $I_{i,j}$ ($i \neq j$) in class c_i which have been *misclassified* into class c_j, and the principal diagonal of CF denotes the total number of instances *correctly* classified. Table 3 shows that most of the classification errors occur when any two ads domains share a similar probability distribution on a considerable number of keywords, e.g., cars and motorcycles, as mentioned earlier.

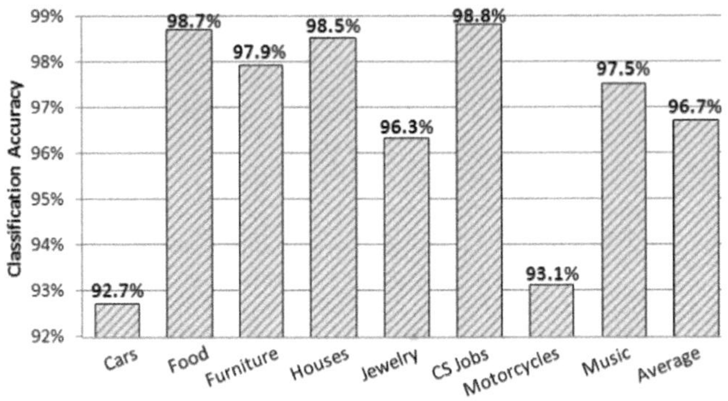

Fig. 2. Classification accuracy of ads in *EData* [25].

Table 3. The confusion matrix generated according to (in)correctly classified ads in *EData* [25].

Ads Domain	Cars	Food	Furniture	Houses	Jewelry	CS Jobs	Motorcycles	Music
Cars	927	1	3	2	0	0	67	0
Food	0	987	3	2	6	1	1	0
Furniture	8	2	979	5	1	0	0	5
Houses	1	1	6	985	1	4	0	2
Jewelry	4	4	5	8	963	5	5	6
CS Jobs	0	1	0	2	2	988	0	7
Motorcycles	58	1	3	3	1	1	931	2
Music	1	3	4	5	3	7	2	975

We have verified the effectiveness of the classifier used by *AdProc* to a greater extent by comparing its classification accuracy with two other well-known classifiers, the multinomial Naïve Bayes (*MNB*) classifier [20] and the SVM extracted from WEKA[14], an open-source collection of machine learning algorithms, based on the 80,000 Craigslist.org and Coupons.com ads in *EData*.

[14] www.cs.waikato.ac.nz/ml/weka/.

MNB follows the premises of the Naïve Bayes classifier (as discussed in Sect. 3.2) in assigning a document to a class. As opposed to the Joint Beta-Binomial Sampling model (JBBSM) introduced in Sect. 3.2, MNB computes the probability of a keyword w_t in a class c_j, $P(w_t|c_j)$, by considering the frequency of keyword occurrence in each document, whereas SVM in [22] is a vector-space-based machine learning method which determines a decision boundary between classes that is maximally far from any point in the training data for classification [19] as defined in Sect. 3.3. Figure 3 shows that the classifier of *AdProc*, i.e., JBBSM, outperforms MNB and SVM in assigning ads in *EData* to their corresponding domain, and the results are statistically significant ($p < 0.05$) based on the Wilcoxon Signed-Rank test [30].

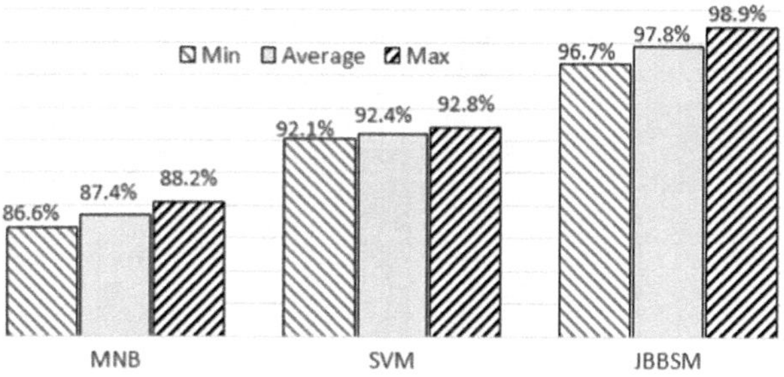

Fig. 3. Classification accuracy obtained by using MNB, SVM, and the classifier of *AdProc*, JBBSM, on *EData* [25].

To further assess the effectiveness of *AdProc* in ads classification, we repeated the conducted experiments two more times, using two new subsets of 80,000 ads uniformly distributed among the eight domains. The overall evaluation of *AdProc* in terms ads classification is shown in Table 4.

Table 4. Averaged accuracy for *AdProc*, as well as the classification approaches used for comparison [25].

Classification Approach	Accuracy
MNB	87.4 ± 0.8
SVM	92.4 ± 0.4
AdProc	97.8 ± 1.1

5.4 Tagging Keywords Based on Types

To assess the effectiveness of the multi-class SVM approach (introduced in Sect. 3.3) on tagging keywords in ads according to their corresponding types, we first created training/test instances using each of the ads in *EData* and for each non-stop, non-numerical keyword in each ad we extracted the corresponding features, as discussed in Sect. 3.3.

As shown in Fig. 4, the overall accuracy of our SVM in assigning Types I-IV tags to keywords in ads is in the 90 percentile. Most of the misclassification errors occur when attribute values that should be assigned a Type I tag are incorrectly labeled as Type II. When none of the keywords in an ad are bolded, italicized, or capitalized, the values assigned to features, such as *Is-Style* or *Is-capitalized*, are the same for keywords of Types I and II, which causes the misclassification. Figure 4 also reveals that Type II attribute values yields the lowest accuracy, since they are often labeled as Type IV due to their proximity in ads.

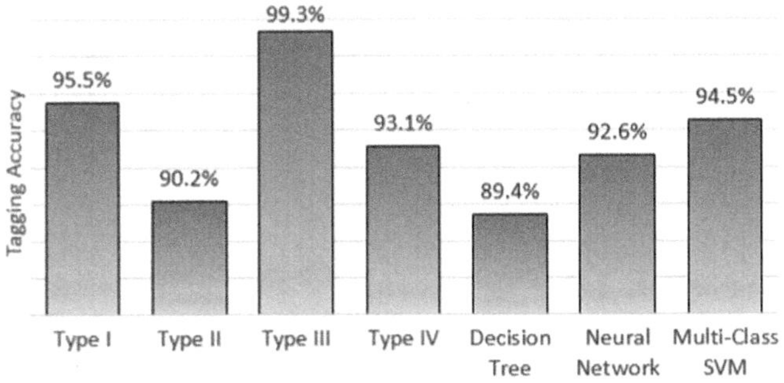

Fig. 4. Performance evaluation using *AdProc*'s multi-class SVM, as well as alternative machine learning approaches, for tagging keywords [25].

To further assess *AdProc*'s tagger, we have compared its performance with two other widely-used approaches, C4.5 decision tree classifier and an artificial neural network, since decision trees and neural networks are frequently adopted to solve machine learning tasks, such as for text classification and labeling [23]. As defined in [23], decision tree learning is an approach used for approximating discrete-valued target functions. In other words, a decision tree classifier categorizes instances, i.e., feature vectors associated with a particular keyword in our case, by organizing them down the tree from the root to some leaf node, which provides the label of the instance, i.e., tag associated with a keyword in our case. Neural network learning provides a robust approach to approximating real-valued, discrete-valued, and vector-valued target functions [23]. The training of a neural net consists of an iterative supervised process in which for each training instance, i.e., feature-vector associated with a particular keyword in our case, the correct class, i.e., keyword Type, is known and thus, it is possible to compare the network's predicted output with the "known" one, and proceed to adjust the weights of the internal nodes of the network so that in future iterations the predicted output classes are closer to the "known" classes and used for comparing the effectiveness of our multi-class SVM approach implemented using WEKA. Figure 4 shows that the multi-class SVM of *AdProc* outperforms the two methods for tagging keywords in ads according to their types, and based on the Wilcoxon Signed-Rank test, the results are statistically significant ($p < 0.04$).

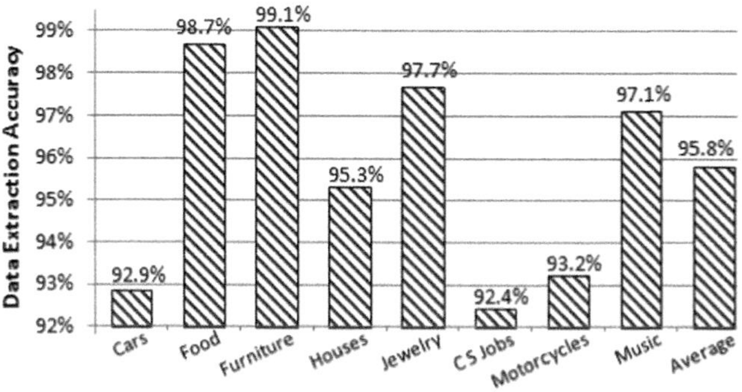

Fig. 5. Accuracy ratios of data extraction computed for the eight domains in our empirical study [25].

5.5 Extracting Ads Data

To determine the effectiveness of the decision-tree based approach (introduced in Sect. 3.4) which assigns non-stop keywords in ads that are valid attribute values to their corresponding DB attributes, we first created training/test instances using the ads in *EData*. In constructing the instances, we considered the (i) domain assigned to each ad in *EData*, (ii) type of each non-stop keyword in the ads, and (iii) features defined in Sect. 3.4. The set of training/test instances includes approximately 135,400 feature vectors, one for each non-stop keyword in *EData* ads.

Figure 5 shows the high accuracy ratio of *AdProc* in assigning valid attribute values to their corresponding DB attributes in different ads domains. On the average, the decision-tree based approach achieves 95.8% accuracy. Based on the conducted empirical study, we have observed that the accuracy ratios for ads domains that contain a *large* number of attributes are *lower* compared with others with a *smaller* number. This is because the *larger* the number of DB attributes, proportionally the *lower* the number of available instances in any training set that describe a particular DB attribute to train the decision tree, which translates into *lower* accuracy ratios in correctly assigning values to the attribute. Furthermore, (i) keywords of Types I and II in the cars and motorcycles domains are often correctly assigned to their corresponding DB attributes, and (ii) keywords of Type IV are not assigned to any DB attribute, as anticipated. However, the overall accuracy of each of these two domains is among the lowest of the eight domains. This is caused by the common (numerical) Type III attribute values which are assigned to incorrect DB attributes with the same or compatible domain(s), e.g., in motorcycles ads, '2000' is assigned to the attribute 'Year', instead of the attribute 'Miles'.

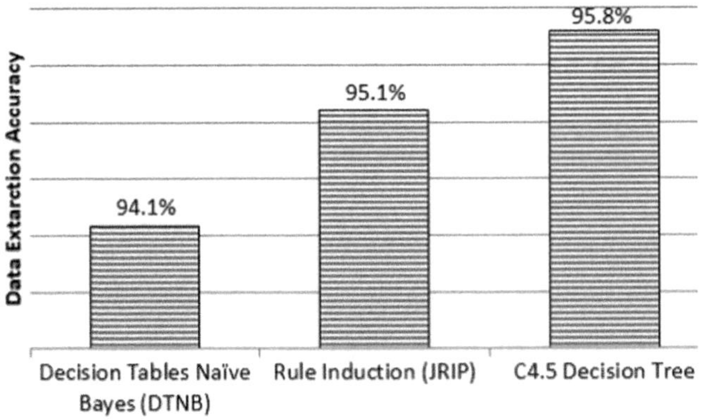

Fig. 6. Performance evaluation using the decision trees of *AdProc* and other machine learning methods for extracting data from ads to create DB records [25].

We have compared the performance of *AdProc*, in terms of extracting data from online ads to populate the DB, with the WEKA implementation of two machine learning approaches: the Decision Tables Naïve Bayes approach (DTNB) in [11] and the Rule Induction approach in [7], denoted JRIP. DTNB is a hybrid method that combines two well-established approaches: decision tables, which define rules that determine to which DB attribute a particular keyword should be assigned, and Naïve Bayes classifiers, which determine the probability of occurrence of a particular rule given a particular DB attribute. Given a word k in an ad, DNTB applies the rule with the highest probability in assigning the DB attribute for which k is a valid attribute value. JRIP, on the other hand, discovers rules that cover or partition the training examples, i.e., the set of feature vectors discussed in Sect. 3.4 in our case. JRIP is a bottom-up method such that given a class, i.e., a DB attribute A, it finds the set of rules that cover all the members of that class, i.e., all the training instances associated with A. Thereafter, JRIP repeats the process of defining rules for each class until all possible classes have been covered, i.e., till rules associated with each of the possible DB attributes are created.

As shown in Fig. 6, the decision-tree based approach of *AdProc* outperforms the alternative approaches for extracting data from online ads. Although the difference in accuracy between JRIP and Decision Tree is less than 1%, *AdProc* is *simpler* in terms of implementation, which has been verified. Even though *AdProc* does not outperform JRIP, it surpasses DTNB and the results are statistical significant ($p < 0.05$) based on the Wilcoxon Signed-Rank test.

In comparing with DTNB and JRIP on extracting data from online ads to populate the underlying DB, Fig. 7 shows that the training and testing time of Decision Tree on *EData* are reduced on an average by 45% and 46%, respectively. These results verify that *AdPorc* is an efficient tool for data extraction.

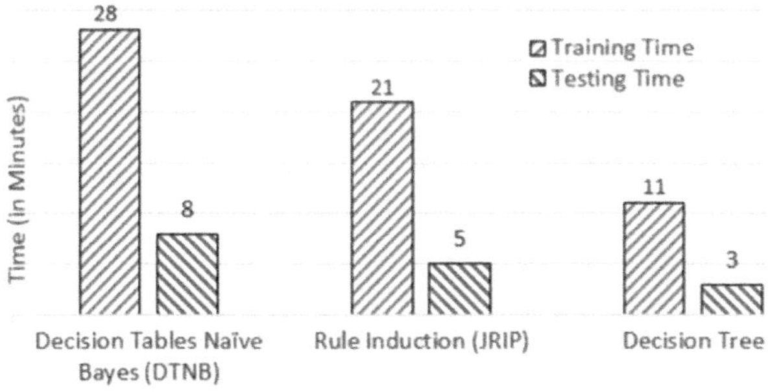

Fig. 7. Training/Testing time of DTNB, JRIP, and decision tree using the instances created for non-stop keywords in *EData* for data extraction [25].

5.6 Accuracy in Populating the DB

To assess the overall performance of *AdProc*, in terms of its *accuracy* in populating the DB with extracted ads data, we first created a new collection of ads, denoted *TData*. *TData* consist of 8,000 Craigslist.org and Coupons.com ads, uniformly distributed among the eight ads domains. A DB record d created by *AdProc* is treated as *incorrect* if (i) at least one valid attribute value in the ad used for creating d is assigned to a wrong DB attribute in d or not assigned to d, or (ii) a Type IV, non-essential attribute, value in an ad is assigned to an attribute in d. The calculated *accuracy ratio* of correct DB records generated by *AdProc* on the ads in *TData* is 80.3%.

Besides measuring the accuracy ratio at the *record* level, we conducted the same evaluation at the DB *attribute* level. In doing so, we determined the portion of attribute values that were correctly assigned to their corresponding DB attributes. The experimental results show that *AdProc* correctly assigned 96.4% ($= \frac{8{,}000-288}{8{,}000}$) of the attribute values in *TData*. As shown in Fig. 8[15], most of the incorrect DB records include a low percentage, i.e., between 10% and 15%, of incorrectly assigned attributes values. In fact, each incorrectly created record includes at least 85% correctly assigned attribute values.

Based on the conducted experiments, we draw the conclusion that *AdProc* is *highly accurate* in assigning keywords in an ad A to its corresponding attribute in the DB record of A (according to the predefined ad schema to which A belongs), since 98.6% of the DB records are either correctly created (i.e., 96.4%) or have at most 15% of invalid attribute values in their DB attributes as shown in Fig. 8.

5.7 Accuracy on Processing Queries

In verifying the accuracy on answers retrieved by *AdProc*, we computed the SRCC values on the answers to each one of the 1,750 Facebook queries in multiple

[15] If the percentage of incorrect attribute values in a given DB record R falls in between the interval of any two percentages shown in Fig. 8, R is counted towards the incorrect number of records rounded to the nearest percentage.

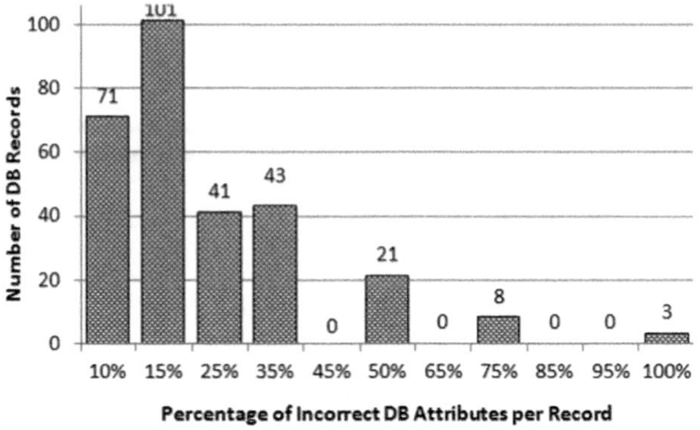

Fig. 8. Error distribution in terms of percentages of attribute values incorrectly assigned to DB attributes in the 288 incorrectly created DB records [25].

ads domains compiled by the appraisers. As shown in Fig. 9, regardless of the ads domain being evaluated, *AdProc* achieves an accuracy for each ads domain that is alike the *gold standard* established by the 350 appraisers. The average SRCC value on the eight domains, which is in the 73%, verifies that the retrieval strategy adopted by *AdProc* is reliable. Figure 9 also includes the *average precision* of the DB records retrieved (up till the 15th) for each one of the 1,750 Facebook queries in the eight ads domains with an *average precision* of 93%.

Fig. 9. Average SRCC values and Precision computed for the eight ads domains [25].

To further measure the performance of *AdProc* in retrieving answers that *match exactly* the constraints specified in users' natural language queries, we processed the 1,750 Facebook user-created queries using *AdProc*. The evaluation metrics for measuring the correctness of retrieving exactly-matched answers to a user query are *precision* (P), *recall* (R), and *F-measure*, where a *correct match* is a retrieved (up till the 15th) DB record that satisfies all the search criteria specified in a query. We measured *precision* based on the correct (up till the 15th) DB records retrieved, and *recall* without restricting the total number of DB records retrieved. Since it can be biased to measure the effectiveness of *AdProc* using *precision* and *recall* alone, we consider the *F-measure* which combines *precision* and *recall* with the same weight assigned to the two measures. *AdProc* achieves 94.2%, 93.3%, and 93.9% for the averaged P, R, and *F-measure*, respectively. We found that most of the test queries yield 100% for *precison* and *recall*, whereas a few yield 0%, i.e., answers are incorrect.

5.8 Accuracy on Ranking Partially-Matched Answers and Evaluation Metrics

In order to objectively measure the accuracy on the rankings of partially-matched answers to user queries, we compared the ranking strategy of *AdProc* with four other ranking approaches, with the most relevant retrieved product ranked first, and so on. We have implemented four baseline approaches on ranking, and compared the ranked results of each approach based on the 1,750 Facebook user-created queries, denoted *Test_Queries*. Since all the ranking methods were given the *same* set of products, i.e., query results, for ranking, the comparison was fair.

To measure the effectiveness of the ranking strategy of *AdProc* on partially-matched answers to queries, we apply two well-known metrics, *average Precision at K* ($P@K$) and *Mean Reciprocal Rank* (MRR) [8]. The average $P@K$ measures the overall user's satisfaction with the top-K ranked answers (generated by *AdProc*) to a particular query which measures the overall user's satisfaction with the top-K answers, whereas MRR is the averaged sum of the reciprocal of the ranking position of the *first* related answer among the top-5 answers, if there is any, or 0, otherwise for each query in *Test_Queries*. Average $P@K$ and MRR evaluate the ranking strategy of *AdProc* and other ranking strategies considered, such that the *higher* relevant answers are positioned in the ranking list, the *higher* their corresponding $P@K$ and MRR scores are.

5.9 Existing Approaches to be Compared

Besides verifying the accuracy of *AdProc* based on *SRCC*, P, R, and *F-measure* for answering natural language queries, we present four query processing approaches to be compared with *AdProc* in terms of ranking partially-matched answers to queries in *Test_Queries* below.

Random Processing [21]**:** shuffles all the retrieved DB records into some potentially new order by using a *random number generator*. In this case, we implemented the number generator using the pseudorandomness from Python's random module, which served as a useful baseline measure.

FAQFinder [3]**:** uses *TF-IDF* for computing the *similarity* of different results to the constraints specified in user queries. Generally, TF is the number of times an ad fulfills some constraint, and IDF is the relative rarity of that constraint being fulfilled by other ads.

Cosine similarity [16]**:** relies on the Vector Space Model (VSM). Each ad is represented as a vector of numeric values, where each value indicates whether the correspondent constraint is satisfied, which is '1' if it is, and '0' otherwise, and the score of the ad is calculated as the cosine similarity between the ad vector and the query vector.

AIMQ [24]**:** depends on attribute-value pairs to generate the associated supertuple of each attribute. A *supertuple* is an inferred DB tuple A that contains a set of attribute values, each of which includes a summary of values in the corresponding table column, and is used for calculating the similarity of categorical attributes. AIMQ determines the similarity between query Q and a DB tuple (i.e., record) A.

Comparison of Performance Evaluation. To avoid bias, the top-5 answers for each one of the 1,750 randomly-selected test queries generated by each of the five natural language query processing approaches were evaluated by the 350 Facebook appraisers. The appraisers were asked to determine the *relevance* of the top-5 partially-matched, retrieved results for the corresponding query. The study was performed between January 21, 2024 and February 22, 2024. The metric $P@1$, $P@5$, and MRR were computed based on the evaluation provided by the 135 Facebook users on the 1,750 test cases, which serve as the *ground truth* for this empirical study.

As shown in Fig. 10, *AdProc* outperforms the other four querying systems based on $P@1$, $P@5$, and MRR, which verifies the *effectiveness* of *AdProc*. The results are *statistically significant* based on the Wilcoxon Signed-Ranks Test ($p < 0.01$).

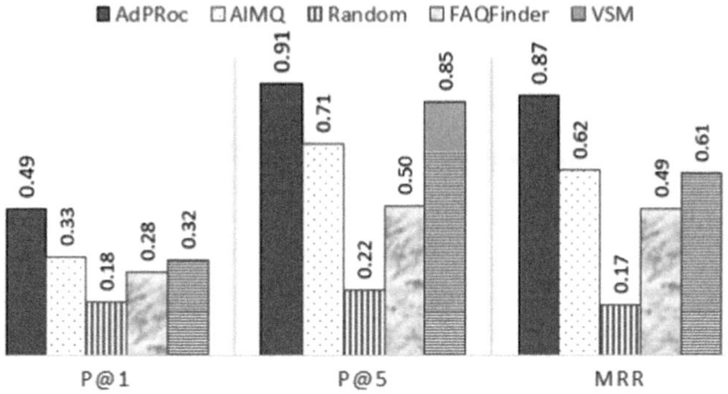

Fig. 10. *Precision@K* (K = 1, 5) and MRR scores on the (top-5) answers achieved by *AdProc* and other query processing approaches for the 1,750 test queries.

Among all the five approaches, the $P@1$, $P@5$, and MRR values for FAQFinder are the *lowest*, except the Random approach, since FAQFinder uses a simple method that simply compares the occurrences of attribute values with constraint values regardless they belong to the same domain. Among all the categories, we observed that the lowest scores on the three measures for *AdProc* occur in the jobs category. For this category, appraisers did not consider the answers based on their similarity to the original query. For example, a Java programmer job is closely related to a C++ programmer job, but the appraisers considered the answers based on which result is more relevant to their own expertise and experience, which is different from one user to another.

5.10 The Efficiency of *AdProc* and Other Ranking Methods

The *efficiency* of *AdProc* and four other ranking approaches are presented in Fig. 11, which shows the average query processing time for the 1,750 test queries obtained from the Facebook survey. *AdProc* outperforms all ranking approaches except the Random ranking strategy (by 0.1 s, on the average). This is due to the fact that the Random approach does not perform any processing or employ any similarity measures, but rather selects records randomly. Unlike the compared ranking methodologies, *AdProc* retrieves exact matches besides partially-matched answers, if needed. Thus, there is an additional time imposed on *AdProc* for retrieving (partially-)matched answers. As shown in Fig. 11, the query processing time for *AdProc* is the fastest compared to FAQFinder, SVM and AIMQ when partially- and exact-matched answers are retrieved by *AdProc*. *AdProc* was implemented using PHP and Javascript and runs on a MacBook Pro with Apple M1 processor and 16 GB of memory.

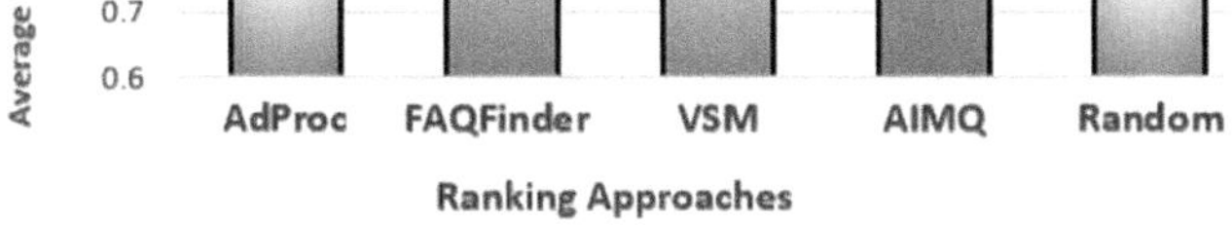

Fig. 11. Average query processing time of *AdProc* and others.

6 Conclusion

We have introduced *AdProc*, a closed domain natural language query processing system on multiple ads domains, which (i) automates the process of *classifying*, *extracting*, and *populating* data from online ads to its underlying database,

(ii) relies on simple probabilistic models to determine the *domain* an ad query belongs, and (iii) generates answers that match the information needs expressed in an ad query. Besides performing exact-match on DB records, *AdProc* also extracts partially-matched DB records. Retrieving partially-matched DB records is desirable, since it is frustrating to the users who look for desired information when there are no exact matches or very few found. Our new holistic partial-matching approach first searches for DB records that satisfy all the constraints specified in a natural language query, which is followed by retrieving partially-matched records by altering one constraint at a time in a sequential order.

Empirical studies conducted on a set of 80,000 online ads show that *AdProc* is highly effective in classifying ads in multiple domains and labeling and extracting their data, with accuracy in the ninety percentile. Furthermore, the approaches adopted by *AdProc* outperform other machine learning approaches (up to 9%) in accomplishing the same task. In addition, a conducted study has verified the effectiveness of *AdProc* in answering natural language queries in multiple ads domains. Furthermore, to objectively measure the accuracy on the rankings of partially-matched answers to user queries, we have compared the ranking strategy of *AdProc* with four other baseline ranking approaches, with the most relevant retrieved product ranked first and so on based on the 1,750 Facebook user created natural language queries. This additional evaluation is based on real users' assessments of the partially-matched DB records which goes beyond the performance analysis conducted based on exact-matched results. The experimental results based on our partial-matched approach demonstrate that it is more effective and efficient than the four baseline ranking strategies, and the results are statistically significant.

References

1. Allison, B.: An improved hierarchical Bayesian model of language for document classification. In: Proceedings of COLING, pp. 25–32 (2008)
2. Bian, J., Liu, Y., Agichtein, E., Zha, H.: A few bad votes too many?: towards robust ranking in social media. In: Proceedings of AIRWeb, pp. 53–60 (2008)
3. Burke, R., Hammond, K., Kulyukin, V., Lytinen, S., Tomuro, N., Schoenberg, S.: Question answering from frequently asked question files: experiences with the FAQ finder system. AI Mag. **18**(2), 57 (1997)
4. Callan, J., Connell, M.: Query-based sampling of text databases. ACM TOIS **19**(2), 97–130 (2001)
5. Chieu, H., Ng, H.: Named entity recognition with a maximum entropy approach. In: Proceedings of Conference on Natural Language Learning, pp. 160–163 (2003)
6. Chouaib, H., Cloppet, F., Tabbone, S.: Generic feature selection and document processing. In: Proceedings of ICDAR, pp. 356–360 (2009)
7. Cohen, W.: Fast and effective rule induction. In: Proceedings of ICML, pp. 115–123 (1995)
8. Croft, W., Metzler, D., Strohman, T.: Search Engines: Information Retrieval in Practice. Addison Wesley (2010)
9. Dayanik, A., Lewis, D., Madigan, D., Menkov, V., Genkin, A.: Constructing informative prior distributions from domain knowledge in text classification. In: Proceedings of SIGIR, pp. 493–500 (2006)

10. Demner-Fushman, D., Lin, J.: Answering clinical questions with knowledge-based and statistical techniques. Comp. Linguist. **33**(1), 63–103 (2007)
11. Hall, M., Frank, E.: Combining Naive Bayes and decision tables. In: Proceedings of Florida Artificial Intelligence Research Society Conference (2008)
12. Jones, B., Kenward, M.: Design and Analysis of Cross-Over Trials, 2nd edn. Chapman and Hall (2003)
13. Khare, R., An, Y.: An empirical study on using hidden Markov model for search interface segmentation. In: Proceedings of ACM CIKM, pp. 17–26 (2009)
14. Ko, J., Nyberg, E., Si, L.: A probabilistic graphical model for joint answer ranking in question answering. In: Proceedings of SIGIR, pp. 343–350 (2007)
15. Koberstein, J., Ng, Y.-K.: Using word clusters to detect similar web documents. In: Lang, J., Lin, F., Wang, J. (eds.) KSEM 2006. LNCS (LNAI), vol. 4092, pp. 215–228. Springer, Heidelberg (2006). https://doi.org/10.1007/11811220_19
16. Li, B., Han, L.: Distance weighted cosine similarity measure for text classification. In: Yin, H., et al. (eds.) IDEAL 2013. LNCS, vol. 8206, pp. 611–618. Springer, Heidelberg (2013). https://doi.org/10.1007/978-3-642-41278-3_74
17. Liu, Y., Lin, Y., Chen, Z.: Using hidden Markov model for information extraction based on multiple templates. In: Proceedings of NLP-KE, pp. 394–399 (2003)
18. Liu, Y., Zheng, Y.: One-against-all multi-class SVM classification using reliability measures. In: Proceedings of IJCNN, pp. 849–854 (2005)
19. Manning, C., Raghavan, P., Schutze, H.: Introduction to Information Retrieval. Cambridge University Press (2008)
20. McCallum, A., Nigam, K.: A comparison of event models for Naive Bayes text classification. In: Proceedings of AAAI Workshop on LTC, pp. 41–48 (1998)
21. Meng, X., Ma, Z., Yan, L.: Answering approximate queries over autonomous web databases. In: Proceedings of WWW, pp. 1021–1030 (2009)
22. Mierswa, I., Wurst, M., Klinkenberg, R., Scholz, M., Euler, T.: YALE: rapid prototyping for complex data mining tasks. In: Proceedings of ACM SIGKDD, pp. 935–940 (2006)
23. Mitchell, T.: Machine Learning. McGraw Hill (1997)
24. Nambiar, U., Kambhampati, S.: Answering imprecise queries over autonomous web databases. In: Proceedings of ICDE, p. 45. IEEE (2006)
25. Ng, Y.K.: Easy to find: a natural language query processing system on advertisements using an automatically populated database. In: Proceedings of the 19th International Conference on Web Information Systems and Technologies (WEBIST), pp. 71–83 (2023)
26. Nguyen, H., Kang, E., Freire, J.: Automatically extracting form labels. In: Proceedings of IEEE ICDE, pp. 1498–1500 (2008)
27. Pera, M., Qumsiyeh, R., Shaikh, M., Ng, Y.K.: Retrieving good, better, and best answers to questions in advertisements. In: Proceedings of ACM WIDM, pp. 3–6 (2009)
28. Radev, D., Fan, W., Qi, H., Wu, H., Grewal, A.: Probabilistic question answering on the web. In: Proceedings of WWW, pp. 408–419 (2002)
29. Rajput, Q., Haider, S.: Use of Bayesian network in information extraction from unstructured data sources. In: Proceedings of WASET, pp. 325–331 (2009)
30. Rey, D., Neuhäuser, M.: Wilcoxon-signed-rank test. In: International Encyclopedia of Statistical Science, pp. 1658–1659. Springer, Cham (2011)
31. Sculley, D., Wachman, G.: Relaxed online SVMs for spam filtering. In: Proceedings of ACM SIGIR, pp. 415–422 (2007)
32. Wang, Z., Luo, X.: A semantic pattern for restricted domain Chinese question answering. In: Proceedings of ICMLC, pp. 1333–1338 (2009)

33. Xu, W., Liu, X., Gong, Y.: Document clustering based on non-negative matrix factorization. In: Proceedings of ACM SIGIR, pp. 267–273 (2003)
34. Xue, G., Dai, W., Yang, Q., Yu, Y.: Topic-bridged PLSA for cross-domain text classification. In: Proceedings of ACM SIGIR, pp. 627–634 (2008)
35. Yang, Y., Pedersen, J.: A comparative study on feature selection in text categorization. In: Proceedings of ICML, pp. 412–420 (1997)
36. Zhang, Z., DaSylva, L., Davidson, C., Lizarralde, G., Nie, J.: Domain-specific QA for the construction sector. In: Proceedings of IR4QA Workshop, pp. 65–71 (2004)

Mapping Business Web Applications for Web Automation

Adrian Sterca(✉), Virginia Niculescu, Alexandru Kiraly, and Darius Bufnea

Department of Computer Science, Babes-Bolyai University, Str. M. Kogalniceanu, Cluj-Napoca, Romania
{adrian.sterca,virginia.niculescu,alexandru.kiraly, darius.bufnea}@ubbcluj.ro

Abstract. This paper presents an automated tool designed to create conceptual maps of business web applications. These maps illustrate, in an abstract and hierarchical manner, the functional navigation and operational paths within the application's user interface (UI). The map presents in a structural way the functionalities of the business web application. Our tool automatically navigates through the UI of the target business web application, mapping UI operations to conceptual operations in the underlying database. The resulting conceptual map is presented in either a graphical or serialized form. This map serves as documentation for the business web application, enabling new users to quickly acquire the knowledge needed to navigate and operate the application's UI screens. A second usage of this conceptual map that is demonstrated in the paper comes in the context of RPA (Robotic Process Automation). Our automated tool is able to execute automatically simple processes on the target business web application, this simple automated processes being described as paths in this hierarchical conceptual map.

Keywords: RPA · Web automation · Web application mapping · DOM · Conceptual operations · Business process execution

1 Introduction

Creating maps that capture the structure of web sites was a very popular idea in the initial days of the world wide web; static websites used to have a tree-like map of the website, describing all or the most important static Html documents of the website. Nowadays this is not common due to the fact that the Html content is generated dynamically from databases. Starting from this classic idea, we propose in this study to create and manage a map for all the functional paths in the UI of a business web application. The targeted applications are the business web applications like CRM (i.e. Customer Relation Management), ERP (i.e. Enterprise Resource Planning) or project management web applications, e.g. Microsoft Dynamics, Atlassian Jira, but generally, any business web application that uses a relational database in the backend and exposes the data in this database through the functionality of the UI (i.e. User Interface). We call this map a conceptual map because is related to the concepts used by the specific web application.

M. Marchiori and F. García Peñalvo (Eds.): WEBIST 2023, LNBIP 543, pp. 133–150, 2025.
https://doi.org/10.1007/978-3-031-89621-7_7

As mentioned above, it would be difficult to create such maps for business web applications due to the possible infinite amount of generated Html content (since the data space of the application is infinite, although the actual data used by the application is, off course, finite).

If we consider all the Html content generated by a business web application, we can see that there are two types of Html content in a business web application:

- Html content used purely for the visual look of the application and for navigation (e.g. menus and menu items, tabs and panels, headers and footers, cover images etc.); this Html content does not use, expose or manipulate the data in the underlying database of the application
- Html content that represents an interface to the data in the underlying database; this Html content is used for showing various entities from the database or for allowing the user to add new content or modify existing content from the database; most Html content generated by a business web application belongs to this second type.

By abstracting over some details, we can consider a business web application to be just a user-friendly interface to the data in a database. For this kind of applications, most of the operations that can be done through the user interface can be reduced to operations on the database. This is because the business functionalities of the application are translated into read or write operations in the database or in SQL terminology, they translate to Select, Insert, Update or Delete operations of various entities in the database.

Having a conceptual structure over the operations that could be done through an application's UI is beneficial for many reasons. We mention two concrete usages of such a conceptual map:

- it could serve as documentation for new users, for allowing new users of the application to see an abstract, birds-eye view of the web application. This can facilitate an easy learning of the functionalities of the business application by new users.
- it could be also useful for advanced RPA (i.e. Robotic Process Automation) robots. Instead of an RPA developer coding various operations on UI controls, an RPA agent can be guided to execute various conceptual operations on the conceptual map. The mapping tool introduced in this paper - a browser extension tool that automatically maps UI controls to properties of entities in the database (i.e. it 'understands' the UI of the application) and maps UI operations to database operations, can also be programmed to manipulate those UI controls (i.e. adding a value to a text input control in the UI, clicking a submit button in the UI etc.) in order to perform the actual conceptual operation from the map (e.g. inserting an 'Account' entity, updating a 'Contact' entity, deleting a 'Resource' entity etc.).

In order to obtain this abstract structure of the website of the application, i.e. the conceptual map, we introduce a tool that is able to navigate through the screens/paths of the target web application and is able to detect the conceptual operations facilitated by the Html content of the application. And it does all of this automatically, without the intervention of the human user.

The central idea of the proposed tool is to map UI operations onto conceptual operations in the corresponding database. The tool automatically translates human user operations on the UI onto conceptual operations in the database.

Since we refer to web applications, the tool comes in the form of a browser extension – a plugin. The output of running our tool on a target business web application is a conceptual map which consists of nodes inter-linked in a tree-like structure, these nodes representing either *generic Html content* or *Html content representing conceptual operations*. A *generic Html content* is just an Html document that contains menus, documentation or any other generic info about the application and does not expose any entity from the database. While an *Html content representing conceptual operations* is an Html code like text inputs, buttons, paragraphs, tables etc. either for displaying properties of an entity from the database (i.e. 'SELECT' the entity) or for allowing the user to add or remove an entity from the database or to alter its properties (i.e. 'INSERT' or 'UPDATE' or 'DELETE' the entity). As a final result, we obtain this conceptual map in a serialized form as JSON and depict it graphically in Html canvas.

We assume in this study that the term *concept* refers to the data type stored in a database table, while *entity* refers to a row/record of a database table. A *concept* always describes a set of *entities*; e.g. an 'Account concept' is stored in an 'Account' table in the database, while each entry/record from this table represents an 'Account entity'.

Since recording all information about entities would lead to a node explosion in the generated conceptual map, for all entities of a specific database concept and each conceptual operation (i.e. CRUD operations on concepts: Create/Insert, Read/Select, Update, Delete) we only store one node in the map. This means that we will have only one node for each concept-operation pair. For example, if we have an 'Account' table in the database and in this table there are three entries, 'Account1', 'Account2' and 'Account3', we only store one single node in our conceptual map for the conceptual operation of 'Update Account'; we don't have three different map nodes for 'Update Account1', 'Update Account2' and 'Update Account3'. For the other conceptual operations: Select, Insert and Delete, things are similarly.

The rest of the paper is structured as follows. Section 2 presents studies related to our work. Section 3 presents the structure of the conceptual map, while Sect. 4 details the algorithm used for building the conceptual map. Following, in Sect. 5 we discuss the method used for detecting conceptual operations in the DOM. Section 6 is devoted to tests and experiments. Section 7 presents a possible usage of the conceptual map of a web application: that is automatic execution of business processes which are described in terms of paths in this hierarchical conceptual map. Section 8 discusses some limitations of our mapping tool. The paper ends with conclusions and future work.

2 Related Work

A related topic is UI screen parsing [17] where a screenshot of a UI is analyzed in order to determine the components it's made from: buttons, menus, labels etc. These components are represented in an hierarchical structure. Screen2Vec [11] uses machine learning autoencoders to represent the layout and UI components of a mobile application screen in a vector of numbers. These studies are performed on a single screen of an applications and consider only mobile applications, not web applications. Our study considers all the screens of a business web application.

[6] uses transformer models Faster RCNN to determine the similarity of various screens of a mobile application. Another approach for determining the similarity

between screens of web applications, not only mobile, is taken in [16]. Authors construct a dataset of web UI screenshots, WebUI and then they use machine learning models with transfer learning to perform screen element detection, screen similarity and screen classification on these screenshots. Similar approaches are described in [14] where authors try to detect UI elements (i.e. buttons, text inputs etc.) from a screenshot of a mobile application in order to generate the UI code for that specific UI.

Another related topic is that of web segmentation which refers to dividing a web page (including the UI of a web application) into constituent segments like: menu items, paragraphs, tables, panels, cards etc. Web segmentation can be useful for information retrieval, archiving, topic extraction, focused crawling, and improving the accessibility of web content in non-visual environments. Generally, there are three approaches for web segmentation: a) a pixel-based one that uses computer vision techniques to determine segments from a screenshot of a web page, b) a DOM-based approach which uses features extracted from the DOM hierarchical structure of the web page, and c) a hybrid approach which combines the previous two.

The VIPS (VIsion-based Page Segmentation) algorithm for segmenting a web document is presented in [4]. The starting point of the VIPS algorithm is separator tags. VIPS uses separator tags in the web document like
and <hr>in order to identify segments. Then it uses heuristics based on the <table>, <tbody>, <tr>, <td>, <p>, <ul>and <li>tags in order to further divide the Html content into segments.

The authors in [5] present another approach to web segmentation. They try to identify objects in the Function-based Object Model (FOM). A FOM object has a function and a category in a web document. The function of an object can be of presentation, navigation, interaction or decoration. Authors use the DBSCAN clustering algorithm in order obtain the web segments.

Another paper [13], introduces a method for recognizing significant structures, such as table-like or list-like structures within a web page. This approach operates at different levels within the DOM tree. Initially, the DOM tree undergoes a cleanup process where unimportant tags are removed, while empirically determined important tags, such as <table>, <tr>, <td>, <ul>, or <div>, are retained. Then, the algorithm proceeds to identify the table-like and list-like structures. The paper by Jeyafreeda et al. [3] examines web page segmentation from the perspective of a clustering problem involving visual elements. The objective is to cluster all visual elements while discovering a predetermined number of clusters, where the elements within each cluster should be visually connected. The study evaluates three clustering algorithms: K-means, F-K-means, and Guided Expansion.

Our present study is connected to RPA (Robotic Process Automation). Robotic Process Automation (RPA) is defined as the application of specific methodologies and technologies that aim to automate repetitive tasks achieved usually by human users [7,8]. RPA frameworks (e.g., UiPath[1] , Automation Anywhere[2] , Blue Prism[3] , Microsoft Power Automate[4], etc.) are designed to develop software robots that improve the busi-

[1] https://www.uipath.com/.
[2] https://www.automationanywhere.com/.
[3] https://www.blueprism.com/.
[4] https://powerautomate.microsoft.com/en-us/.

ness environment in various ways. RPA refers to those tools that operate on the user interface (UI) aiming to perform automation tasks using an "outside-in" approach. The information systems are kept unchanged, compared to the traditional workflow technology, that allows the improvement using an "inside-out" approach [1]. In order to describe business processes in a way that can later be executed automatically by programs, RPA platforms like UiPath, Power Automate, Automation Anywhere etc. operate (i.e. create automated business processes) in the following way: RPA developers identify UI (i.e. User Interface) components of a software application like buttons, text input controls, dropdown lists and tables, and then customize activities by writing code snippets in a programming language in order to act on these UI controls (e.g. click the selected button, write data in the text input, select all data from a table or a dropdown list etc.); this code that references the selected UI controls forms the automated business process which can be executed many times later with different input parameters. Other alternative to the aforementioned commercial RPA platform do exist and they are usually based on PBD (i.e. Program-By-Demonstration) methodologies [2,9,10].

A work somewhat similar to ours is [12]. In this paper, authors build a topic hierarchy of a website starting from the homepage of the website and continuing in a tree-like structure in which the vertices and edges correspond to web pages and hyperlinks. But our work deals with a much complex topic of mapping the web content generated by a business web application and the map is a conceptual map (it does not include actual web pages).

3 The Structure of the Map

The conceptual map describes in an abstract, conceptual way, the use cases (functionalities) of the business web application. The structure of the map has a form of a tree, and the map is a navigational map meaning that it describes in a compact way all possible navigation paths in the UI of the web application. The elements of the map are the following:

- *Clickable elements*
- *Conceptual operation DOM parts*
- *Web pages*
- *Links*

Conceptual operation DOM parts and *Web pages* are the nodes of the map, *clickable elements* are attributes of the nodes and *links* are the edges of the map.

3.1 Clickable Elements

The clickable elements are Html elements which can be clicked by the user in the browser. The effect of such a click is the sending of a complete HTTP request or an XHR/Fetch API HTTP request whose response would replace totally or partially the currently loaded Html content in the browser window/tab. These *clickable elements* create *links* in the conceptual map of the application between two *Web pages* of the application. From a technical point of view, *clickable elements* are the following Html elements: <button>, <input type="submit">, <a>or any other Html element that has an *onclick* attribute or has a click event listener added on it. From a technical point of view, the discovery of the Html elements that have a click event listener from the current document is realised through overwriting the *addEventListener()* method on the *Element.prototype* object in the browser.

The identification of each *clickable element* in the browser is done by assigning an unique signature to it (i.e. unique to the current Html document) – the XPATH of that *clickable element* in the current DOM (i.e. Document Object Model).

3.2 Conceptual Operation DOM Parts

A *conceptual operation DOM part* is that part of the DOM in a business web application that allows the human user to execute a *conceptual operation* on the database. We denote by a *conceptual operation* any CRUD operation (i.e. Create, Read, Update, Delete) on a concept or entity from the underlying database of the web application.

For example, we may consider the operations on the concept *Account*:

- *Update* operation – usually the web application would provide an Html form with labels and inputs for each field of the *Account* entity.
- *Create* operation – similarly, the web application will usually construct also a form with labels and inputs for all the fields of the *Account* concept.
- *Read* operation – a similar DOM part is shown, but this time the text inputs are read-only.
- *Delete* operation – usually the web application requires just a key of the entity (like the name of the entity) from the user.

It's worth mentioning that *conceptual operation DOM parts* are always the leaves of the tree-like structure of the conceptual map of the web application.

3.3 Web Pages

From a technical point of view, an *web page* represents an Html document that is loaded in the browser. In the map of the web application, the *web pages* are nodes of the map. A *Web page* contains *clickable elements*. Actually, the set of all *clickable elements* from a *web page* form a signature of that *web page*; as mentioned before, each *clickable element* is identified by its XPATH in the current document. On the map, *web pages* may lead to other *web pages* or to *conceptual operation DOM part* nodes. The *web pages* are linked together in the map by *links*.

Web pages could be seen as container nodes which contain a set of *clickable elements*, each *clickable element* connecting the current *web page* with another *web page* or with a *conceptual operation DOM part*.

An actual Html document may contain an Html section corresponding to a *conceptual operation DOM part*, still this *conceptual operation DOM part* is not considered to be part of the *web page* corresponding to this Html document. This *conceptual operation DOM part* will form a separate node in the conceptual map connected through a *link* to the original *web page*.

3.4 Links

The *links* are oriented connections between *web pages* or between a *web page* and a *conceptual operation DOM part* in the conceptual map. A *link* unites two *web pages* in the map if by clicking on a *clickable element* of the first *web page*, we navigate to the second *web page*. They form the backbone of the map of the web application.

4 Building the Map

In order to construct the map we apply a depth-first traversal algorithm on the entire web application. The various web pages of the application are traversed through clickable elements. The algorithm for constructing the map of the web application is described in code listing 1. In the first lines of the algorithm, lines 1–5, initializations of various structures used in the algorithm are performed: the stack, the root of the map and the current $WebPage$ node. The algorithm assumes that the starting Html document of the web application is already loaded in the browser. The algorithm initializes a new *WebPage* node from this Html document in line 4, node that will also be the root of the conceptual map. As mentioned before, a *WebPage* node of the map is identified by a list of *clickable elements* and also it has a set of *links* pointing towards other *WebPage* nodes or *Conceptual operation DOM parts* (in line 4, *links* are initialized with an empty set).

The lines 6–25 define the main cycle of the iterative depth-first traversal algorithm. The $Stack$ stores all the *WebPages* that need to be traversed/visited. A *WebPage* is considered visited and removed from the stack when all its *clickable elements* were visited by our tool. On lines 7–10 we take the first *WebPage* from the top of the stack, get it's first unvisited (i.e. unclicked) *clickable element* and navigate through that path by clicking on this unvisited *clickable element* (on line 10). Following, if all *clickable elements* from the *WebPage* node from the top of the stack were already visited, we remove this *WebPage* from the top of the stack (line 12).

Next, on line 14, we try to determine if this new Html document that was loaded in the browser (after executing line 10) represents a generic *WebPage* or a *conceptual operation DOM part* leaf node. If we have found a conceptual operation in the currently loaded Html document (this is performed by the $DetectConceptualOperation()$ algorithm which is detailed in a subsequent code listing), we add this *conceptual operation DOM part* to the map on lines 15–17. Otherwise, in lines 19–23 we add a new, generic *WebPage*, $newWebPage$ to the map. We don't forget to link this $newWebPage$ to the previous *WebPage* and to add it to the stack in order to be visited later.

5 Detecting Concepts and Conceptual Operations in the DOM

We mentioned in Sect. 1, that in order to avoid a node explosion problem in the generated conceptual map, we only store one node in the map for all entities of a specific database concept and each conceptual operation (i.e. Select, Insert, Update, Delete); i.e. we have only a single map node for the whole concept, and not one map node for each entity. The considered conceptual operations are the 4 basic SQL operations: Select,

The MapBuilding algorithm is:

```
Stack = []
MapRoot = []
clickableElements = GetClickableElements()
MapRoot = {WebPage : clickableElements, Links : []}
Stack.push(MapRoot)
while (Stack is not empty) do
   WebPage = Stack.top()
   ClickElem = GetFirstUnvisitedClickableElement(WebPage)
   ClickElem.state = visited
   Navigate(ClickElem)
   if VisitedAllClickableElements(WebPage) == True then
      Stack.pop()
   end if
   conceptualOperation = DetectConceptualOperation()
   if conceptualOperation! = NULL then
      WebPage.Links.add({ClickableElement : ClickElem, WebPage :
      conceptualOperation})
      { This is a leaf node of the map }
   else
      { This is a new WebPage node of the map }
      clickableElements = GetClickableElements()
      newWebPage = {WebPage : clickableElements, Links : []}
      WebPage.Links.add({ClickableElement : ClickElem, WebPage : newWebPage})
      Stack.push(newWebPage)
   end if
end while
```

Algorithm 1. The algorithm that constructs the map of the web application.

Insert, Update, Delete, and an additional one: *SelectAll*, which represents the operation of selecting many entities, not necessary all, of the same concept. This is due to the fact that a popular UI pattern for CRM and ERP web applications is to show a list of entities (i.e. a *SelectAll* operation) from where the human user can choose which entity to update or delete.

The algorithm depicted in listing 2 is applied for the detection of an operation on a concept in the current DOM (i.e. Document Object Model). After each UI change generated by a click event, the *diffDOM* (i.e. the difference DOM) is computed. For efficiency reasons, the algorithm is run on this difference DOM, not on the whole DOM loaded in the browser; the *diffDOM* is just the part of the DOM that is changed after the click event and the previous run of $DetectConceptualOperation$.

The algorithm begins with the function $TableOfEntitiesDetected(diffDOM)$ which tries to detect any tables in the *diff DOM*. Tables are important because they may indicate a SelectAll operation on a concept. We can not present here the body of this function due to space constraints, but in essence, the algorithm for detecting a table tries to horizontally align leaf html tags containing only text into horizontal row clusters and then tries to vertically align these horizontal rows into a table. After detecting a table, the algorithm tries to detect if the entities of a concept are rendered in that table through function $DetectConceptInTable(diffDOM)$. Again, we omit the details here, but simply said, the concept is determined by searching its attributes in the table cells of the table header or, if these are not found (e.g. the table may not have a header), the concept is determined by inspecting (i.e. triggering click events) the entities from the first two rows of the table; detecting an entity is done similarly to the way function $FindOneConcept(diffDOM)$ determines that an entity (of that concept) is rendered in the $diffDOM$: several text attributes (labels) of a concept from the Data Model are found in the $diffDOM$ - more than 75% of the labels of a concept are found.

Next, if there was no *SelectAll* or *Delete* operation on a concept/entity detected, the algorithm tries to find text input elements/tags (i.e. "<input type=text >,<select >,<textarea >") associated to the text labels (i.e. the text attributes of the found entity). We do this in the $FindInputFieldsForTextLabels()$ function. The associated text input tag is searched in the neighborhood of the label, more specifically, the associated input tag is searched in the South-East quarter of a circle with the center in the middle of the label - the closest such text input tag from this South-East quarter is associated to the label. If no such input html tags were detected, that the operation is *Select* on an entity of the detected concept. If, on the other hand, there was text input html elements found in the *diffDOM*, but they are empty (i.e. don't contain values), the conceptual operation would be *Insert*. Otherwise, the conceptual operation is *Update* on an entity of the detected concept.

The DetectConceptualOperation algorithm is:

```
if TableOfEntitiesDetected(diffDOM) == TRUE then
    DetectedConcept = DetectConceptInTable(diffDOM)
    return {Concept : DetectedConcept, Operation : SELECTAll}
else
    if (DetectedConcept = FindOneConcept(diffDOM))! = NULL) then
        {Several text attributes (labels) of a Concept from the Data Model are found in the DOM}
        if (The previous click event was triggered on a tag with the caption "DELETE") then
            return {Concept : DetectedConcept, Operation : DELETE}
        end if

        InputFieldsFound = FindInputFieldsForTextLabels()
        if InputFieldsFound == FALSE then
            return {Concept : DetectedConcept, Operation : SELECT}
        else
            if (Found Text Input elements are empty (i.e. don't contain values)) then
                return {Concept : DetectedConcept, Operation : INSERT}
            else
                return {Concept : DetectedConcept, Operation : UPDATE}
            end if
        end if
    end if
end if
```

Algorithm 2. The conceptual operation detection in DOM algorithm.

5.1 Configuration Settings for the Plugin

In order to use the plugin for a specific application we need to configure the plugin correspondingly, with the following settings:

- the URL of the target web application together with access credentials
- the Data Model of the database used by the target web application

Primary access to the database used by the target web application is not required in order to use the plugin because the Data Model configuration does not need to match

```
DataModel = {
    "Account" : ["Account Name", "Phone", "Fax", "Website", "Parent Account",
    "Ticker Symbol", "Address", "Primary Contact", "Description", "Industry",
    "SIC Code", "Ownership"],

    "Contact" : ["Full Name", "Job Title", "Account Name", "Email",
    "Business Phone", "Mobile Phone", "Fax", "Preferred Method of Contact",
    "Address", "Gender", "Marital Status", "Birthday", "Spouse/Partner Name",
    "Anniversary", "Personal Notes", "Company", "Originating Lead",
    "Last Campaign Date", "Marketing Materials", "Contact Method", "Email",
    "Bulk Email", "Phone", "Fax", "Mail"],

    "ForeignKeys" : [ { ForeignKey : "Company", ForeignTable : "Contact",
    PrimaryKey : "Account Name", PrimaryTable : "Account" } ]
        ...
};
```

exactly the structure of the database used by the application, but it should match the text labels used for each concept of the database on the UI of the application.

A small snippet from a data model example used for the Microsoft Dynamics 2016 CRM application is given above in JSON format.

This data model describes the concept *Account* with its associated attributes and the concept *Contact* with its associated attributes. It also describes that the *Company* attribute from *Contact* is a foreign key and refers to the primary key *Account Name* from the *Account* concept. As specified before, these attributes' names don't have to match exactly the attributes of the database tables, but instead they have to match the text labels for the respective entities on the UI of the target web application. The foreign key relation is used by the plugin to identify entities that are linked with the currently identified entity (i.e. entities that are values for a property of the currently identified entity).

6 Experiments

The implementation of our web automation tool is based on common web design principles (like the fact that an input field is always placed in the web UI either on the right or below or in the south-east part of its corresponding text label), so it should work correctly on any business web applications. In addition, the plugin uses general templates for the conceptual operations which are not specific to a particular web application, it should, in principle, work with any typical business web application that uses a relational database. For validation we chose two commercial business web application – Microsoft Dynamics 2016 CRM[5] and Atlassian Jira[6].

The validation is based on tests that evaluate the ability of our web automation tool to automatically discover conceptual operations on the UI by going through all use cases/functionalities of the aforementioned business applications. The test cases are defined based on all use cases that involve (i.e. whose outcome is) a CRUD operation in the backend database. The use cases that involve only UI modifications, without reading

[5] https://learn.microsoft.com/en-us/lifecycle/products/dynamics-crm-2016-dynamics-365.

[6] https://www.atlassian.com/software/jira.

from or changing the data in the database are ignored. So, we ignored use cases that do not operate on the database (they operate just on the data that is shown in HTML documents in the browser) like for example: print current HTML document, draw reports, email a link/current document, import from Excel into the browser window, etc. For the considered use cases, we tested whether our plugin is able to automatically detect the main conceptual operations. The results are summarized in Table 1.

Table 1. Automatic discovery of conceptual operations on the User Interface [15].

Target application	Total number of use cases (i.e. conceptual operations)	Number of conceptual operations automatically discovered	Percentage of conceptual operations automatically discovered
Microsoft Dynamics 2016 CRM	102	81	79%
Atlassian Jira	39	26	67%

In the Table 1 the first column specifies the tested web application, the second column specifies the number of use cases (i.e. potential conceptual operations), and the third column presents the absolute number of conceptual operations detected. The fourth column presents the percentage of the number of automatically detected conceptual operations in the total number of potential conceptual operations considered (i.e. the second column value). The results emphasize an automated detection of 79% and respectively 67% of the available conceptual operations in the considered business web applications. The difference is given by the operations which affect the database but they involve just partial CRUD operations, i.e. operations that only change a single field or a couple of fields from a database entity (e.g. the operation "Fulfill Order" which only changes a field of the "Order" entity and does not perform a full-fledged Update operation of all fields).

Figure 1 presents a partial conceptual map generated for the business web application Microsoft Dynamics 2016 CRM. Due to the space constraints we could not place the complete map here.

From the performance point of view, the evaluation reflects the fact that the time required by our web automation tool to build a complete map of a business web application is similar to the time required by a human user to navigate through all the use cases of the application.

Figure 2 shows a screenshot of our plugin associating text inputs to labels in Atlassian Jira in order to detect a *conceptual operation DOM part*. You can see that each text input in the DOM is correctly associated (through the green line) to its text label (i.e. the text label is also a property of the 'Issue' concept in Jira).

The full code of our browser plugin is available at
https://github.com/KiralyCraft/WAPlugin.

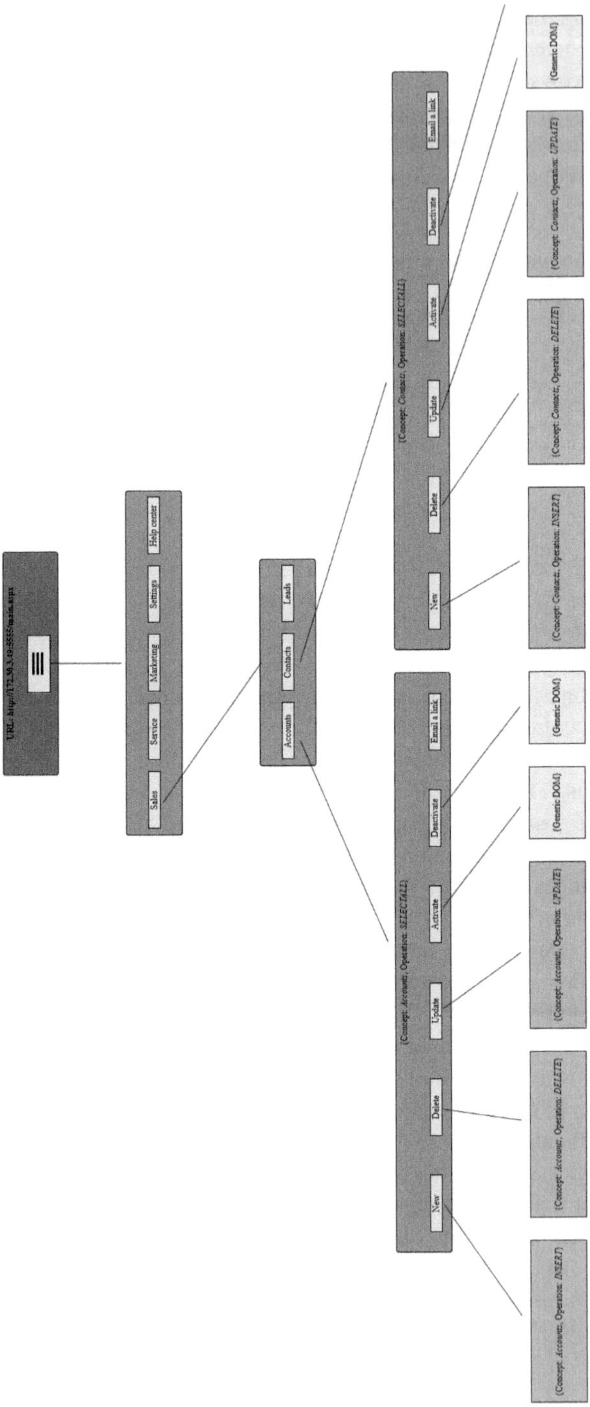

Fig. 1. Partial map of Microsoft Dynamics 2016 CRM [15].

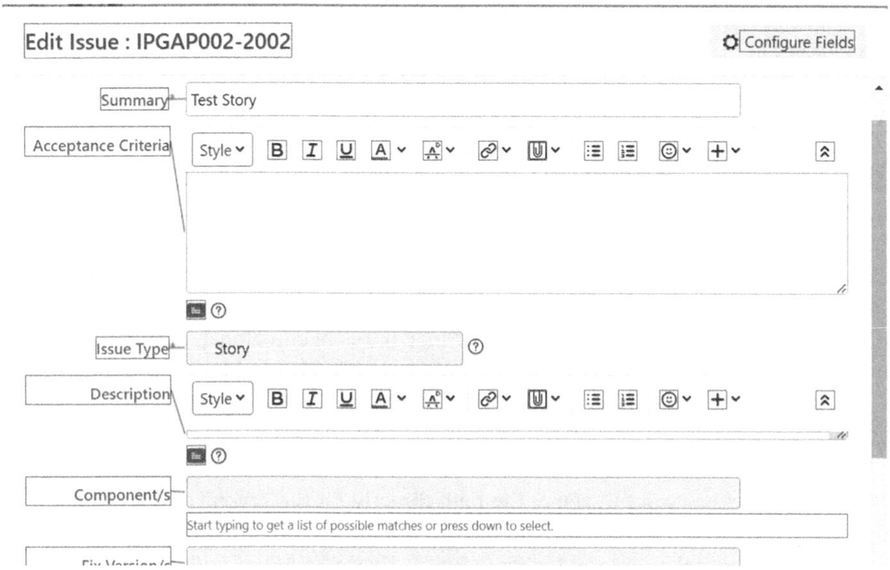

Fig. 2. Associating test inputs to labels in Atlassian Jira.

7 Using the Map in RPA Context. Automatic Execution of Paths in the Map

The conceptual map can be used by an RPA agent which can automatically execute paths in this conceptual map; our browser plugin can be used as such RPA agent. After the browser plugin maps the target web application and constructs the hierarchy of web pages and conceptual nodes parts, the human user can start using the map and executing paths from the map. The human user will choose the edge nodes of this path (i.e. starting node and the end node), will specify possible parameters for a conceptual operation and then the RPA agent can execute that specific path in the map by navigating through the target web application as instructed by the chosen path in the map and if the last node in the path is a conceptual operation (Select, Insert, Update, Delete), it will use the parameters input by the user in order to perform this conceptual operation.

For example, if we consider the map in Fig. 1, the human user may choose the *WebPage* identified by the clickable elements set {*Accounts, Contacts, Leads*} as the starting node of the map path and then he/she can choose the *conceptual operation DOM part* node {Concept: Account, Operation: INSERT} as the end node of the map path. Assuming the current web page of the target web application currently loaded in the browser is the {*Accounts, Contacts, Leads*} *WebPage* (the browser plugin will verify if this is the web page currently loaded in the browser - by verifying the signature of the web page), the plugin starts going down the map path, clicking on the corresponding clickable elements, until the currently loaded web page is the end node of the path.

This functionality of automatically executing paths of the hierarchical map can help the human user use the application even if he/she does not know how to use the target web application; the human user only needs to understand the map of the target web application.

7.1 Selecting a Path in the Conceptual Map

The user can select a path in the conceptual map (i.e. selects the start node and the end node of the path) either using the graphical representation of the map as provided by our browser plugin (as seen in Fig. 1) or by searching the two end nodes using textual search. The textual search of the nodes of the map is described in the next subsection. In order to select a node on the graphical representation of the map, the human user only needs to click this node on the map. Since our conceptual map is a tree, there is an unique downward path between the selected start node and end node.

7.2 Searching Nodes in the Map

If the user does not want to select the path directly on the graphical representation of the map, he/she can specify this path using a textual search mechanism. Let's consider again the conceptual map example from Fig. 1. Every *WebPage* node from the map can have text labels attached to them for easy identification (these labels are not visible in Fig. 1). After the web application is mapped, the user can attach these text labels to each node for easy identification of that node. When the user wants to specify a map path for the plugin to execute, he/she can do it very easily in the following way:

```
Path = { StartNode: {Label: "Sales WebPage"},
         EndNode: {Label: "Account INSERT"} }
```

In the above listing, the two labels are textual labels attached by the user to the nodes of the map, after the map is constructed. If the user did not attach any textual labels to the map's nodes, he/she can still specify the start and end node of the path by specifying the conceptual operation for a *conceptual operation DOM part* node (e.g. {Concept: Account, Operation: INSERT}) or he can specify a sublist of all clickable elements that identify a *WebPage*. In the following example, the start node is a *WebPage* node from the map example from Fig. 1 specified by a three clickable elements sublist: *"Sales", "Service", "Marketing"*. This clickable elements sublist identifies second *WebPage* node from the left.

```
Path = { StartNode: {ClickableElements: ["Sales", "Service", "Marketing"]},
         EndNode: {Concept: "Account", Operation: "INSERT"} }
```

In this example, *"Sales", "Service", "Marketing"* are the captions of the respective clickable elements (which are actually buttons in the Microsoft Dynamics 2016 CRM application). This sublist of clickable elements identify uniquely a *WebPage* node in the small conceptual map presented in Fig. 1 - the second node from the left in the map (because no other *WebPage* node in this small map contains these three clickable elements). Please note that this method of specifying a *WebPage* node by a sublist of clickable elements may fail when this sublist of clickable elements is small - the longer this list, the better; it can also be less practical for complex applications, for example, some web pages in the Dynamics 2016 CRM application can contain more than 30 clickable elements; specifying all of them involves more work than just selecting the respective node in the graphical map. Also, another problem can be that some clickable elements do not have textual captions, for example, the delete button may not have any textual caption and it can contain just a recycle bin icon. In this case, we can not specify the button with a caption. Please note that even if the button does not have a textual caption, the button is still identifiable in the map as the signature of a clickable element is its XPATH, not its caption. An example of a web page from Ms Dynamics 2016 CRM with clickable elements highlighted in red is visible in Fig. 3.

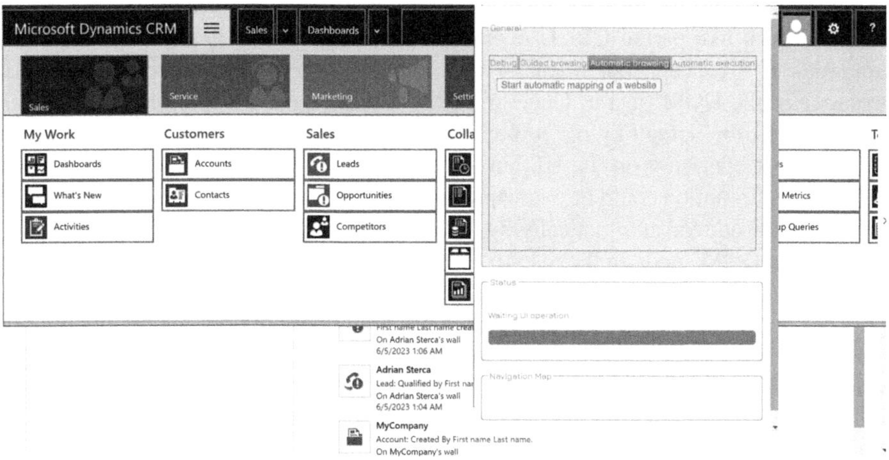

Fig. 3. A web page example from Ms Dynamics 2016 CRM with clickable elements highlighted.

7.3 Executing a Path in the Map

Since the conceptual map defines the links between various web pages of the application and these web pages are ultimately defined by sets of clickable elements which are identified by XPATHs in the DOM, it is not hard for the browser plugin tool to select a clickable element based on its XPATH and then trigger a click event on that element which makes the browser navigate to a different web page. In this way, all what the browser plugin must do in order to execute a path in the map is follow the path, trigger click events on the necessary clickable elements on the path and identify the currently loaded web page in the browser (as either a *WebPage* or a *conceptual operation DOM part*) along the way. If the browser plugin navigated to a *conceptual operation DOM part* node, then it can assign input parameters from the user to various input fields of this *conceptual operation DOM part* - the input fields from the DOM are automatically detected and associated to fields in the Data Model as part of Algorithm 2 described above.

8 Limitations

Our goal was to define a general mapping tool that should work with any business web application that uses a relational database in the backend. For achieving this goal we based the plugin definition on general templates for conceptual operations which are not specific to a particular web application, and the plugin implementation on common web design principles (like the fact that an input field is always placed in the web UI either on the right or below or in the south-east part of its corresponding text label). The limitations to our approach are related to the fact that our mapping tool detects a conceptual operation in the DOM by mapping text labels in the DOM (that represent properties of a concept from the underlying database of the application) to text inputs. For operations like partial CRUD operations that only modify a single or just a few properties of a concept, our mapping tool would not be able to detect a conceptual operation in the DOM (just a 'Generic DOM' would be detected). The operation will still be labeled by default by our tool with the text found on the *clickable element* that triggered that operation on the UI, but the effect on the UI would still be a 'Generic DOM'. This limitation could be overcame by using a manual verification and adaptation of the result produced automatically by the plugin. A human user is allowed to change the 'Generic DOM' node in the map to *conceptual operation DOM part* node, and this way increasing the accuracy of the result. As our experiments demonstrated these could not represents more than 20%–30% of all conceptual operations.

Another limitation is due to the fact that a path in our current conceptual map ends with a *Conceptual operation DOM part* node that is always a leaf node. This means that our mapping tool cannot map use cases where several entities are altered in a sequence. But we plan to resolve this issue in future work by removing the rule which states that a *Conceptual operation DOM part* node is always a leaf node and including cycle detection in our algorithm.

9 Conclusions

The presented automated tool can be used to construct conceptual maps of business web applications, and this way to automatically extract information about the conceptual operations that could be executed inside that application. This conceptual map depicts in an abstract, hierarchical way the possible navigation and operational paths in the UI (i.e. User Interface) of a web application. The result is a map that is produced in either serialized, JSON format or in a graphical form in Html. The tool was tested on two commercial web applications and the tests validated the fact that our tool is able to construct a conceptual map for those applications.

Obtaining such a map offers important advantages especially in the RPA tools' development. The map could be used as a foundation for automatizing the operations allowed to be executed through the web UI of an application. We have implemented in our tool an RPA functionality, i.e. automatic execution of business processes which we describe as paths in this hierarchical conceptual map. The human user can specify in several ways the start node and the end node of a path in the map (representing a simple business process), he/she can provide parameters for conceptual operations in this path, and our automation tool is able to execute this business process automatically.

The detection of the conceptual operations could be further improved so that to work also on partial CRUD operations (operations that read/update only few attributes of a concept).

Disclosure of Interests. The authors have no competing interests.

References

1. van der Aalst, W., Bichler, M., Heinzl, A.: Robotic process automation. Bus. Inf. Syst. Eng. **60**(4), 269–272 (2018). https://doi.org/10.1007/s12599-018-0542-4
2. Agostinelli, S., Lupia, M., Marrella, A., Mecella, M.: Automated generation of executable RPA scripts from user interface logs, pp. 116–131 (2020). https://doi.org/10.1007/978-3-030-58779-6_8
3. Andrew, J.J., Ferrari, S., Maurel, F., Dias, G., Giguet, E.: Web page segmentation for non visual skimming. In: The 33rd Pacific Asia Conference on Language. Information and Computation (PACLIC 33), pp. 423–431. Hakodate, hal-02309625 (2019)
4. Cai, D., Yu, S., Wen, J.-R., Ma, W.-Y.: Extracting content structure for web pages based on visual representation. In: Zhou, X., Orlowska, M.E., Zhang, Y. (eds.) APWeb 2003. LNCS, vol. 2642, pp. 406–417. Springer, Heidelberg (2003). https://doi.org/10.1007/3-540-36901-5_42
5. Chen, J., Zhou, B., Shi, J., Zhang, H., Fengwu, Q.: Function-based object model towards website adaptation. In: Proceedings of the 10th International Conference on World Wide Web, pp. 587–596. Association for Computing Machinery, New York (2001)
6. Feiz, S., Wu, J., Zhang, X., Swearngin, A., Barik, T., Nichols, J.: Understanding screen relationships from screenshots of smartphone applications. In: 27th International Conference on Intelligent User Interfaces, pp. 447–458, IUI 2022, Association for Computing Machinery, New York (2022). https://doi.org/10.1145/3490099.3511109
7. Hofmann, P., Samp, C., Urbach, N.: Robotic process automation. Electron. Markets **30**(1), 99–106 (2020). https://doi.org/10.1007/s12525-019-00365, https://ideas.repec.org/a/spr/elmark/v30y2020i1d10.1007_s12525-019-00365-8.html
8. Primer, A.: IRPA: introduction to robotic process automation (2015). https://irpaai.com/wp-content/uploads/2015/05/Robotic-Process-Automation-June2015.pdf
9. Leno, V., Deviatykh, S., Polyvyanyy, A., Rosa, M.L., Dumas, M., Maggi, F.M.: Robidium: automated synthesis of robotic process automation scripts from UI logs. In: Proceedings of the Best Dissertation Award, Doctoral Consortium, and Demonstration & Resources Track at BPM 2020 co-located with the 18th International Conference on Business Process Management (BPM 2020). Sevilla, 13–18 Sep 2020, vol. 2673, pp. 102–106, CEUR-WS.org (2020). https://ceur-ws.org/Vol-2673/paperDR08.pdf
10. Li, T.J.J., Azaria, A., Myers, B.A.: Sugilite: creating multimodal smartphone automation by demonstration. In: Proceedings of the 2017 CHI Conference on Human Factors in Computing Systems, pp. 6038–6049. CHI 2017. Association for Computing Machinery, New York (2017). https://doi.org/10.1145/3025453.3025483

11. Li, T.J.J., Popowski, L., Mitchell, T., Myers, B.A.: Screen2vec: semantic embedding of gui screens and gui components. In: Proceedings of the 2021 CHI Conference on Human Factors in Computing Systems, CHI 2021. Association for Computing Machinery, New York (2021). https://doi.org/10.1145/3411764.3445049
12. Liu, N., Yang, C.: Mining web site's topic hierarchy. In: Special Interest Tracks and Posters of the 14th International Conference on World Wide Web, pp. 980–981. WWW 2005. Association for Computing Machinery, New York (2005). https://doi.org/10.1145/1062745.1062828
13. Mantratzis, C., Cassidy, S.: DOM-based XHTML document structure analysis separating content from navigation elements. In: International Conference on Computational Intelligence for Modelling, pp. 632–637. Control and Automation and International Conference on Intelligent Agents, Web Technologies and Internet Commerce (CIMCA-IAWTIC 2006). Vienna (2005)
14. Nguyen, T.A., Csallner, C.: Reverse engineering mobile application user interfaces with remaui. In: Proceedings of the 30th IEEE/ACM International Conference on Automated Software Engineering, pp. 248–259. ASE 2015. IEEE Press (2015). https://doi.org/10.1109/ASE.2015.32
15. Sterca, A., Niculescu, V., Kiraly, A., Bufnea, D.: Automatic mapping of business web applications. In: García-Peñalvo, F.J., Marchiori, M. (eds.) Proceedings of the 19th International Conference on Web Information Systems and Technologies, WEBIST 2023, Rome, 15–17 Nov 2023, pp. 437–445. SCITEPRESS (2023). https://doi.org/10.5220/0012212700003584
16. Wu, J., Wang, S., Shen, S., Peng, Y.H., Nichols, J., Bigham, J.P.: WebUI: a dataset for enhancing visual UI understanding with web semantics. In: Proceedings of the 2023 CHI Conference on Human Factors in Computing Systems, CHI 2023. Association for Computing Machinery, New York (2023). https://doi.org/10.1145/3544548.3581158
17. Wu, J., Zhang, X., Nichols, J., Bigham, J.P.: Screen parsing: towards reverse engineering of UI models from screenshots. In: 34th Annual ACM Symposium on User Interface Software and Technology (UIST 2021), pp. 470–483. New York (2021). https://doi.org/10.1145/3472749.3474763

On the Construction of Text-to-SQL Tools Based on Large Language Models for Real-World Relational Databases

Eduardo Nascimento[1,2], Gustavo Coelho[1], Lucas Feijó[1], Yenier Izquierdo[1], Grettel García[1], Aiko Oliveira[1,2], João Pinheiro[2], Antony Seabra[2], Antonio Furtado[2], Luiz Paes Leme[3], Melissa Lemos[1,2], and Marco Casanova[1,2(✉)]

[1] Instituto Tecgraf, PUC-Rio, Rio de Janeiro, RJ 22451-900, Brazil
{rogerrsn,gustavocoelho,lucasfeijo,ytorres, ggarcia,melissa}@tecgraf.puc-rio.br

[2] Departamento de Informática, PUC-Rio, Rio de Janeiro, RJ 22451-900, Brazil
{aramalho,jpinheiro,amedeiros,furtado,casanova}@inf.puc-rio.br

[3] Instituto de Computação, UFF, Niterói, RJ 24210-310, Brazil
lapaesleme@ic.uff.br

Abstract. A strategy to construct natural language database interfaces is to use Large Language Models (LLMs) to translate the end-user questions into SQL queries. Such interfaces will be called LLM-based text-to-SQL tools. This article analyses the limitations and proposes solutions to improve the performance of LLM-based text-to-SQL tools for real-world relational databases that have large, complex schemas often expressed in terms different from those adopted by end-users to formulate their questions. The article considers implementations based on Prompt Engineering, including Retrieval-Augmented Generation, and LLM fine-tuning. Finally, it describes experiments that analyze the accuracy of some implementations on two benchmarks built upon databases with large schemas.

Keywords: Text-to-SQL · Large language models · Prompt engineering · Retrieval-augmented generation · RAG · Fine-tuning · Relational databases

1 Introduction

A Large Language Model (LLM) [16] is a language model based on a deep neural network architecture with a very large number of parameters, trained on large quantities of unlabeled text, using self-supervised or semi-supervised learning. LLMs are a transformative technology, enabling developers to build novel applications. LLMs came to the foreground with the announcement of conversational interfaces that are able to generate high-quality textual answers.

Natural language database interfaces (NLIDBs) allow users to access databases using questions formulated in natural language (NL) [1]. A strategy for constructing NLIDBs is to use LLMs to translate the NL questions into SQL queries. Such interfaces will be called *LLM-based text-to-SQL tools*.

One may classify a text-to-SQL tool into:

M. Marchiori and F. García Peñalvo (Eds.): WEBIST 2023, LNBIP 543, pp. 151–167, 2025.
https://doi.org/10.1007/978-3-031-89621-7_8

- *General-purpose*, when the tool is designed to work with any database.
- *Domain-specific*, when the tool is designed to work with any database of a given domain, such as Health, Energy, or Cultural Heritage, for which a Conceptual Reference Model (CRM) or a similar artifact exists.
- *Database-specific*, when the tool is designed to work with a specific database based on its schema, data, and associated documentation.

Real-world relational databases (RW-RDBs) are especially challenging for text-to-SQL tools, since they feature:

C1. A relational schema that has a large number of tables, columns per table, and foreign keys – which may not fit in the prompt area and leads to queries with many joins, which are difficult to synthesize.
C2. A relational schema whose table and column names are often different from the terms the users adopt to formulate their NL questions.
C3. A complex data semantics; for example, some data values may encode enumerated domains – again, the terms the users adopt to formulate their NL questions may have to be mapped to this internal semantics.
C4. Ambiguous metadata and data, which influences the behavior of the text-to-SQL tool, leading to unexpected results.

The article is an extended and revised version of [26]. The main contribution of this article is to analyze the limitations of general-purpose LLM-based text-to-SQL tools and propose how to improve their performance when they are specialized to an RW-RDB. The article considers implementations based on Prompt Engineering, including Retrieval-Augmented Generation (RAG), or developed by fine-tuning and running privately an LLM. Finally, it analyzes some implementations on two benchmarks built upon databases with large schemas[1]. The contributions should broadly apply to an application domain for which a Conceptual Reference Model (CRM) indeed exists, which will then play the role of the RW-RDB relational schema and associated documentation. However, the domain-specific scenario is outside the scope of the article.

The article is organized as follows. Section 2 reviews the background and related work. Section 3 discusses implementations based on Prompt Engineering. Section 4 covers implementations based on RAG. Section 5 addresses implementations based on fine-tuning and running privately an LLM. Section 6 presents the experiments. Finally, Sect. 7 contains the conclusions.

2 Background and Related Work

2.1 Large Language Models

The approaches involved in Natural Language Processing (NLP) tasks have undergone profound changes over the past decades. [16] classifies NLP approaches in roughly

[1] For the interested reader, several implementations that use the Mondial benchmark are available at https://github.com/dudursn/text_to_sql_chatgpt_real_world/.

four eras. From 1950 to 1969, the first era was characterized by simple rule-based models, achieving very limited results due to the lack of computation, data, and knowledge of human language structure. The second era, from 1970 to 1992, benefited from the rapid development of linguistic theories, allowing the creation of a new generation of hand-built systems from knowledge-based artificial intelligence models. The third era, from 1993 to 2012, saw a shift from knowledge-based to machine-learning models, when digital text became widely available, providing enough data to allow some level of language understanding by using mainly supervised machine-learning methods. The last era, from 2013 to the present, extended the use of machine learning, allowing a more generalized language understanding by using self-supervised learning and deep learning-based models to represent words with dense vectors. Within this context, the first Large Language Model (LLM) was successfully trained in 2018 by exposing a model to an extremely large quantity of text, representing an enormous amount of language knowledge.

LLMs follow the *transformer* neural network architecture [12, 27, 28]. Manning [16] contains a readable introduction to transformers. The central intuition behind transformers is the use of the *attention* mechanism, where the representation of a given position is computed based on a weighted combination from others. The result is a much more complex setup when compared to simple neural networks. Based on this architecture, the self-supervised strategy for LLMs usually involves masking words in the text and training the model to predict the missing words from the outer context. After this process is extensively repeated, the model learns generalized syntactic structures of sentences.

As a consequence of their generalization capacity, LLMs perform well in a wide variety of NLP tasks, to the point of shifting the focus of NLP research away from the previous paradigm of training specialized supervised models for each task. An LLM can be redeployed to a particular NLP task with a small number of further instructions.

However, an LLM reflects the data it was trained with. In particular, an LLM suffers from the "temporal generalization problem" – capturing facts that change over time – and the "factual grounding problem" – capturing specific facts. To circumvent these limitations, the user may *fine-tune* the LLM, that is, retrain it with more examples, or he may adopt *few-shot learning*, that is, add a few examples in a dialog interaction so that the model can capture what the user is trying to do and generate a plausible completion.

Chat Models (CMs) must, in turn [22]: avoid unacceptable answers, including avoiding incorrect or made-up facts, nonsensical answers, and biased or toxic text; seek useful user dialogues, that is, the language modeling objective must be aligned with the objective "follow the user's instructions helpfully and safely". Putting it in more explicit terms, a Chat Model must be [2]: *helpful* (they should help the user solve their task), *honest* (they shouldn't fabricate information or mislead the user), and *harmless* (they should not cause physical, psychological, or social harm to people or the environment).

Table 1 summarizes some of the major LLMs announced up to February 2024. A more extensive list can be found at the Awesome-LLM Web Site[2].

[2] https://github.com/Hannibal046/Awesome-LLM.

Table 1. Some major LLMs announced up to February 2024.

Date Announced	LLM Name	Type	Comment
Nov. 30th, 2022	GPT-3.5	CM	used in ChatGPT
Feb. 23rd, 2023	LLaMA	LLM	developed by Meta
Mar. 14th, 2023	GPT-4	CM	used in ChatGPT Plus
May 10th, 2023	PaLM 2	CM	used in Bard
Jul. 18th, 2023	LLaMA 2	LLM	developed by Meta
Aug. 24th, 2023	Code LLaMA	LLM	LLaMA 2 for coding
Nov. 06th, 2023	GPT-4-turbo	CM	major upgrades to GPT
Dec. 06th, 2023	Gemini	CM	Bard renamed to Gemini
Jan. 29th, 2024	Code LLaMA instruct 70B	CM	fine-tuned for NL instructions
Feb. 16th, 2024	Gemini 1.5	CM	uses a 1M Context Window

2.2 Natural Language Interfaces for Databases

Briefly, Affolters et al. [1] offers a comparative survey of 24 NL database interfaces, classified as keyword-, pattern-, parsing- and grammar-based. As lessons learned, the authors indicate that, for simple questions, keyword-based systems are enough, whereas for complex questions, an NL database interface needs to apply some parsing to identify structural dependencies, and that grammar-based systems are the most powerful, but depend on manually designed rules.

Diefenbach et al. [6] assumes that an NL query can be correctly interpreted considering only the semantics of the words. For example, the NL query "Give me actors born in Berlin" is reformulated as a keyword question "Berlin, actors, born in", that is, the semantics of the words "Berlin", "actors", and "born" suffice to deduce the intention of the user. The authors proceed to map the NL query into one of four templates, which limits the usefulness of the process. Interestingly, however, they show that the process has good performance on several RDF datasets from the QALD benchmarks in different languages.

Recent surveys of conversational interfaces (software-based dialogue systems or chatbots) can be found in [3, 17]. Such systems offer users the possibility to interact with databases using techniques similar to human conversation; they accept queries typically formulated in NL, return answers also typically expressed in NL, and support persistent conversational contexts, thereby generating user-computer dialogues.

2.3 Text-to-SQL Benchmarks

The Spider – Yale Semantic Parsing and Text-to-SQL Challenge [30] defines 200 datasets, covering 138 domains, for training and testing text-to-SQL tools. For each database, Spider lists 20–50 hand-written NL questions and their SQL translations. An NL question Q, with an SQL translation S, is classified as easy, medium, hard, and extra-hard, where the difficulty is based on the number of SQL constructs of S – `GROUP`

`BY`, `ORDER BY`, `INTERSECT`, nested sub-queries, column selections, and aggregators – so that Q is considered harder if S contains more SQL constructs.

Most databases in Spider have small schemas: the largest five databases have between 16 and 25 tables, and about half of the databases have schemas with five tables or fewer. Also, all Spider NL questions are phrased in terms used in the database schemas. These two limitations considerably simplify the text-to-SQL task. Therefore, the results reported in the Spider leaderboard are biased toward databases with small schemas and NL questions written in the schema vocabulary, which is not what one finds in real-world databases.

Spider has two interesting variations. Spider-Syn [7] is used to test how well text-to-SQL tools handle synonym substitution, and Spider-DK [8] addresses testing how well text-to-SQL tools deal with domain knowledge.

BIRD – BIg Bench for LaRge-scale Database Grounded Text-to-SQL Evaluation [14] is a large-scale cross-domain text-to-SQL benchmark in English. The dataset contains 12,751 text-to-SQL data pairs and 95 databases with a total size of 33.4 GB across 37 domains. BIRD addresses real-world applications by exploring three additional challenges: dealing with large and messy database values, external knowledge inference, and optimizing SQL execution efficiency. However, BIRD still does not have many databases with large schemas: of the 73 databases in the training dataset, only two have more than 25 tables, and, of the 11 databases used for development, the largest one has only 13 tables.

WikiSQL [31] has 80,654 NL sentences and SQL annotations of 24,241 tables. Each query in WikiSQL is limited to the same table and does not contain complex operations such as sorting and grouping.

SQL-Eval [25] is a framework[3] that evaluates the correctness of text-to-SQL strategies, created during the development of the SQLCoder.

The `sql-create-context`[4] dataset also addresses the text-to-SQL task, and was built from WikiSQL and Spider. It contains 78,577 examples of NL queries, SQL `CREATE TABLE` statements and SQL Queries answering the questions. The `CREATE TABLE` statement provides context for the LLMs without having to provide actual rows of data.

Despite the availability of these benchmarks for the text-to-SQL, and inspired by them, Sect. 6 describes two benchmarks that feature databases with large schemas. The first benchmark is based on a real-world database, a set of views, specified as proposed in [21], and a set of 100 test NL questions, that mimic those posed by real users, and their ground truth SQL translations. The second benchmark[5] is based on Mondial, a database that stores geographic data and is openly available, and a set of 100 test NL questions.

[3] Available at https://github.com/defog-ai/sql-eval.

[4] https://huggingface.co/datasets/b-mc2/sql-create-context.

[5] Available at https://github.com/dudursn/text_to_sql_chatgpt_real_world/.

2.4 Text-to-SQL Tools

The Spider Web site[6] publishes a leaderboard with the best-performing text-to-SQL tools. At the time of this writing, the top 5 tools achieved an accuracy that ranged from an impressive 85.3% to 91.2% Four tools use GPT-4, as their names imply. The three tools that provide detailed documentation have an elaborate first prompt that tries to select the tables and columns that best matches the NL question. This first prompt is, therefore, prone to failure if the database schema induces a vocabulary that is disconnected from the NL question terms. This failure cannot be easily fixed by even more elaborate prompts that try to match the schema and the NL question vocabularies, as argued in [21].

The BIRD Web site[7] also publishes a leaderboard with the best-performing tools. At the time of this writing, out of the top 5 tools, two use GPT-4, one uses CodeS-15B, one CodeS-7B, and one is not documented. The sixth and seventh tools also use GPT-4, appear in the Spider leaderboard, and are well-documented. MAC-SQL [29] is a recent addition to this list and features an LLM-based multi-agent collaborative framework.

The Awesome Text2SQL Web site[8] lists the best-performing text-to-SQL tools on WikiSQL, Spider (Exact Match and Exact Execution) and BIRD (Valid Efficiency Score and Execution Accuracy).

Other text-to-SQL tools include the Defog SQLCoder. The "60 Top AI Text To SQL Bot Tools" Web site[9] lists other AI tools for text-to-SQL.

The DB-GPT-Hub[10] is a project exploring how to use LLMs for text-to-SQL. The project contains data collection, data pre-processing, model selection and building, and fine-tuning weights, including LLaMA-2, and evaluating several LLMs fine-tuned for text-to-SQL.

Finally, LangChain[11] is a generic framework that offers several predefined strategies to build and run SQL queries based on NL prompts. Section 6 uses LangChain in the experiments.

2.5 Retrieval-Augmented Generation – RAG

Retrieval-Augmented Generation (RAG), introduced in [13], is a strategy to incorporate data from external sources before proceeding to the generation phase. This process ensures that the responses are grounded in retrieved evidence, thereby significantly enhancing the accuracy and relevance of the output.

There is an extensive literature on RAG. Recent references include a RAG technique for an LLM-based Text-to-SQL framework involving sample-aware prompting and a dynamic revision chain [10]. A RAG technique is used in [24] to retrieve the table and column descriptions from a metadata store to ensure that the NL question is related to the right tables and columns. A recent survey can be found in [9], encompassing

[6] https://yale-lily.github.io/spider.

[7] https://bird-bench.github.io.

[8] https://github.com/eosphoros-ai/Awesome-Text2SQL.

[9] https://topai.tools/s/Text-to-SQL-bot.

[10] https://github.com/eosphoros-ai/DB-GPT-Hub.

[11] https://python.langchain.com.

the *Naive RAG*, the *Advanced RAG*, and the *Modular RAG*. Advanced RAG refined the indexing approach using techniques such as sliding window, fine-grained segmentation, and metadata and introduced various methods to optimize retrieval. The modular RAG structure integrates multiple methods to enhance functional modules, such as incorporating a search module for similarity retrieval and applying a fine-tuning approach in the retriever.

3 Prompt Engineering for Text-to-SQL

The simplest way to create an LLM-based text-to-SQL tool would be to design Prompts that implement three steps:

S1. (Schema Linking) Given an NL question Q, ask the LLM to retrieve a description of the relation tables and columns, creating a context C.
S2. (SQL Translation) Ask the LLM to translate Q into an SQL query S under the context C.
S3. (SQL Execution) Execute S.

The same approach also applies to RDF datasets with RDF schemas to translate NL queries into SPARQL. Several LLMs have good performance for these tasks, including GPT-3.5 and GTP-4 [15]. As mentioned in Sect. 2.4, the Awesome Text2SQL Web site[12] lists the best-performing text-to-SQL tools on WikiSQL, Spider (Exact Match and Exact Execution) and BIRD (Valid Efficiency Score and Execution Accuracy), most of which use sophisticated Prompt Engineering strategies. The experiments in Sect. 6 use LangChain SQLQueryChain.

LangChain predefined chains[13] and agents considerably simplify the text-to-SQL task. *SQLQueryChain* receives the database schema as a prompt and convert an NL question into an SQL statement. The schema is automatically extracted from the database, and the prompt is customized with some tips. Then, the instruction establishes the desired format for the model's output. Lastly, the complete database schema and the natural language query are provided.

SQLDatabaseSequentialChain is a more sophisticated version that employs a sequential chaining of prompts. The first prompt, the *decider chain prompt*, identifies tables related to the user's query without providing the database schema. The second prompt is identical to the prompt in the SQLQueryChain, but only the involved tables are provided for the LLM to generate the SQL statement.

Lastly, *SQLAgent* uses the ReAct strategy to create SQL statements. For each instance, it invokes a chain of "reasonings" that involve an "Action" (such as "I need to list the tables in this database"), an "Observation" (the tables), and a "Thought" ("I should access details only from table X"). This reasoning process iterates, recovering from errors by running a generated SQL query, capturing the trace, and regenerating it until the desired result is achieved. In all these alternative chains, it is possible to include samples of database instances.

[12] https://github.com/eosphoros-ai/Awesome-Text2SQL.
[13] https://python.langchain.com/docs/use_cases/sql/prompting.

Despite their performance on Spider and BIRD, some of the most popular benchmarks for the text-to-SQL task, general-purpose LLM-based text-to-SQL tools face considerable challenges when applied to RW-RDBs, as reported in [18, 19]. Schema linking is especially challenging, since the vocabulary induced by the relational schema and the vocabulary of the NL questions may be difficult to match (Challenge C2), and the matching may be ambiguous (Challenge C4). SQL translation is hindered by the RW-RDB large schema (Challenge C1) and by the fact that some column values, or combinations of column values, hide semantic information that does not directly correspond to end-user terms (Challenge C3). For example, NL questions may use "overdue order" which translates to "`Maintenance_Order.Status = 1`"; this translation is in the database documentation but it is not visible to the LLM.

As a more amusing example, consider the NL query "Where is Khwarazm?", which asks where the oasis region "Khwarazm" is located (the region where Muhammad ibn Musa al-Khwarizmi lived – the mathematician from whose name the term "algorithm" is derived), but omits the detail that "Khwarazm" is the name of a region (the answer is "Uzbekistan" and "Turkmenistan"). Interestingly, ChatGPT can infer the missing information and synthesize a correct SQL query. Note that passing the entire database as context would not be feasible due to current prompt size limits.

To circumvent some of these challenges, an approach would be to use a database keyword search tool to mediate access to the database, as in [20]. Initially, an LLM would extract keywords from the input NL question. Then, the tool would match the keywords with schema and data terms and create a single "universal" table by querying the database. Finally, the LLM would translate the original NL query into an SQL query over this table (the LLM would also be prompted with the matching between the NL question terms and the columns of the table). This approach would transfer the vocabulary matching problem (including schema linking) to an external database keyword search tool.

But other approaches are possible. For example, as argued in [21], the construction of a database-specific text-to-SQL tool can be greatly facilitated by a database description that provides:

- *LLM-friendly views* that map (fragments of) the database schema to terms close to the terms users frequently adopt and that try to predefine frequently used joins.
- *LLM-friendly data descriptions* of the database values.

LLM-friendly views provide a better description of the database to the LLM and can be implemented within the database. Therefore, they shift the solution from Prompt Engineering to database design. LLM-friendly data descriptions refer to constructs that try to capture the data semantics. Section 6.2 describe experiments that use data samples in the LLM prompt, as mentioned above in the context of LangChain, and explore the use of Retrieval-Augmented Generation [13], discussed in the next section.

4 RAG for Text-to-SQL

An LLM-based text-to-SQL tool that takes advantage of RAG would be designed roughly as follows:

Pre-processing

P1. (Sample Generation) Create a *synthetic dataset* E of pairs (Q_i, S_i), where Q_i is an NL question and S_i is the corresponding SQL query.
P2. (Embedding) Embed each NL question Q_i into a vector V_i and store V_i in a vector store along with the pair (Q_i, S_i).

Processing

S1. (Sample Retrieval) Given an NL question Q, embed Q in a vector V_Q; retrieve the top-k vectors V_i, $i \in [1, k]$, most similar to V_Q; create a context C with the pairs (Q_i, S_i), $i \in [1, k]$.
S2. (Schema Linking) Ask the LLM to retrieve a description of the relation tables and columns, passing the context C; add these descriptions to C.
S3. (SQL Translation) Ask the LLM to translate Q into an SQL query S under the context C.
S4. (SQL Execution) Execute S.

The key step is the generation of the synthetic dataset E in such a way as to improve the performance of the Schema Linking and SQL Translation steps. The technique introduced in [4] provides a robust strategy to construct database-specific synthetic datasets, especially when it comes to conveying the data semantics of the database to the LLM. Using the same notation as above, the generation of E is based on an algorithm that samples the database and its schema and associated documentation, and calls an LLM to create an NL question Q_i from the sampled data and to translate Q_i into S_i. Roughly, by varying how the sampling works, the pairs in E provide SQL examples illustrating how the database schema is structured, how the user's language maps to the database schema, and how NL language constructions map to data values. Therefore, the dataset E is specific to the database, but the algorithm is generic and applicable to any database.

A *single-attribute synthetic dataset* is generated by sampling just single attributes, whereas the *multi-attribute synthetic dataset* is generated by sampling multiple attributes. Therefore, the single-attribute syntactic dataset will contain only pairs (Q_i, S_i), where Q_i is a *simple NL question* and S_i is an SQL query over a single table, with no join clauses, and a `WHERE` clause with one filter over a single column.

Section 6.2 describes experiments with single- and multi-attribute synthetic datasets combined with LLM-friendly views. These experiments test if RAG adds information not conveyed by the LLM-friendly views, thereby leading to a better text-to-SQL prompt strategy.

Finally, note that a standardized domain may have a Conceptual Reference Model with taxonomies that reflect the user's vocabulary, specific data types with measurement units, etc. These metadata can be explored to construct a domain-specific synthetic dataset, which can then be reused for any database pertaining to the domain.

5 Fine-Tuning and Running Privately an LLM for Text-to-SQL

A text-to-SQL tool may be based on fine-tuning an LLM specifically for this task as follows:

Pre-processing

F1. (Training Dataset Generation) Create a *training dataset* T of pairs (Q_i, S_i), where Q_i is an NL question and S_i is the corresponding SQL query.
F2. (LLM Selection) Select an LLM $\mathcal{L}$ for fine-tuning.
F2. (LLM Fine-tuning) Fine-tune $\mathcal{L}$ for the task-to-SQL task using T.

Processing

S1. (Schema Linking) Given an NL question Q, ask $\mathcal{L}$ to retrieve a description of the relation tables and columns, creating a context C.
S2. (SQL Translation) Ask $\mathcal{L}$ to translate Q into an SQL query S under the context C.
S3. (SQL Execution) Execute S.

The pre-processing steps pose difficult challenges. Creating a training dataset requires generating a large number of pairs (Q_i, S_i) using, for example, crowdsourcing, as in Dolly[14], or self-instruct, that is, using an LLM to expand a small set of manually created pairs, as in Alpaca[15]. Fine-tuning an LLM may use LoRA–Low-Rank Adaptation [11] or QLoRA–Quantized Low-Rank Adaptation [5], or use a specific fine-tuning strategy [23].

The Defog SQLCoder[16] is an example of a general-purpose LLM-based text-to-SQL tool created as above. The latest model, sqlcoder-70b-alpha, features 70B parameters and was fine-tuned on a base StarCoder model on more than 20,000 human-curated questions, classified as in Spider, based on ten different schemas. The training dataset[17] consisted of prompt-completion pairs, encompassing several schemas with varying difficulty levels, whereas the evaluation dataset featured questions from novel schemas. The use of complex schemas, with 4–20 tables, challenged the model. The fine-tuning process occurred in two stages: the base model was first refined using easy and medium questions and then further fine-tuned on hard and extra-hard questions to yield the final model. The accuracy of the sqlcoder-70b-alpha model on the Defog's dataset achieved 93.0%, whereas GPT-4 (Feb. 5, 2024) achieved 86%.

LLM fine-tuning for the database-specific text-to-SQL task may follow strategies similar to fine-tuning an LLM in general, except that the training dataset should be created for the target database. In fact, the automated process described in Sect. 4 to create a synthetic dataset for RAG may be re-targeted to create a training dataset.

Finally, the question of running privately an LLM is relevant for text-to-SQL since enterprises frequently require that sensitive data remain in-house. An approach would be to select an LLM $\mathcal{L}$ with an appropriate license[18] fine-tune $\mathcal{L}$ for the text-to-SQL task, and run $\mathcal{L}$ privately.

---

[14] https://huggingface.co/databricks/dolly-v2-12b.
[15] https://crfm.stanford.edu/2023/03/13/alpaca.html.
[16] https://github.com/defog-ai/sqlcoder?tab=readme-ov-file.
[17] https://defog.ai/blog/open-sourcing-sqlcoder/.
[18] A comprehensive list can be found at https://github.com/Hannibal046/Awesome-LLM,.

6 Experiments

6.1 Benchmark Datasets Adopted

Following [19], the primary benchmark dataset adopted in this article is based on an RW-RDB, referred to as *IndDB* for privacy reasons, with a set of 100 NL questions and their translations to SQL.

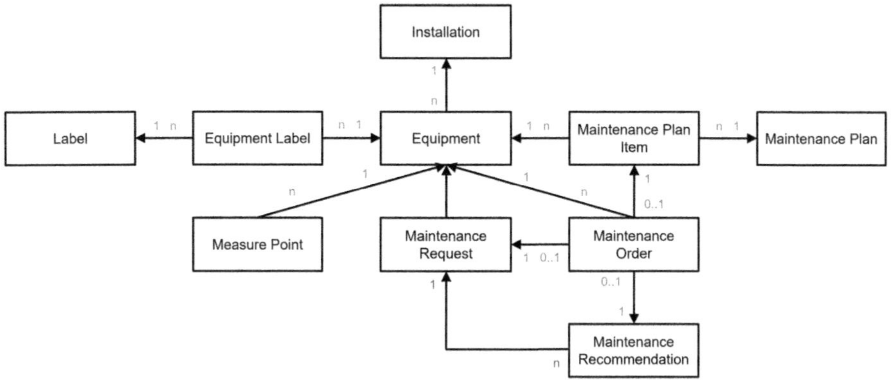

Fig. 1. The referential dependency diagram of a simplified version of the RW-RDB.

The IndDB schema (see Fig. 1) has 27 tables, 585 columns, and 30 foreign keys (some multi-column); the largest table has 81 columns. Table and column names follow an internal naming convention, which is very different from the end-user vocabulary. Thus, based on the discussion in Sect. 3, a set of *LLM-friendly views* was defined in such a way that the views: (1) remap the relational schema names to end users' terms; (2) predefine joins that follow foreign keys.

The set of NL questions contains 33 simple, 34 medium, and 33 complex questions. This classification corresponds to the easy, medium, and hard classes used in the Spider benchmark (extra-hard questions were not considered) (see Table 2 for examples). As in the Spider benchmark, the difficulty is based on the number of SQL constructs, so that queries that contain more SQL constructs (GROUP BY, ORDER BY, INTERSECT, nested subqueries, column selections, and aggregators) are considered to be harder.

Since IndDB is proprietary and is not openly available, the experiments cannot be replicated. Thus, as in [19], this article also considers a benchmark[19] based on Mondial, a database that stores geographic data and is openly available. The Mondial schema[20] has 46 tables, with a total of 184 columns and 49 foreign keys, some of which are multi-column. The benchmark has a set of 100 NL questions and their translations to SQL, divided into 34 simple, 33 medium, and 33 complex questions.

19 Available at https://github.com/dudursn/text_to_sql_chatgpt_real_world/.

20 https://www.dbis.informatik.uni-goettingen.de/Mondial/.

Table 2. A sample of the NL questions and their translations.

NL Question	Ground Truth SQL Query Table	Type
Show all orders with type MO05	SELECT id FROM vw_maintenance_order WHERE type = 'MO05'	simple
Which IBX15 installation recommendations are expired?	SELECT id FROM vw_maintenance_recommendation WHERE expired ='true' AND installation_name = 'IBX-15'	medium
Which IBX-67 installation label has the most approved maintenance requests with a priority greater than 6?	SELECT el.label_id FROM vw_maintenance_request Mr JOIN vw_equipment_label el ON mr.equipment_id = el.equipment_id WHERE mr.installation_name = 'IBX-67' AND mr.situation = 'Approved' AND mr.priority >6 GROUP BY el.label_id ORDER BY COUNT(mr.id) DESC FETCH FIRST 1 ROWS ONLY	complex

Finally, to help assess the difficulty of the NL questions in each benchmark, Table 3 shows basic statistics of the sets of SQL queries, where #cols refers to the number of columns of the target clause and the other columns refer to the number of joins, filters, and aggregations that occur anywhere in the query, including any nested query. Note that, on average, the medium and complex SQL queries for IndDB have roughly twice the number of joins and filters as the medium and complex queries for Mondial. Therefore, the set of medium and complex NL questions of IndDB are more challenging for text-to-SQL tools than those of Mondial.

Table 3. Basic statistics of the sets of queries.

		IndDB				Mondial			
	Query Type	#cols.	#joins	#filters	#aggr	#cols.	#joins	#filters	#aggr
Average	complex	2,00	3,03	2,55	0,39	1,09	1,41	1,65	0,38
	medium	1,44	2,47	1,68	0,29	1,18	0,79	0,85	0,30
	simple	1,30	0,21	0,82	0,15	1,33	0,42	0,73	0,12
Maximum	complex	19	5	4	2	2	4	5	1
	medium	9	5	3	2	2	3	1	2
	simple	6	2	2	1	4	3	2	1

6.2 Results for the Industrial Database

The experiments used the *accuracy* of a given text-to-SQL tool over a benchmark, defined as the number of correct predicted SQL queries divided by the total number of SQL queries, as usual.

All experiments were based on LangChain SQLQueryChain with GPT-4, summarized in Sect. 3. This implementation was adopted because it proved less expensive and had an accuracy comparable with much more complex text-to-SQL tools [19].

The experiments involving RAG used the strategy described in Sect. 4 to create a multi-attribute synthetic dataset with 46,215 pairs, and a single-attribute synthetic dataset with 24,861 pairs, both on top of the LLM-friendly views.

Table 4 shows the results obtained with the different configurations. Columns under #Correct Predicted Queries indicate the number of simple, medium, or complex SQL queries correctly synthesized; columns under Accuracy, the accuracy results for simple, medium, or complex SQL queries.

Table 4. Results for the different setup configurations – IndDB.

#Line	Configuration*	#Corrrect Predicted Queries				Accuracy			
		Simple	Medium	Complex	Total	Simple	Medium	Complex	Total
1	Relational Schema	22	11	8	41	0,67	0,33	0,23	0,41
2	LLM-friendly Views Only	30	25	19	74	0,91	0,76	0,56	0,74
3	RAG (multi-attribute) Only	26	7	9	42	0,79	0,21	0,26	0,42
4	RAG (multi-attribute) + LLM-friendly Views	31	26	19	76	0,94	0,79	0,56	0,76
5	RAG (single-attribute) + LLM-friendly Views	32	24	23	79	0,97	0,73	0,68	**0,79**

(*) All experiments used LangChain SQLQueryChain with GPT-4.

The lines of Table 4 show the accuracy results of the various configurations:

Line 1: A poor total accuracy was achieved by directly using the relational schema, for the reasons anticipated in Sect. 1, especially the problem that the relational schema of IndDB has table and column names very different from the terms used in the benchmark NL questions. This configuration was taken as the baseline.

Line 2: A substantially better total accuracy was reached by using LLM-friendly views, exactly because this configuration prompts the LLM with table and column names much more similar to those observed in the NL questions of the benchmark, and also because the LLM-friendly views simplify the synthesis of some SQL queries by pre-computing frequently used joins.

Line 3: A poor total accuracy was again obtained by using RAG without schema information, with the multi-attribute synthetic dataset, retrieving the top-15 most similar vectors under cosine similarity. These results suggest that, even retrieving 15 samples, the multi-attribute synthetic dataset alone was not sufficient to capture the necessary schema information to translate the NL questions to SQL queries, especially for medium and complex NL questions, whose accuracy was especially poor.

Line 4: The total accuracy was improved by a small margin (with respect to Line 2) by combining LLM-friendly views and RAG, with the multi-attribute synthetic dataset, retrieving the top-8 most similar vectors under cosine similarity.

Line 5: The best total accuracy was obtained by combining LLM-friendly views and RAG, but with the single-attribute synthetic dataset, retrieving the top-8 most similar vectors under cosine similarity. For simple NL questions, this configuration achieved a 97% accuracy, the best result. For complex NL questions, it achieved a 68% accuracy, also the best result; this represents a significant improvement over the second-best result for complex NL questions (Lines 2 and 4), which achieved a 56% accuracy. However, for medium NL questions, the configuration had a lower accuracy than the second-best configuration (Line 4).

In summary, the experiments with IndDB suggested that:

- The relational schema is not an appropriate description of IndDB for an LLM.
- LLM-friendly views that describe IndDB in end-user terms permit achieving a reasonable accuracy.
- RAG with the single-attribute synthetic dataset helped achieve a non-trivial improvement on the results obtained using only LLM-friendly views.

6.3 Results for Mondial

Table 5 shows the accuracy results for two configurations using the Mondial benchmark.

Line 1 repeats the accuracy results obtained using C3 with GPT-4 over the relational schema (without RAG). This configuration was taken as the baseline since it achieved the best total accuracy in the experiments reported in [19].

Line 2 shows the accuracy results for a configuration based on SQLQueryChain with GPT-4 over the relational schema with RAG, using a multi-attribute synthetic dataset with 60,000 pairs, created on top of the Mondial relational schema and associated documentation. Note that Mondial precludes the use of LLM-friendly views since the relation table and column names are close to the geographic object class names, such as Country, City, etc. The adoption of RAG with the relational schema substantially improved the accuracy for medium and complex NL questions. But, for simple NL questions, the multiple similar examples confused the LLM.

The experiments with Mondial therefore corroborated that RAG indeed helps improve total accuracy for a database with a large schema, and in particular the accuracy for medium and complex NL questions.

For display Table with Figure:

Table 5. Results for the different setup configurations – Mondial.

#Line	Configuration	#Corrrect Predicted Queries				Accuracy			
		Simple	Medium	Complex	Total	Simple	Medium	Complex	Total
1	C3 with GPT-4 over the relational schema	29	25	24	78	0,88	0,76	0,71	0,78
2	SQLQueryChain with GPT-4 over the relational schema using RAG	28	31	27	86	0,85	0,94	0,79	**0,86**

7 Conclusions

This article argued that LLMs can be profitably used to construct natural language database interfaces that compile NL questions into SQL queries. The main contribution of the article was to analyze the limitations of general-purpose LLM-based text-to-SQL tools and propose how to improve their performance when they are specialized to RW-RDBs. The article considered implementations based on Prompt Engineering, including Retrieval-Augmented Generation, and fine-tuning and running privately an LLM. Sections 3, 4, and 5 briefly addressed these approaches, whereas Sect. 6 analyzed some implementations on two benchmarks built upon databases with large schemas.

The take-home lessons can be summarized as follows: (1) The relational schema is not an appropriate description of an RW-RDB for an LLM; (2) The use of LLM-friendly views to describe an RW-RDB in end-user terms to the LLM considerably improves accuracy; such views are reasonably simple to define, provided the database is well-documented; (3) The use of RAG helps achieve non-trivial improvements on the results obtained using only LLM-friendly views; RAG captures metadata and data semantics but requires the construction of synthetic datasets.

As future work, we plan to continue to explore how to construct an LLM-based text-to-SQL tool for an RW-RDB such that the tool can be run privately and either use an LLM fine-tuned specifically for the RW-RDB, or be based on RAG with a synthetic dataset built for the RW-RDB.

Acknowledgments. This work was partly funded by FAPERJ under grants E-26/200.834/2021, by CAPES under grant 88881.134081/2016-01 and 88882.164913/2010-01, and by CNPq under grant 305.587/2021-8.

References

1. Affolter, K., Stockinger, K., Bernstein, A.: A comparative survey of recent natural language interfaces for databases. VLDB J. **28**(5), 793–819 (2019). https://doi.org/10.1007/s00778-019-00567-8
2. Askell, A., et al.: A general language assistant as a laboratory for alignment. arXiv (2021). https://doi.org/10.48550/arXiv.2112.00861
3. Caldarini, G., Jaf, S., McGarry, K.: A literature survey of recent advances in chatbots. Information **13**(1) (2022). https://doi.org/10.3390/info13010041, https://www.mdpi.com/2078-2489/13/1/41
4. Coelho, G.M.C., et al.: Improving the Accuracy of Text-to-SQL Tools Based on Large Language Models for Real-World Relational Databases. In: Strauss, C., Amagasa, T., Manco, G.,

Kotsis, G., Tjoa, A.M., Khalil, I. (eds.) Database and Expert Systems Applications. DEXA 2024. Lecture Notes in Computer Science, vol. 14910, pp. 93–107. Springer, Cham (2024). https://doi.org/10.1007/978-3-031-68309-1_8
5. Dettmers, T., Pagnoni, A., Holtzman, A., Zettlemoyer, L.: Qlora: Efficient finetuning of quantized LLMs. In: Advance in Neural Information Processing System, vol. 36, pp. 10088–10115 (2023)
6. Diefenbach, D., Both, A., Singh, K., Maret, P.: Towards a question answering system over the semantic web. Semant. Web **11**(3), 421–439 (2020). https://doi.org/10.3233/SW-190343
7. Gan, Y., et al.: Towards robustness of text-to-SQL models against synonym substitution. CoRR **abs/2106.01065** (2021). https://doi.org/10.48550/arXiv.2106.01065
8. Gan, Y., Chen, X., Purver, M.: Exploring underexplored limitations of cross-domain text-to-SQL generalization. In: Proceedings of the 2021 Conference on Empirical Methods in Natural Language Processing, pp. 8926–8931 (2021). https://doi.org/10.18653/v1/2021.emnlp-main.702
9. Gao, Y., et al.: Retrieval-augmented generation for large language models: a survey. arXiv preprint (2024). https://doi.org/10.48550/arXiv.2312.10997
10. Guo, C., et al.: Retrieval-augmented GPT-3.5-based text-to-SQL framework with sample-aware prompting and dynamic revision chain. arXiv preprint (2023). https://doi.org/10.48550/arXiv.2307.05074
11. Hu, E.J., et al.: Lora: low-rank adaptation of large language models. arXiv (2021). https://doi.org/10.48550/arXiv.2106.09685
12. Jurafsky, D., Martin, J.H.: Speech and Language Processing: An Introduction to Natural Language Processing, Computational Linguistics, and Speech Recognition. Prentice Hall (2023). third Edition draft
13. Lewis, P., et al.: Retrieval-augmented generation for knowledge-intensive NLP tasks. In: Larochelle, H., Ranzato, M., Hadsell, R., Balcan, M., Lin, H. (eds.) Advances in Neural Information Processing Systems, vol. 33, pp. 9459–9474. Curran Associates, Inc. (2020). https://api.semanticscholar.org/CorpusID:218869575
14. Li, J., et al.: Can LLM already serve as a database interface? A big bench for large-scale database grounded text-to-SQLs. arXiv preprint (2023). https://doi.org/10.48550/arXiv.2305.03111
15. Liu, A., Hu, X., Wen, L., Yu, P.S.: A comprehensive evaluation of ChatGPT's zero-shot text-to-SQL capability. arXiv preprint (2023). arXiv:2303.13547
16. Manning, C.D.: Human language understanding & reasoning. Daedalus **151**(2), 127–138 (2022). https://doi.org/10.1162/daed_a_01905
17. Motger, Q., Franch, X., Marco, J.: Software-based dialogue systems: survey, taxonomy, and challenges. ACM Comput. Surv. **55**(5), 1–42 (2022). https://doi.org/10.1145/3527450
18. Nascimento, E.R.: Querying Databases with Natural Language: The use of Large Language Models for Text-to-SQL tasks. Master's thesis, Dissertation presented to the Graduate Program in Informatics, PUC-Rio (2024)
19. Nascimento, E.R., et al.: Text-to-SQL meets the real-world. In: Proceedings of the 26th International Conference on Enterprise Information Systems, vol. 1, pp. 61–72. ICEIS; SciTePress (2024). ISBN 978-989-758-692-7. https://doi.org/10.5220/0012555200003690
20. Nascimento, E.R., et al.: A family of natural language interfaces for databases based on ChatGPT and LangChain (short paper). In: Companion Proceedings of the 42nd International Conference on Conceptual Modeling: Posters and Demos co-located with ER 2023. Lisbon, 06–09 November 2023, CEUR Workshop Proceedings, vol. 3618 (2023). https://ceur-ws.org/Vol-3618/pd_paper_1.pdf
21. Nascimento, E.R., et al.: My database user is a large language model. In: Proceedings of the 26th International Conference on Enterprise Information Systems, vol. 1, pp.

800–806. ICEIS; SciTePress (2024). ISBN 978-989-758-692-7. https://doi.org/10.5220/0012697700003690

22. Ouyang, L., et al.: Training language models to follow instructions with human feedback. In: Proceedings of the Advances in Neural Information Processing Systems, NeurIPS 2022, vol. 35 (2022). https://doi.org/10.48550/ARXIV.2203.02155, online: 25 May 2023
23. Pal, A., Karkhanis, D., Dooley, S., Roberts, M., Naidu, S., White, C.: Smaug: fixing failure modes of preference optimisation with DPO-positive. arXiv preprint (2024). https://doi.org/10.48550/arXiv.2402.13228
24. Panda, S., Gozluklu, B.: Build a robust text-to-SQL solution generating complex queries, self-correcting, and querying diverse data sources. AWS Mach. Learn. Blog (2024)
25. Ping, W.J.: Open-sourcing sqleval: our framework for evaluating LLM-generated SQL (2023). https://defog.ai/blog/open-sourcing-sqleval/
26. Pinheiro., J., et al.: On the construction of database interfaces based on large language models. In: Proceedings of the 19th International Conference on Web Information Systems and Technologies - WEBIST, pp. 373–380. INSTICC, SciTePress (2023). https://doi.org/10.5220/0012204000003584
27. Tunstall, L., von Werra, L., Wolf, T.: Natural Language Processing with Transformers. Revised Inc, O'Reilly Media (2022)
28. Vaswani, A., et al.: Attention is all you need. In: Proceedings of the Advances in Neural Information Processing Systems (NIPS 2017), vol. 30 (2017). https://doi.org/10.48550/arXiv.1706.03762
29. Wang, B., et al.: Mac-SQL: A multi-agent collaborative framework for text-to-SQL. arXiv preprint (2024). https://doi.org/10.48550/arXiv.2312.11242
30. Yu, T., et al.: Spider: a large-scale human-labeled dataset for complex and cross-domain semantic parsing and text-to-SQL task. In: Proceedings of the 2018 Conference on Empirical Methods in Natural Language Processing, pp. 3911–3921 (2018). https://doi.org/10.18653/v1/D18-1425
31. Zhong, V., Xiong, C., Socher, R.: Seq2sql: generating structured queries from natural language using reinforcement learning. CoRR **abs/1709.00103** (2017). http://arxiv.org/abs/1709.00103

Refining Community Boundaries in Bee Swarm Optimization for Enhanced Community Detection in Social Networks

Narimene Dakiche(✉), Karima Benatchba, and Fatima Benbouzid-Si Tayeb

Laboratoire des Methodes de Conception de Systèmes (LMCS), Ecole nationale Supérieure d'Informatique (ESI), 68M - 16270 Oued Smar, Alger, Algeria
{an_dakiche,k_benatchba,f_sitayeb}@esi.dz

Abstract. Community detection in social networks plays a crucial role in understanding the underlying structure and dynamics of complex systems. In this paper, we present a novel approach for enhancing community detection by refining community boundaries within the framework of Bee Swarm Optimization (BSO). Our method focuses on identifying and contracting the contour of communities, where nodes exhibit greater external degrees than internal degrees, thus enhancing the coherence and granularity of detected communities. Through BSO, we iteratively explore the solution space, employing a contour-based contraction operation to refine community structures. Experimental evaluations on fifteen real-word networks demonstrate the effectiveness of our approach, revealing significant improvements in modularity compared to traditional and state-of-the-art methods.

Keywords: Social networks · Community detection · Bee swam optimization · Modularity

1 Introduction

The field of network science has advanced significantly in modeling complex real-world systems, with one prominent feature being the presence of community structure. Various algorithms for community detection have emerged in recent years, aiming to unveil structural properties and collective behaviors within networks [1]. Community detection involves partitioning a network into groups of interconnected nodes based on their structural attributes [22]. This task is recognized as an optimization problem, known to be NP-Hard, as it seeks to maximize diverse objective functions [4]. Among the metrics used to evaluate the quality of network partitions, modularity stands out as the most commonly employed [16], despite its resolution limit problem in certain realistic scenarios [11]. Consequently, numerous community detection algorithms have been proposed with the aim of optimizing modularity.

Swarm Intelligence encompasses the study and design of computationally intelligent interactive solutions inspired by the collective behavior of natural swarms, such as birds, fish, and ants. Many swarm-based optimization algorithms have been adapted to address the community detection problem. In our work, we are specifically interested

M. Marchiori and F. García Peñalvo (Eds.): WEBIST 2023, LNBIP 543, pp. 168–178, 2025.
https://doi.org/10.1007/978-3-031-89621-7_9

in algorithms derived from the behavior of bees. In [15], the authors used Artificial Bee Colony (ABC) optimization to detect communities in networks, aiming to maximize the modularity measure. Their approach employed a string-based representation for food sources (community structures), assuming a predefined number of communities. Their findings demonstrated the effectiveness of ABC for community detection tasks. In a similar vein, in [14], the authors employed ABC with improved efficiency for community detection. They adopted a locus-based adjacency encoding scheme to represent food sources, enabling automatic determination of the number of communities. By employing various popular community quality measures as objective functions, their study highlighted modularity optimization as yielding superior community structures. In [21], the authors proposed enhancements to the ABC algorithm tailored for community detection. Their modifications included dynamic adjustments to the number of initial food sources and search scope, aimed at mitigating local optimization pitfalls. Experimental evaluations showcased the superiority of their algorithm over traditional ABC methods, particularly in complex networks. Furthermore, in [13], the authors introduced a heuristic ABC-based algorithm for community detection. Their approach involved redefining the search process of ABC and utilizing an agglomerate probability of neighboring communities as a heuristic function. This heuristic-driven approach aimed to improve the effectiveness and efficiency of community detection using ABC. In [8], the authors proposed a hybrid Artificial Bee Colony algorithm with Simulated Annealing (ABC-SA) to address the community detection problem. The SA algorithm enhanced the exploitation by searching the most promising regions located by ABC algorithm. Their successful results showed the effectiveness and efficiency of the hybridization between ABC and SA algorithms. Another inspiration from the foraging behavior of real bees called Bee Swarm Optimization (BSO [9]) has been used in [3]. In their algorithm, the authors take modularity Q as objective function and choose a string based schema to represent the solutions. No explicit local search was adopted to visit the solutions' neighborhood except for a flipping algorithm to determine the search area. Later, the authors of [7] adapted the BSO algorithm for community detection problem in which they proposed to use the network knowledge to select the most appropriate neighbor, while exploring each bee's neighborhood. Precisely, they used Pearson's correlation as structural information to guide the solutions' construction. Their algorithm outperformed other existing swarm-based methods, delivering higher-quality results.

In this paper, we propose a novel technique to explore the bee's neighborhood. The main idea consists of refining community structures by contracting a community through the removal of nodes with weaker connections. Specifically, focusing on the contour of the community, where nodes exhibit greater external degrees than internal degrees, aligns with the intuition of identifying boundary nodes that are less tightly connected internally and might contribute less to the coherence of the community.

The paper is structured as follows: Sect. 2 outlines community detection as an optimization problem. In Sect. 3, we introduce the general algorithm of BSO and discuss its application for community detection. Section 4 presents the results of our proposed method, BSODCS, and compares them with those obtained from other methods. Finally, in Sect. 5, we provide concluding remarks and suggest directions for future research.

2 Community Detection Problem

A complex network can be represented as a graph $G = (V, E)$, where V denotes the set of nodes and E denotes the set of edges. The graph consists of $N = |V|$ nodes and $m = |E|$ edges. To describe the network, an adjacency matrix A of size $N \times N$ can be used. Each element A_{ij} in the matrix corresponds to the presence ($A_{ij} = 1$) or absence ($A_{ij} = 0$) of an edge between the ith and j^{th} nodes. The final community structure $P = \{C_1, C_2, ..., C_k\}$ of the network is defined, where each $C_l (l = 1, 2, ..., k)$ is a subset of V, and k represents the total number of communities.

Community detection as an optimization problem requires the use of an objective function to guide the search for optimal solutions. The objective function acts as the "steering wheel" in the process, leading to favorable solutions. The most widely used and well-known quality function for evaluating the quality of a network partition is modularity Q [16]. In the context of a graph G with n nodes and m edges, modularity is computed in the following manner:

$$Q = \frac{1}{2m} \Sigma_{ij}^{n} \left(A_{ij} - \frac{d_i \times d_j}{2m} \right) \delta(C_i, C_j) \tag{1}$$

where A is the corresponding adjacency matrix, $d_i(d_j)$ denotes the degree of node $i(j)$, $C_i(C_j)$ is the community to which node $i(j)$ belongs to. $\delta(C_i, C_j) = 1$, if $C_i = C_j$, otherwise, 0. A high modularity indicates that there is a dense concentration of connections within nodes of the same community, while connections between nodes from different communities are sparse [10].

3 Proposed BSODCS Algorithm

The BSO algorithm, introduced as a metaheuristic in [9], is inspired by the foraging behavior of real bees. It operates as a search process employing a population of artificial bees to solve optimization problems. The process begins with a designated bee, BeeInit, responsible for generating the initial solution known as the reference one (RefSolution). Then, additional solutions are derived from RefSolution using a flipping strategy, which defines the search space. After that, each solution is assigned to a bee, acting as a starting point for exploring its local neighborhood. BSO has been adapted for static community detection problem and reached better results compared to other algorithms from the literature [7].

In contrast to this previous work, we propose in this paper a new technique for exploring the neighborhood of bees. Otherwise, we maintain the locus-based adjacency encoding scheme to represent the solutions, which has the advantage of enabling the algorithm to detect communities' number automatically. Besides, we use the modularity as objective function to detect communities. The whole process is given in Algorithm 1, and subsequent subsections provide detailed explanations of its different steps.

Algorithm 1. BSO for detecting community structure [7].

```
Input: A network G = (V, E)
Output: Community structure of the network C = {C_1, C_2, ..., C_k}
Initialize the algorithm' parameters: N (number of bees), itermax (maximum number of iterations), flip (search
  space creation parameter), and limit (abandonment criterion)
  // Initial reference solution, see subsection 3.2
RefSolution ← GenerateInitialSolution();
Set TabuList = newlist()
Set iter = 0
while iter < itermax do
   //Save RefSolutio in Tabu list
   TabuList ← TabuList + RefSolutio;
   // Determine search space from RefSolution, see subsection 3.3
   for i ← 1 to N do
      ApplyFlippingStrategies;
   end for
   Set nbChances = limit
   // Neighborhood exploration (N), see subsection 3.4
   for i ← 1 to N do
      NeighborBee ← GenerateNeighborBee(Bee[i]);
      // Save best solution in Dance table
      DanceTable ← DanceTable + BestSol(Bee[i], NeighborEBee)
   end for
   // Select the new RefSolution, see subsection 3.5
   if BestModularity(DanceTable) > BestModularity(TabuList) then
      RefSolution ← BestModularity(DanceTable);
      nbChances ← limit;
   else
      if nbChances > 0 then
         nbChances ← nbChances − 1;
         //Intensification
         RefSolution ← RefSolution;
      else
         //Diversification
         RefSolution ← GenerateInitialSolution();
      end if
   end if
   iter ← iter + 1;
end while
BestModularity ← BestModularity(TabuList)
Decode the optimal solution to obtain the communities of network G, C = {C_1, C_2, ..., C_k}
```

3.1 Solution Encoding

The algorithm uses a locus-based adjacency representation [19], automatically detecting the number of communities. In this scheme, each solution represents a set of communities. Each individual g in the population, for a network with n nodes, is denoted as $\langle g_1, g_2, ..., g_n \rangle$, with each g_i representing a node and taking a randomly selected allele value j from adjacent nodes. This representation effectively extracts community structures by decoding connected components of the graph, ensuring nodes within the same component belong to the same community. This decoding is highly effective and efficient, as demonstrated in [6]. For illustration, see Fig. 1 for a visual representation.

3.2 Initial Reference Solution

Initially, the algorithm generates a solution to represent the BeeInit's food source. This solution is generated randomly or via a heuristic. For community detection problem, a randomly generated solution may not be feasible. Indeed, it could contain a value

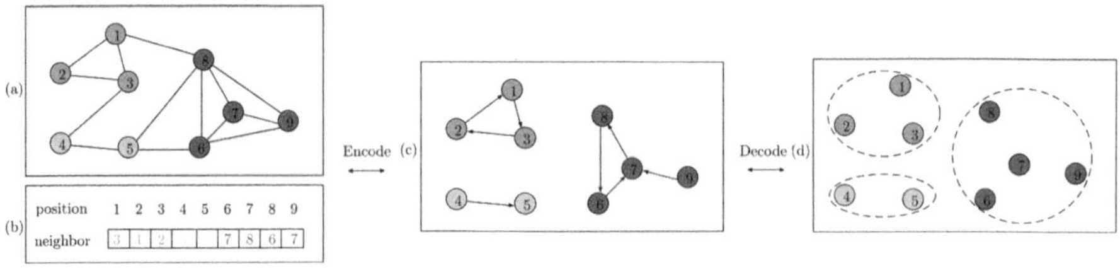

Fig. 1. Example of a solution representation [7].

j in the ith position, but no edge (i, j) exists in the network. Therefore, in order to obtain a solution with certain quality, for each node i, a value j in the ith position is randomly chosen among its neighbor nodes. This means that the edge (i, j) exists in the network. This strategy provides a good starting point for the algorithm and improves the convergence of the method.

3.3 Search Space

The search space comprises N solutions, each assigned to a bee for local search. Solutions are generated by flipping bits from the reference solution (Refsolution), with flip determining the number of flips per solution. Low flip values may trap in local optima, while high values risk deviating too far. To maintain diversity, two strategies are used [18]: one flips bits with a gap of flip bits, the other flips contiguous bits. For example, in Fig. 2, with N = 9 and flip = 3, the first strategy flips positions (1,4,7), (2,5,8), and (3,6,9), while the second flips (1,2,3), (4,5,6), and (7,8,9). Positions are then altered by randomly selecting a neighbor from each node's list of neighbors.

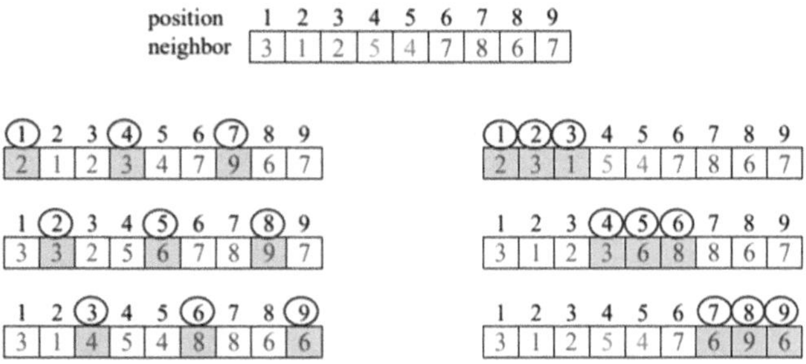

Fig. 2. (left) solutions generated by the first strategy, (right) solutions generated by the second strategy [7].

If the number of generated solutions is insufficient (less than the defined number of bees), a random approach can be used. These strategies aim to maintain distinct solutions while exploring the search space effectively.

3.4 Neighborhood Procedure

The bees explore their neighborhood in search of potentially better solutions. To do so, we propose a method to refine community structures. The first step consists in identifying boundary nodes within a community, which have more connections outside than inside it. Once identified, these nodes are removed to contract the community, and transferred to other communities. This process eliminates nodes with weaker internal connections, enhancing internal cohesion, while essential nodes preserving internal structure are retained. The method is applied iteratively in the evolutionary process, refining community structure with each iteration and improving coherence and integrity.

Formally, the proposed method aims to transfer the boundary nodes of the communities within the network in a manner that increases the modularity value. In Fig. 3 (left partition), nodes 1 and 6 indicated in yellow in community A represent boundary nodes. Each one is transferred to a neighboring community using Pearson's correlation. If the modularity value increases after this transfer process, the new solution is accepted by the bee. If not, the transfer is canceled, and the relevant border node remains in its community. In this example, the modularity of the initial partition is 0.467, which improved to 0.5878 after transferring nodes 1 and 6 to communities B and C, respectively. Thus, the corresponding bee retains the new solution, presenting the improved structure (right partition).

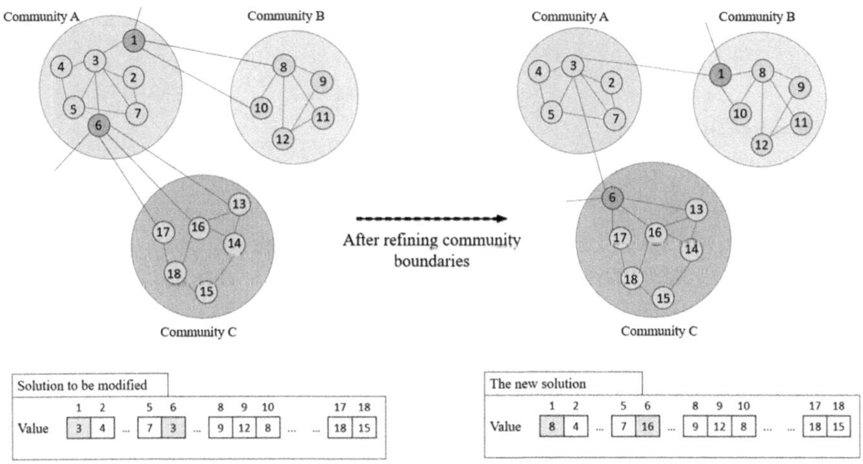

Fig. 3. Community contour-based neighbor procedure.

3.5 Selection of Reference Solution

The selection of the reference solution, for next iteration, initially relies on the modularity of solutions in the dance table. However, if the swarm observes a lack of progress in terms of modularity over a certain period of time, it incorporates a restart strategy. This restart enables the swarm to escape from a potentially stagnant region and explore alternative regions in the search space. By combining both quality and diversity considerations, the swarm ensures a more comprehensive and effective search process.

4 Experimental Results

In this section, our algorithm was tested on a selection of fifteen real-world networks of different sizes. Given the fact that no ground truth is available for these networks, we used modularity as a quality measure to assess the resulting community structures. We compared the results with six algorithms from the literature, namely GN [12], CNM [5], Leiden [20], GaNet [17], BCD [2], and ABC [8]. The choice of these algorithms is motivated as follows: GN and CNM are two classical methods for community detection. They are based on well-established principles and have been widely used and studied in the literature. Leiden is considered a successful modern method for community detection built upon the Louvain method. GaNet, BCD, and ABC can be classified as metaheuristic algorithms for community detection. They offer novel optimization approaches inspired by principles from different fields, such as the process of natural selection and genetics, brainstorm optimization, and swarm intelligence, respectively. In what follows, we present the used datasets and discuss the results.

4.1 Datasets

The networks used in this study are sourced from diverse domains, exhibiting varying scales and degree distributions. Additionally, in the case of weighted networks, we standardized the edge weights to 1, converting them into unweighted networks. Table 1 provides a summary of these networks, encompassing information such as their node and edge counts, as well as their average degree.

4.2 Results Discussion

The performance of BSODCS, like most metaheuristics, is heavily influenced by its parameter values. In our experiments, we determined the optimal parameter values through extensive runs using various parameter combinations. Consequently, we set the parameters as follows: $flip = 5$, $limit = 5$, the number of bees $N = 50$, and the number of iterations 100.

Table 2 presents the modularities obtained from various algorithms, including Girvan-Newman (GN), Clauset-Newman-Moore (CNM), Leiden, GaNet, BCD, ABC, and BSODCS, across different real-world network datasets. The comparison of these results reveals distinctive performance trends among the algorithms evaluated. While Girvan-Newman demonstrates moderate performance across different networks, it falls short compared to others in certain cases, such as the Zachary karate club and Collaboration GRQC networks. Similarly, the Clauset-Newman-Moore algorithm exhibits competitive performance but lacks consistency across datasets. In contrast, Leiden consistently demonstrates strong performance, particularly evident in networks such as Contiguous USA and Facebook (NIPS). However, the standout performer across various datasets appears to be BSODCS, consistently achieving the highest modularity scores across different network topologies. This is particularly evident in networks like Zachary karate club, Contiguous USA, Dolphins, and American football, where BSODCS consistently achieves the highest modularity scores compared to other algorithms. These findings suggest that while Leiden shows robust performance, BSODCS

Table 1. Real networks information.

Network	No. of nodes	No. of edges	Average degree
Zachary karate club [a]	34	78	4,59
Contiguous USA [a]	49	107	4,37
Dolphins [a]	62	159	5,13
Les Misérables [a]	77	254	6,59
Political books [a]	105	441	8,40
David Copperfield [a]	112	425	7,59
American football [a]	115	613	10,66
Jazz musicians [a]	198	2742	27,70
Neural [b]	297	2162	14,56
08blocks [b]	300	584	3,89
Metabolic [a]	453	2040	9,01
Polblogs [b]	1490	16718	27,32
NetScience [b]	1589	2742	3,45
Facebook (NIPS) [a]	2888	2981	2,06
Collaboration GRQC [b]	5242	14496	5.53

[a] http://konect.uni-koblenz.de/networks/.
[b] http://networkrepository.com/networks.

Table 2. Comparison of modularity results on real-world networks.

Network	GN	CNM	Leiden	GaNet	BCD	ABC	BSODCS
Zachary karate club	0.4013	0.3807	**0,4198**	0,3940	**0.4198**	**0.4198**	**0,4198**
Contiguous USA	0.5958	0.5824	**0,5970**	0,5481	0.5882	0.5935	**0,5970**
Dolphins	0.5194	0.4955	0,5165	0,4562	0.5253	**0.5285**	**0.5285**
Les Misérables	0.5381	0.5006	0,5594	0,5062	0.5430	0.5594	**0,5600**
Political books	0.5168	0.5020	0,5253	0,4632	0.5189	0.5116	**0.5262**
David Copperfield	0.0805	0.2947	0,3071	0,2011	0.2231	0.2933	**0.3072**
American football	0.5996	0.5497	0,6043	0,3853	0.5146	0.6009	**0,6046**
Jazz musicians	0.4051	0.4389	**0,4451**	0,2589	0.3125	0.4384	0,3931
Neural	0.3010	0.3728	**0,4032**	0,1988	0.2319	0.3137	0.3599
08blocks	0.8599	**0.8750**	**0,8750**	0,7594	0.8104	**0.8750**	**0,8750**
Metabolic	0.4048	0.4172	**0,4442**	0,2771	0.3128	0.3664	0,3503
Polblogs	0.4180	0.4270	0.4269	0.4220	0.3642	0.3702	**0.4271**
NetScience	0.9579	0.9551	**0.9593**	–	0.9006	0.9086	0,9092
Facebook (NIPS)	0.8079	**0.8087**	**0.8087**	–	0.8086	**0.8087**	**0.8087**
Collaboration GRQC	**0.8492**	0.8119	0.8418	–	0.7054	0.7028	0,7061

emerges as the most effective algorithm for achieving high modularity across diverse network structures. Its consistent performance across various datasets highlights its potential as a reliable choice for community detection tasks.

Figure 4 illustrates the interconnected relationships among characters in Victor Hugo's 19th-century novel, *Les Misérables*. Within this network, 77 nodes represent the characters, while 254 links signify their co-appearance in scenes. This visualization highlights the community structure, revealing six distinct communities identified by our algorithm BSODCS, achieving the highest modularity of $Q = 0.5600$.

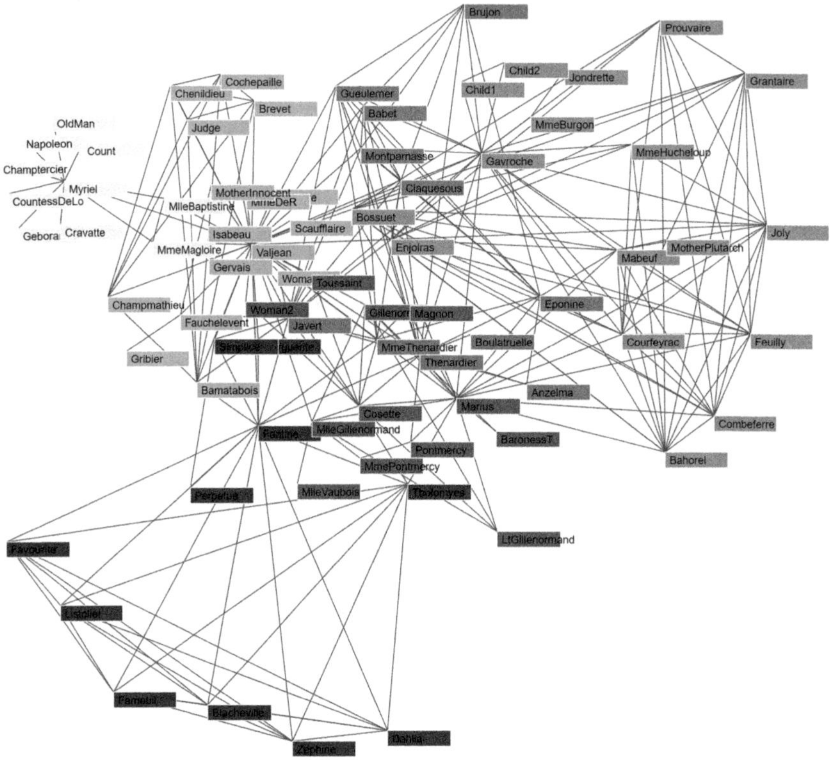

Fig. 4. The community structure of Les Misérables network found by BSODCS algorithm (Q = 0.5600).

5 Conclusion

In this paper, we have introduced a novel approach for refining community structures in complex networks by focusing on the contraction of communities through the removal of boundary nodes with weaker connections. Our approach, which targets nodes with greater external degrees than internal degrees, aims to enhance the coherence and integrity of identified communities within the framework of a Bee Swarm Optimization (BSO) algorithm. By iteratively removing boundary nodes, our method systematically refines the boundaries of communities, resulting in more compact and cohesive structures. This approach, integrated into the BSO algorithm, aligns with the intuition of identifying and addressing the periphery of communities to improve their overall functionality.

Through empirical evaluations on benchmark datasets, we have demonstrated the effectiveness of our approach in enhancing community detection performance within the BSO framework. Our results showcase improvements in community quality metrics, highlighting the efficacy of targeted boundary node removal for community refinement. Looking ahead, there are several avenues for future research. Further exploration could focus on refining the criteria for identifying boundary nodes and optimizing the community contraction process within the BSO framework. Additionally, the integration of our approach with other optimization techniques or algorithms could lead to further enhancements in community detection performance.

References

1. Bedi, P., Sharma, C.: Community detection in social networks. Wiley Interdisc. Rev. Data Min. Knowl. Discov. **6**(3), 115–135 (2016)
2. Beldi, Z., Bessedik, M.: A new brainstorming based algorithm for the community detection problem. In: 2019 IEEE Congress on Evolutionary Computation (CEC), pp. 2958–2965 (2019). https://doi.org/10.1109/CEC.2019.8789897
3. Belkhiri, Y., Kamel, N., Drias, H., Yahiaoui, S.: Bee swarm optimization for community detection in complex network. In: Recent Advances in Information Systems and Technologies, vo. 2 5, pp. 73–85. Springer (2017)
4. Brandes, U., et al.: On modularity clustering. IEEE Trans. Knowl. Data Eng. **20**(2), 172–188 (2007)
5. Clauset, A., Newman, M.E.J., Moore, C.: Finding community structure in very large networks. Phys. Rev. E **70**(6) (2004). https://doi.org/10.1103/physreve.70.066111
6. Cormen, T.H., Leiserson, C.E., Rivest, R.L., Stein, C.: Introduction to Algorithms 2 edn. (2001)
7. Dakiche, N.: BSODCS: bee swarm optimization for detecting community structure. In: International Conference on Web Information Systems and Technologies (2023). https://api.semanticscholar.org/CorpusID:265361939
8. Dakiche, N., Benatchba, K., Tayeb, F.B.S., Slimani, Y., Brahmi, M.A.: A hybrid artificial bee colony algorithm with simulated annealing for enhanced community detection in social networks. In: 2022 IEEE/ACM International Conference on Advances in Social Networks Analysis and Mining (ASONAM), pp. 104–108. IEEE (2022)
9. Drias, H., Sadeg, S., Yahi, S.: Cooperative bees swarm for solving the maximum weighted satisfiability problem. In: International Work-Conference on Artificial Neural Networks, pp. 318–325. Springer (2005)
10. Fortunato, S.: Community detection in graphs. Phys. Rep. **486**(3–5), 75–174 (2010)
11. Fortunato, S., Barthélemy, M.: Resolution limit in community detection. Proc. Natl. Acad. Sci. **104**(1), 36–41 (2007)
12. Girvan, M., Newman, M.E.: Community structure in social and biological networks. Proc. Natl. Acad. Sci. **99**(12), 7821–7826 (2002)

13. Guo, Y., Li, X., Tang, Y., Li, J., et al.: Heuristic artificial bee colony algorithm for uncovering community in complex networks. Math. Probl. Eng. **2017** (2017)
14. Hafez, A.I., Zawbaa, H.M., Hassanien, A.E., Fahmy, A.A.: Networks community detection using artificial bee colony swarm optimization. In: Proceedings of the Fifth International Conference on Innovations in Bio-Inspired Computing and Applications IBICA 2014, pp. 229–239. Springer (2014)
15. Naser, A., Alshattnawi, S.: An artificial bee colony (ABC) algorithm for efficient partitioning of social networks. Int. J. Intell. Inf. Technol. (IJIIT) **10**(4), 24–39 (2014)
16. Newman, M.E., Girvan, M.: Finding and evaluating community structure in networks. Phys. Rev. E **69**(2), 026113 (2004)
17. Pizzuti, C.: Ga-net: A genetic algorithm for community detection in social networks. In: International Conference on Parallel Problem Solving from Nature, pp. 1081–1090. Springer (2008)
18. Sadeg, S., Hamdad, L., Benatchba, K., Habbas, Z.: BSO-FS: bee swarm optimization for feature selection in classification. In: Rojas, I., Joya, G., Catala, A. (eds.) Advances in Computational Intelligence, pp. 387–399. Springer, Cham (2015)
19. Shi, C., Wang, Y., Wu, B., Zhong, C.: A new genetic algorithm for community detection. In: Zhou, J. (ed.) Complex Sciences, pp. 1298–1309. Springer, Heidelberg (2009)
20. Traag, V.A., Waltman, L., van Eck, N.J.: From louvain to leiden: guaranteeing well-connected communities. Sci. Rep. **9**(1) (2019). https://doi.org/10.1038/s41598-019-41695-z
21. Wang, Z., Zhao, X., Wen, P., Xue, J., Hu, C.: Community detection in complex networks using improved artificial bee colony algorithm. In: 2016 International Conference on Communications, Information Management and Network Security, pp. 283–288. Atlantis Press (2016)
22. Yang, J., McAuley, J., Leskovec, J.: Community detection in networks with node attributes. In: 2013 IEEE 13th International Conference on Data Mining, pp. 1151–1156. IEEE (2013)

Digital Citizen Rights: Accessibility of E-Government Websites in Italy

Maria Claudia Buzzi[1](✉), Marina Buzzi[1], Giuseppe Della Penna[2], Barbara Leporini[3,4], and Francesca Ricci[2]

[1] CNR-IIT, via Moruzzi 1, 56124 Pisa, Italy
marina.buzzi@iit.cnr.it

[2] Department of Information Engineering, Computer Science and Mathematics, University of L'Aquila, L'Aquila, Italy

[3] Department of Computer Science, University of Pisa, via Fibonacci, Pisa, Italy

[4] CNR-ISTI, via Moruzzi 1, Pisa, Italy

Abstract. Nowadays, digital services pervade all our life. Accessible services can greatly empower the personal autonomy of people with disabilities. In Europe, accessible and usable interfaces for assuring a satisfying interaction with applications and services created by Public Administrations is a right for any citizen, regardless of any functioning or ability. In this paper, starting from experience of people with disabilities, the accessibility of Italian PA services has been investigated to evaluate progresses in a time range of six years. Results suggest that PA services have increased but no substantial improvement in terms of accessibility has been recorded. Simplified interaction and increased efficiency are still lacking, even if global user satisfaction seems to have slightly improved. However, the virtuous cycle of accessibility improvement is steadily in progress thanks to tools and resources made available by eGovernment agencies with the voluntary contribution of the Italian designer and development community, and this will hopefully lead to further improvements in the coming years.

Keywords: Human computer interaction · Web interfaces · Accessibility · People with disabilities · e-Government

1 Introduction

In most cases, the Internet is an opportunity for people with disabilities: accessible Public Administration (PA) online services can be performed autonomously, without the need for an accompanying person, greatly empowering personal autonomy. Autonomy is one of the most important conditions for people with disability and therefore it is very important to progressively increase the portfolio of available services. Investigating the main reasons why users who could access such services avoid using them could provide interesting information on barriers or problems to be solved.

This paper extends [11], a study on accessibility of e-Government services from the perspective of people with disability to assess if and how accessibility improved

M. Marchiori and F. García Peñalvo (Eds.): WEBIST 2023, LNBIP 543, pp. 179–207, 2025.
https://doi.org/10.1007/978-3-031-89621-7_10

over time. Specifically, a comparison between 2023 and 2017 has been introduced in (Buzzi et al. 2023) exploiting a questionnaire extended with additional questions and semi-structured interviews.

This paper extension provides a complete background on accessibility, the European and Italian directives and governance policies about Digital Citizen Rights. It also includes and discusses the complete details of the 2023 versus 2017 comparison, and finally discusses the study results with respect to official analyses performed by the Italian Government on its PA websites. Findings showed that PA services have increased but no substantial improvement in terms of accessibility has been recorded. Simplified interaction and increased efficiency are still lacking, even if global user satisfaction seems to have slightly improved.

The paper is organized in six sections. After discussing the background in Sect. 2, the related work is introduced in Sect. 3. The study and its results are then illustrated and discussed in Sect. 4. Next, Sect. 5 compares the current results with the 2017 study and assess the coherence of such results with official studies of the Italian Government. Finally, Sect. 6 outlines some concluding considerations and remarks.

2 Background

Users accessing online resources and sites may do so in different ways and devices, and thus may have different needs. For example, some people may use desktop computers, laptops, tablets or smartphones to access online information. In addition, there is a significant proportion of users who rely on assistive technologies to interact with digital content. These technologies, ranging from screen readers to special input devices, need to well interact with the user interface, i.e. web sites and pages. However, it is important that a user can easily interact with the interface, reach the contents and complete their task in a reasonable time and with as little effort as possible. It is usual to consider this for the ordinary user, while it is not kept in mind for the user with disabilities or who may have specific needs. This can then lead to interaction problems even for those web pages that are considered accessible. In the following few examples of possible interaction issues will be presented.

2.1 User Interaction and Assistive Technology

Users with disabilities may interact with web pages and services through assistive technologies exploiting also navigation techniques to interact in a simpler and more natural way, for instance users reading on a smartphone may need to zoom in, change the color of the text or background, or may need to interact in a simpler and more natural way. Furthermore, assistive technologies such as screen readers and magnifiers must interpret the content of the user interface in order to properly and appropriately deliver content to the user who cannot see or is partially sighted. For instance, the screen reader will have to announce when a link is a menu item, or which are the possible button choices the user can make, or which are the main areas or sections on a page. Let us also not forget those parts of the page that update very frequently and are very difficult, and sometimes almost impossible, to capture via screen reader. These are just a few examples of possible user interactions, from which it emerges how important it is that the user interface is well designed and developed.

A screen reader is a software installed in the system able to interface between the user (blind/visually impaired) and the operating system and its applications. It interprets the contents of the interface (text, links, buttons, text boxes, drop-down menus, etc.) to announce to the user the type of content and object on which the focus is. For the detection of interface components to occur correctly, it is necessary that the developer has used and applied accessibility features, and thus has used the appropriate attributes and elements at the user interface level. For example, most screen readers, when accessing a web page, interpret the HTML code of the page, announcing the elements in the order in which they are written (entered) on the page. Therefore, the correct order of the contents and elements of a web page is very important to guarantee a correct sequence of the contents to be announced to the blind person.

A magnifier is an assistive technology used by partially sighted persons to better read the contents of the user interface by allowing the user to magnify the mouse pointer, increasing the font size or changing the background and foreground colours. If the system focus is not linked to the mouse pointer, this assistive technology does not work properly.

Another interaction issue may occur for a user interacting via keyboard. If every element of the interface has not been programmed to receive the system focus, only the mouse will have to be used to activate it. Thus, if a web page does not have any links or buttons programmed in a standard or accessible way granting focus, the user will not be able to interact via the Tab or Enter key. For a user who must interact via pointing systems, it is essential that the control elements of the interface are well separated, distinguishable and clickable. Also, if subtitles have not been provided in a video, or a script has not been associated with it, or if it does not provide an alternative form of speech and audio content to comprehension, the user who cannot hear will not be able to enjoy the content.

All these are just some examples of interaction with the web page by people who may have difficulties or use assistive technology to give an idea about the potential encountered problems.

2.2 Usability and Accessibility

It is important that a user can easily interact with the interface, can reach the contents and complete their task in a reasonable time and with as little effort as possible. That is why when designing and developing, we should not limit this only to the basic concept of accessibility but go beyond it by considering how the user achieves the objective of navigating the website or service. It is usual to consider this for the ordinary user, while it is not so for the user with disabilities or specific needs. This can then lead to interaction problems even for those web pages that are considered accessible. If there are further accessibility problems, then it becomes even more complicated.

Usually, accessibility refers to the minimum requirements that allow everyone, regardless of ability, to access content and more generally the user interface. However, there is no concern about how these contents are reachable and usable while they greatly impact on the user experience. There is a tendency to consider the concepts of accessibility and usability as separate, with usability suitable only for non-disabled users. Conversely, usability is a crucial aspect that must be considered when designing for making content technically accessible but also easy to use removing perceived barriers and difficulties in navigation.

Leporini et al. [29] proposed a combination of usability and accessibility in the web, identifying the relationships between these two concepts, in particular considering usability criteria for accessible Web sites: the three properties 'efficiency, effectiveness and satisfaction' were related to the concept of accessibility, and especially of interaction/navigation with the web interface. Many studies investigated the relationship between accessibility and usability for the web [9, 21, 37]. As a result, a number of criteria have been proposed in the literature to decline the concept of usability with respect to the accessibility of web pages for various types of users [30, 33].

2.3 Digital Citizen Rights

eGovernment services, such the Public Administration digital services addressed in this paper, represent a great opportunity to increase and simplify citizen relationship with governance structures [39]. However, the European digital rights and principles [17, 18] clearly indicate that such a transition from traditional to digital services must take into consideration some basic principles: one of these is *solidarity and inclusion*, which explicitly addresses the previously described usability and accessibility issues.

Specifically, in Italy such principles have been transposed in the Digital Citizenship Rights Guide [15], which is now considered a mandatory guideline for the design of PA online services, and explicitly states that every citizen has the right to access online services in a simple, safe, and fast way.

Therefore, investigating such aspects of PA digital services, especially through the eyes of citizens with disabilities that mostly suffer from poorly designed applications and interfaces, is strictly required for the (mandatory) fulfillment of the European Declaration on Digital Rights and Principles as well as of National laws and guidelines.

Legislation is a way for inducing changes in society. The last few decades have been characterized by a legislative effort for improving accessibility of public services and digital products in Europe. In 2016, the EU Commission published the European Directive (EU) 2016/2102 of the European Parliament and of the Council on the accessibility of websites and mobile applications of public sector bodies. Two years later the Commission Implementing Decision (EU) 2018/1523 established a model accessibility statement in accordance with Directive (EU) 2016/2102. Furthermore in 2018 the European Directive (EU) 2019/882 (Accessibility Act) was approved. Lastly, the Commission delegated ETSI to define a technical specification for accessible products and services: the EN 301 549 v.3.2.1 (Accessibility requirements for ICT products and services).

These directives have a significant impact fueling the internal market of accessible ICT products and removing barriers to their free circulation. Member States applied these directives at national level. They are applied not only to the Public Bodies but also to companies and legal entities offering services to the public through websites or mobile applications, with an average revenue in the last three years greater than 500 million euros.

2.4 The Italian Context

Many websites of the Italian Public Administration are still not compliant with the minimum level of accessibility requirements [19]. Previous studies investigated the accessibility of PA services in Italy, mostly exploiting automatic tools for assessing the degree of accessibility of PA web sites [10, 41]. Unfortunately, the ability of automated test tools to identify problems is limited because they cannot detect all types of WCAG accessibility errors [3, 42]. A comparison of ten popular website accessibility tests in 2017 found 37% of problems (errors and warnings) at best and 17% at worst [15].

Considering the limits of automatic evaluation of web accessibility (due to their inability to check all aspects of accessibility automatically), our study investigates the current perceived accessibility level of e-Government services in Italy from the point of view of people with disabilities, who can highlight issues and problems not detected by automatic validators. A similar work was performed in 2017 investigating Italian public administration accessibility via an online survey answered by 68 people with disabilities accessing PA services in Italy [12]. Results from the sample highlighted the need to improve service accessibility and usability, and the request for increasing their number and set of functions. Our study thus follows the same methodology of Buzzi et al. [12], investigating the accessibility of e-Government services in Italy in 2023 from the point of view of people with disabilities. Seventy-six users participated in an online survey, and the collected answers have been further refined through semi-structured interviews on a subset of participants. To make the results further comparable, the present study only extends the number of questions of the previous one to collect more data (adding three new questions related to usability and support) but keeps the original ones unchanged. Moreover, have a similar number of participants (68 vs. 76) and both recruited mainly blind and visually impaired (62% in the 2019 study and 85.5% in this study).

To better investigate and understand differences and similarities between the Italian situation in 2017 and 2023, in this section we briefly summarize main results in the 2017 study [12].

Regarding accessibility issues encountered by the 2017 sample, a first point concerns the participants' experience in accessing the online PA services. To better understand collected data, it is important to bear in mind that unskilled people having problems accessing online services often blame the failure on themselves – as unable to act correctly in a digital environment – instead of considering the cause as due to poor usability design.

Most users (88%) declared they have experienced some issues: 68% of participants had problems sometimes and 20% many times, as shown in Fig. 1.

Concerning the ability to successfully completing the service, only 11% of the participants declared to always successfully complete a service while less than one third only often. More than half of the sample (i.e. 51%) instead was successful only sometimes, and two participants never. Three participants never tried to access online services, in fact for users with visual/hearing impairment it is very difficult to access online services, since they are unable to use the screen reader but can still exploit braille devices as assistive technology (Fig. 2).

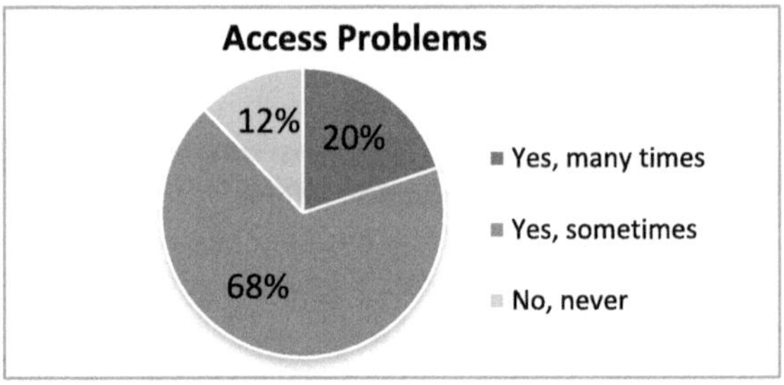

Fig. 1. Problems detected when accessing online PA Services.

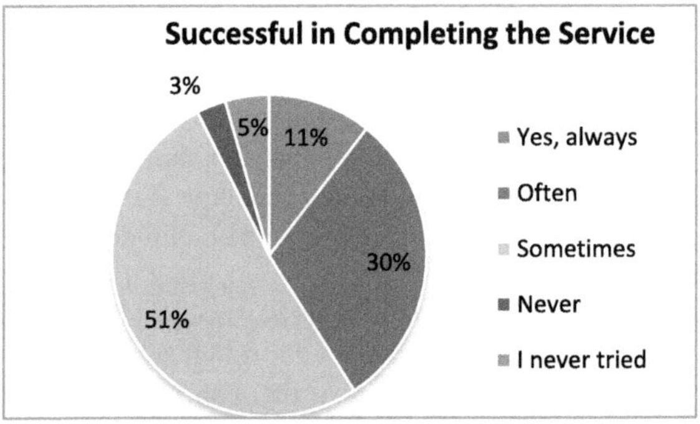

Fig. 2. Completing online PA Services.

The interaction was evaluated as quite simple in most cases by 46% participants, and always simple by 9%, while big difficulties still are detected by 9% (always difficult) and 36% participants found it quite difficult (Fig. 3).

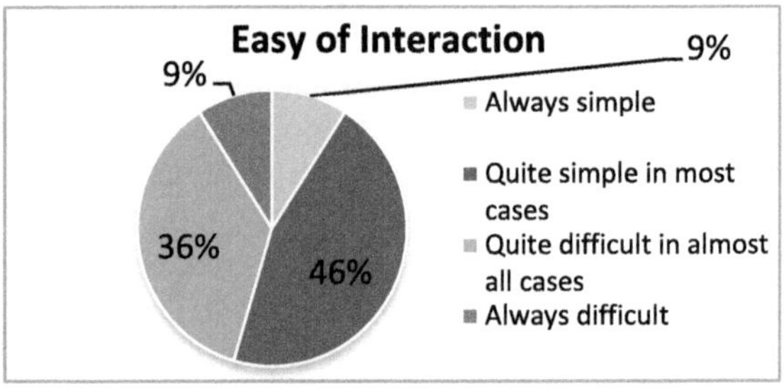

Fig. 3. Interaction in accessing online PA Services.

Despite the very small sample, result indicated that the Italian PAs should improve their services to better guarantee web accessibility and usability for all. A very frequent issue encountered was related to the difficulty of finding the information or service the user was looking for. For this kind of issue, a more usable reorganization and presentation of web content could be beneficial to improving user orientation and usability overall.

After more than 6 years we have decided to investigate if and how the perception of accessibility of PA services by people with disability has changed over time.

3 Related Work

Technology may play a crucial role in promoting the participation and inclusion of people with disabilities in society [18]. Accessibility of websites and services is important for everyone, but it is crucial for people with disabilities, since their interaction may require more time and cognitive effort and may encounter more obstacles [45]. If not designed with accessibility in mind, the Internet may increase the marginalization of people with disability in society [25].

Although accessibility to any type of website is essential to allow full equity for people with disabilities, guaranteeing access to the services provided by Public Administration is especially delicate: if the eGovernment is not accessible, users with disabilities may suffer discrimination [8]. Different kinds of barriers due to competences and skills, geographic position and resources, income, education, and special needs, create the so-called "digital divide" and without urgent countermeasures, in the future this gap could even increase [38]. Several countries have published laws to favor and guarantee equal rights for all to access digital resources, web sites and services. These laws are mostly based on accessibility web standards defined by the World Wide Web Consortium (W3C), the Web Content Accessibility Guidelines (WCAG), which evolved for over 25 years to better incorporate usability into accessibility guidelines [43, 44].

E-Government accessibility is a widely investigated topic [5], due to the importance to ensure equal opportunities for all citizens. Attention to accessibility seems also bound to the transparency of public administrations [4]. Alcaraz-Quiles et al. investigated relations between transparency, accessibility, and usability of Spanish Regional Government websites, showing that the transparency of analyzed websites is inversely related to accessibility since without visibility retrieving information can require considerable time [6].

The assessment of the degree of accessibility of e-Government services has been a subject of investigation all over the world. Galvez and Youngblood [20] analysed more than one hundred national and local e-government website sites in Rhode, using a combination of code inspection, heuristic evaluation, and automated analysis. Results suggest that best-practice templates may be helpful in improving usability, accessibility, and mobile readiness, while it is critical for designers to receive training and for governments to monitor Web sites compliance with standards.

Paul [35] presented the evaluation of the accessibility of Indian e-government websites using a sample of 65 websites of various ministries based on the WCAG 2.1 standard, finding most e-government websites do not meet Level A conformance with WCAG 2.1.

Al-Sakran and Alsudairi [7] investigated how well the Saudi mobile e-government websites comply with usability standards and accessibility guidelines recommended in the WCAG. Websites assessments were conducted using manual evaluation and complemented by different automated analysis tools. The findings revealed usability and accessibility problems that affect the performance of government websites. Several recommendations for improving the usability and accessibility of websites in Saudi Arabia were also suggested.

Agrawal et al. [2] investigated the usability, accessibility, and mobile readiness of Indian government websites. The analysis covered 164 government websites delivering e-services. The evaluation of websites compliance with the WCAG 2.0 was done on various quality parameters using automatic online tools. The results indicate that many websites had low usability: most of the website does not follow WCAG 2.0 guidelines, while none was usable and fully accessible on mobile devices.

Siquiera et al. [40] investigated whether public prosecutors' websites are following web accessibility guidelines. The authors evaluated the websites of each of the 27 states of Brazil using the WCAG 2.0. Results indicated that the websites violated between 16 and 33 different success criteria out of 61 criteria.

Referring to European countries, Kous et al. [26] performed an accessibility evaluation (compliance with Standard EN 301549) of 189 Slovenian municipalities' websites using an automatic tool (Achecker). The results of the statistical analysis showed that the website home pages in 2018 (after the adoption of European Standard EN 301 549) have significantly improved compared to 2017.

Inal et al [23] analyzed the accessibility of 356 municipality websites in Norway using automatic tools. Results show that none of the municipality websites meet an acceptable level of compliance. The websites violate an average of 40 checkpoints, of which many are Level A criteria that represent the minimum level of satisfaction of checkpoints [23].

The situation in Romania in terms of web accessibility has been depicted by two key institutions in the field of administering social services and benefits in Romania – County Directorates of Social Work and Child Protection (DGASPC) as well as County Pensions Houses (CJP). The assessment grid was represented by the Web Content Accessibility Guidelines (WCAG) 2.1 requirements and techniques, mainly regarding Level A, minimum level of conformance. Results underlined significant discrepancies on implementing various accessibility characteristics among similar institutional bodies and the needs to take further steps on improved accessibility [1].

Recently, Lynn et al. [31] examined the web accessibility of local authority websites in the Republic of Ireland specifically referring the Web Accessibility Directive, Results confirmed that most local government websites examined present significant accessibility issues.

Similar studies were performed also in Italy. Barricelli et al. [10] exploited automatic tools for assessing the degree of accessibility of PA web sites in Italy, showing that on more than 8000 websites of the Italian municipalities only 1% fully conformed to the Italian legislation that required PA to implement a subset of the WCAG guidelines. This issue was confirmed by several successive studies. Buzzi et al. [12] created an online survey answered by 68 people with disabilities accessing PA services in Italy. Results from the sample highlighted the need to improve service accessibility and usability, and the request for increasing their number and set of functions. Furthermore, Valtolina and Fratus [41] investigated accessibility issues of local government websites in Italy proposing an evaluation strategy via two validators (AChecker and VaMolà) and analysing 7,713 homepages against WCAG 2.0 and Italian Law recommendations. Results confirmed a low degree of web accessibility of municipal websites.

However, most studies exploited only automated analysis tools (validators). A recent systematic literature review on automated tool utilization in web accessibility research from 2002 to 2021 showed an increasing trend in its use year by year [32]. Considering the limits of automatic evaluation of web accessibility (due to their inability to check all aspects of accessibility automatically) [3, 42], our study investigates the current degree of accessibility of Italian PA services from the point of view of users with disabilities, who can highlight actual problems via an online survey further deepened through semi-structured interviews.

4 The Study

4.1 Methodology

Although recent years have shown an increasing attention at legislative, academic, and social levels, concerning accessibility of public services, many websites of the Italian Public Administration are still not compliant with the minimum level of accessibility requirements [19].

This study investigates the accessibility of digital services provided by Italian PAs for people with disabilities. To this aim, the study applies two methodologies for gathering participants' feedback: an online survey and semi-structured interviews.

The survey was administered via Google Forms, which supports screen readers and keyboard shortcuts offering a good degree of accessibility for screen reading users [28]. The questions focus on interaction aspects such as effectiveness in successfully completing the service, usability of the interface, overall user satisfaction.

To recruit survey participants, the main Italian organizations of people with disabilities (visual, motor, hearing, intellectual) were asked to distribute an invitation letter (with the survey link) to their associates. This assures the authors do not have any direct contact with participants.

On the other hand, the purpose of the semi-structured interviews was to better understand the main accessibility problems identified by the survey. Some participants accepted to participate. The interviews were designed to reveal the participants' personal narratives, leaving them free to express their opinions, considerations, and suggestions that could not emerge in closed questions. Interviews were conducted via a video conferencing tool (Google Meet) or by phone (preferred by visual impaired users) by two of the authors: one conducted the interview and the other one annotated answers, comments, and observations. After each interview, both authors checked and integrated the annotated answers.

4.2 Questions

The questionnaire used for the online survey is reported below. For each question, we report the descriptive text presented to the user and the set of possible single- or multiple-choice answers. In the paper, we will use the numbers listed here to refer to the single questions.

Section 1: Sample Characterization

Q1. Gender
Single choice: Male/Female/I don't want to specify.
Q2. Age
Single choice: 18–29/30–39/40–49/50–59/60–69/$\geq$70.
Q3. Occupation/Job
Single choice: Student/Employee/Professional/Unemployed/Retired.
Q4. Disability
Multiple choice: Visual/Motor/Hearing/Intellectual/Behavioural.
Q5. Internet usage – How often do you use the internet?
Single choice: Almost Never/Sometimes/Weekly/Daily.
Q6. Which online PA services do you access most frequently?
Multiple choice: Health/Municipality/Job/Pension/Tax/Justice/Official Documents/Finance/Economy/School & Education.

Section 2: Interaction with e-Government services

Q7. Access problems – Have you ever had problems accessing the PA's online services?
Single choice: Yes, often/Yes, sometimes/No, never.
Q8. Effectiveness of PA service interaction – Were you able to get what you wanted and to complete the requested services?
Single choice: Yes, always/Often/Sometimes/Never/I've never tried.
Q9. Ease of the interaction
Single choice: Always simple/Simple (in most cases)/Difficult (It was quite difficult in most cases)/Very difficult (No, it was difficult in all cases and sometimes I was unable to complete the service).
Q10. User satisfaction
Single choice: Very unsatisfied/Unsatisfied/Neutral/Satisfied/Very Satisfied.
Q11. Documentation and help – Rate documentation and help (manuals, interface messages, contacts, etc.)
Single choice: Totally dissatisfied/Dissatisfied/Neutral/Satisfied/Totally satisfied.
Q12. Errors – Were you able to manage it and move forward if you made mistakes? (clear messages and interaction mechanisms).
Single choice: I made mistakes and could not correct them/I made mistakes and could hardly correct them/I don't know/I made mistakes, but I could easily correct them.
Q13. User Interface customization – Could the interface be adapted to your needs? (e.g. increase font size, change contrast, increase line spacing).
Single choice: No, in general/Only in some sites/I don't know/Yes, generally/I have not tried.
Q14. Aspects to be improved urgently.
Single choice: None, everything is fine/Response times/Ease of use/Increase the number of online services to avoid face-to-face services/Others.

Section 3: Authentication & citizen rights.

Q15. Do you have SPID credentials?
Single choice: I don't know what they are/I know what SPID is, but I don't own it/I own it, but I don't use it/I own it and I use it.

Q16. Did you know that every citizen has the right to easily access all PA services?
Single choice: Yes/No.
Q17. Did you know that AgID (Agency for digital Italy) is the authority in charge of receiving reports on difficulties in accessing PA services?
Single choice: No, I didn't know/Yes, but I didn't report it/Yes, and I made some reports.
Q18. Each Administration must identify a Digital Public Advocate. Would you like to participate by sending them your personal suggestions?
Single choice: No, I'm not interested/No, I wouldn't be able/Yes, I would like to participate/I don't have time to do it.

On the other hand, the semi-structured interviews consisted in four questions about difficulties, effectiveness, efficiency, user satisfaction and possible improvements:

IQ1. What obstacles have you encountered in accessing PA Web sites?
IQ2. Were you able to do what you wanted, or did you need help?
IQ3. Was the interaction challenging, did it require effort and/or a lot of time?
IQ4. Were you satisfied or what would you like to improve?

4.3 Participants

A total of 76 participants filled out the online survey: 50 males and 26 females. Participants are aged between 18 and more than 70 years: 2 (2.6%) participants in the 18–29 range, 8 (10.5%) in the 30–39 range, 20 (26.3%) in the 40–49 range, 30 (39.5%) aged 50–59, 8 (10.5%) aged 60–69, and 8 (10.5%) participants were 70 or older as shown in Fig. 4.

All participants indicated their gender: 2/3 are males (50 out of 76) and 1/3 are female (25 out of 76).

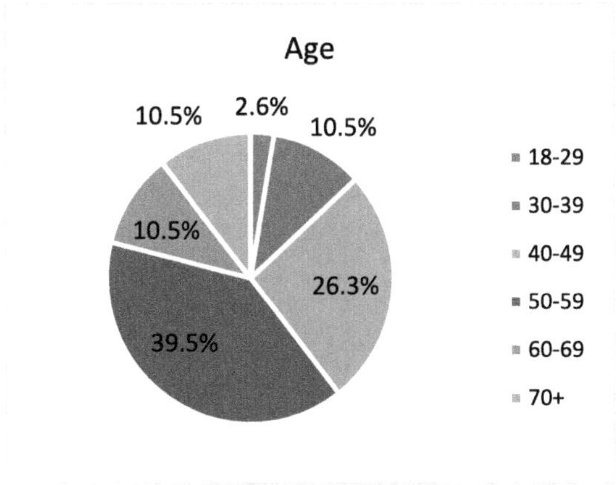

Fig. 4. Sample age.

Regarding the type of disability, most participants are visually impaired (65 out of 76), followed by people with motor disability (8 out of 76) and three people having a mild cognitive impairment. Lastly, two hearing impaired people who communicate through the Italian sign language (signists) and one with a behavioral disability complete the sample. Note that some participants had multiple disabilities.

Among these participants, a total of 8 accepted to take part in the semi-structured interviews: 3 visually impaired (2 blind and one visually impaired that navigates via screen reader), 2 motor impaired, and 3 old persons with mild cognitive and motor impairment.

Clearly, the sample composition suffers from an imbalance in the representation of disability types, with 81.25% of the overall participants being individuals with visual disabilities. However, it should be taken into consideration that several studies have shown that users with visual impairment, and more specifically blind users, encounter greater difficulties than users with other disabilities when interacting with websites. Visually impaired users searching the Web content took an average of 2.5 times longer than sighted users [13]. An efficiency gap was also confirmed by [24]: when blind people executed a search task, they took twice as long as sighted users to explore search results and three times as long to explore the corresponding web pages. Petrie et al. in 2004 performed an accessibility test of one hundred websites with users having visual, motor and perceptual disabilities. Results showed a mean task success rate of 76% which falls to 53% considering only the totally blind (the lowest score of all categories). Besides, regarding user satisfaction, blind users confirmed the lowest score of all the user categories [36]. Overall, blind users interacting on the Web lose, on average, 30.4% of time due to frustrating situations, such as page layout causing confusing screen reader feedback, conflict between screen reader and application, poorly designed/unlabeled forms, no alternative text for pictures, misleading links, inaccessible PDFs, and screen reader crashes [27]. For these reasons, we believe this study's sample including a high percentage of visually impaired users, who experience the greatest difficulties in web navigation, can offer interesting insights on the accessibility of PA services.

The distribution of participants by their occupation/job (Fig. 5) depicts that more than half of participants (~54%) are employees of a PA (Italian law requires PAs to recruit a percentage of people with disabilities). Twenty-three participants (30.3%) are retired, two participants are students (2.6%), five participants (6.6%) are practitioners and five unemployed people complete the sample.

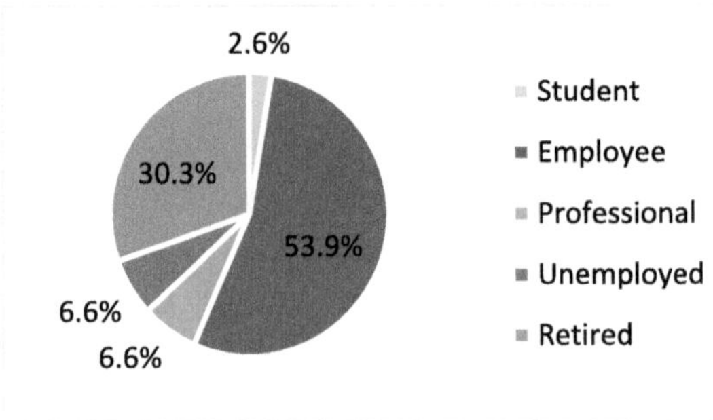

Fig. 5. Participant occupation/job.

Finally, concerning the frequency of Internet use, most (68 out of 76) participants use the Internet daily, three weekly and two sometimes (Fig. 6). Three participants declared they use the Internet rarely. Two of them are 70+ aged and one 50+.

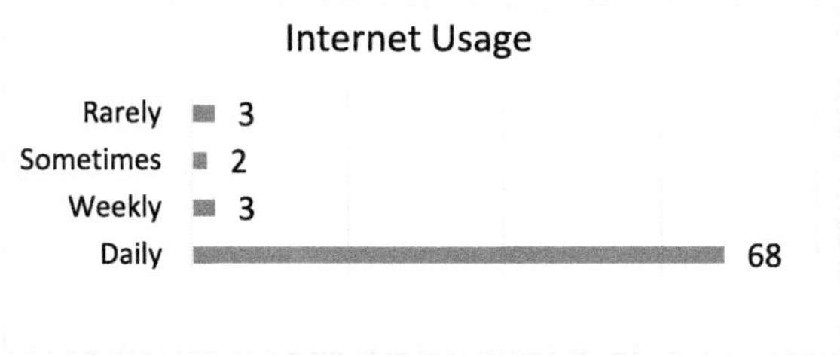

Fig. 6. Sample characterization: Participants' Internet use (Q5).

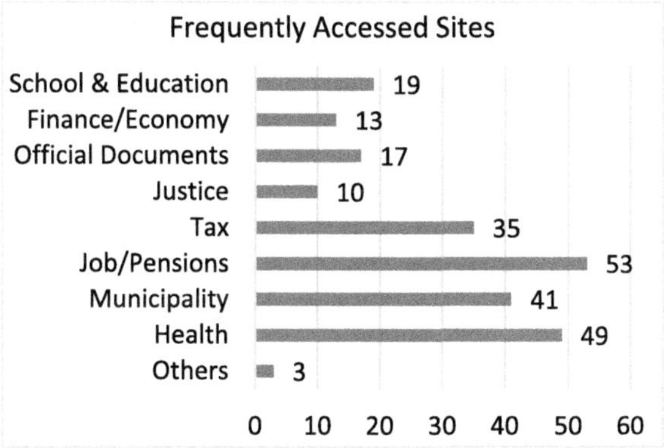

Fig. 7. Frequently accessed PA services (Q6).

Participants interact with several services of both local and central public administrations (Fig. 7). Most frequently accessed services are job or pension, Health, Local Administrations, and taxation.

4.4 Online Survey Results

Results are presented by analyzing the answers to sections 2 and 3 of the questionnaire.

Accessibility and Usability. Section 2 contains eight questions related to participants' usability experience when interacting with e-Government services. Answers to question Q7: "Have you had problems accessing the PA's online services?" offer indications on the access issues. It is remarkable that 89% of the sample reported access problems ('Sometimes' for 46 out of 76 users, and 'Often' for 22). Only 11% of the users (8 out of 76) never had an access problem (Fig. 8).

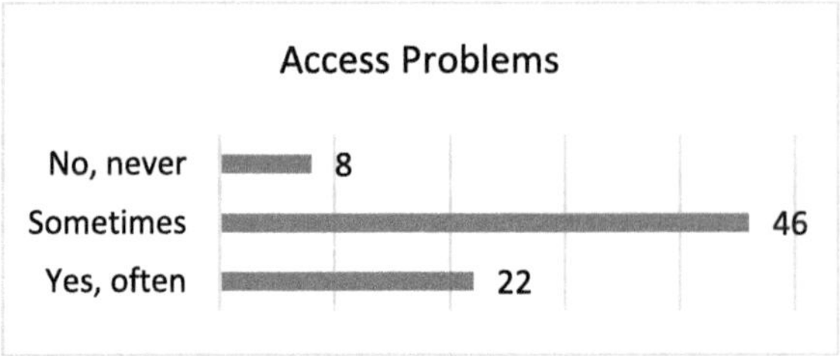

Fig. 8. Frequency of problems reported/encountered by participants when accessing e-Government services (Q7).

Regarding the effectiveness of interaction with PA services (Q8), more than half of the participants reported they were able to complete the requested services always (6 users, 8%) or often (37 users, 49.3%), while 27 users (36%) only sometimes and 3 never, as shown in Fig. 9.

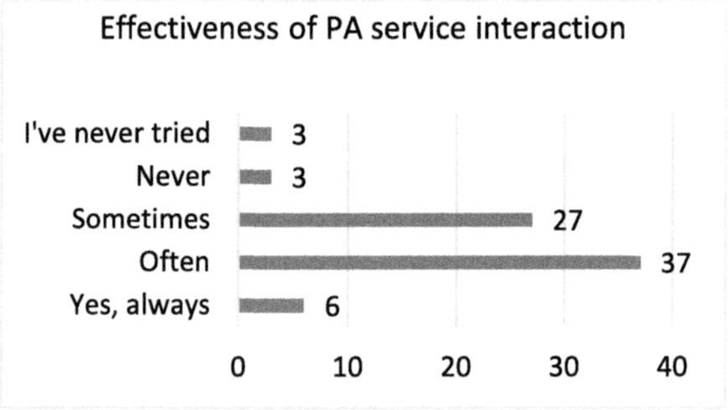

Fig. 9. Effectiveness of PA service interaction (Q8).

Considering how frequently the interaction with e-Government services was easy (Q9), most participant were positive: 4 users (5.3%) found it always simple and 37 users (48.7%) found it simple in most cases, but it is also relevant the percentage of participants that found it difficult in most cases (32.9%, 25 users) or very difficult in all cases (13.2%, 10 users), as shown in Fig. 10.

User satisfaction (Q10) reported a consistent number of participants (the majority) with a neutral position (27 users, 35.5%), as shown in Fig. 11. Positive rating is 30.3% (satisfied, 23 users) and 7.9% (very satisfied, 6 users), while negative rating is 15.8% unsatisfied (12 users) and 10.5% very unsatisfied (8 users).

For evaluating the PA website documentation and support (documentation and manuals, pop-up messages, etc.) a 5-item Likert scale (1 = *Totally dissatisfied*, to 5 = *Truly satisfied*) has been administered (Q11). Most users, 31 out 76, (i.e. 40.8%) chose the neutral option (*neither satisfied nor dissatisfied*), while the percentage of satisfied (22.4% *satisfied* and 7.9 *very satisfied*) is only slightly higher than the one of dissatisfied (19.7% *somewhat dissatisfied* and 9.2% *totally dissatisfied*), as shown in Fig. 12.

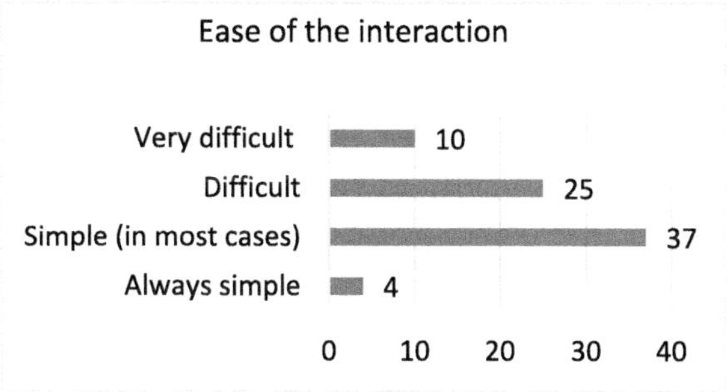

Fig. 10. Ease of the Interaction with PA services (Q9).

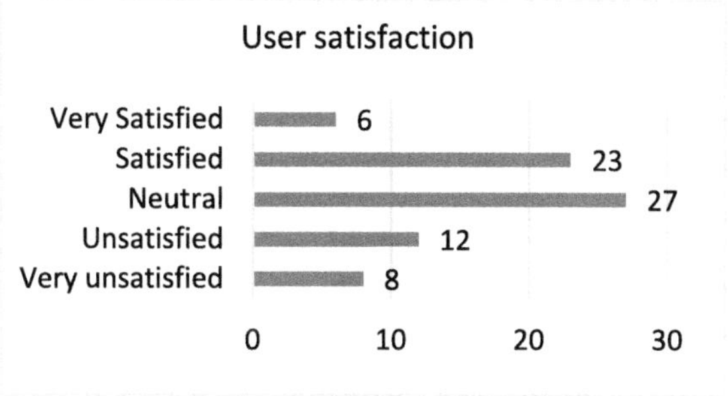

Fig. 11. User satisfaction about the Interaction with PA service (Q10).

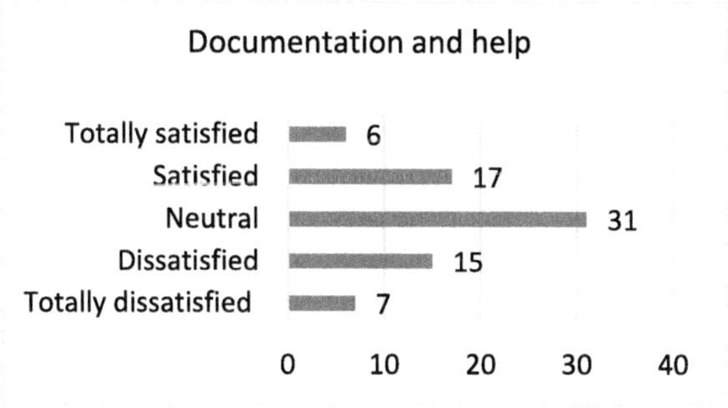

Fig. 12. User satisfaction about documentation and help (Q11).

Another question (Q12) was about errors: if you made mistakes, were you able to fix them easily (and were the error messages clear and helpful)?

This question aims to understand if the recovery of errors is provided through clear messages and easy interaction mechanisms. Most users (40%, 30 users) reported that in case of errors they could hardly correct them, reaching half of the sample if we also consider the ones who said they could not correct errors (12%, 9 users). The neutral option (I don't know, 17.3%, 13 users), "I made mistakes, but I could easily correct them"

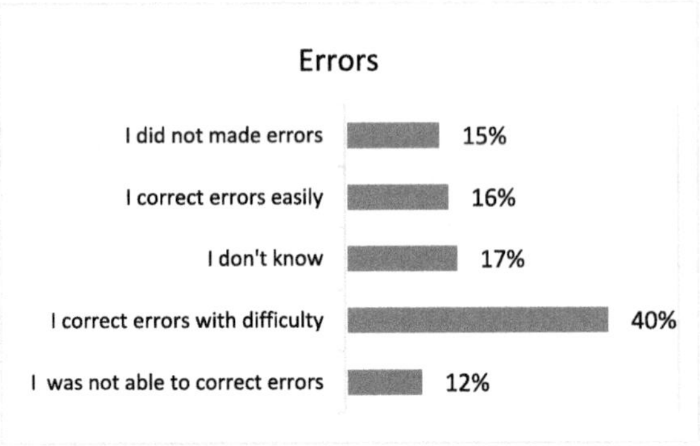

Fig. 13. Errors visibility and easy of recovery (Q12).

(16%, 12 users) and "I don't made mistakes" (14.7%, 11 users) complete the answers (Fig. 13). Visibility of the error cause is a very important principle, and for people with disabilities this can imply to be perceptible with different sensorial channels.

Concerning the ability to customize or adapt the interface to the user's needs (Q13), one-fifth of the sample (20.3%) was *unable* to adapt the interface to their needs, 12.2% were *able* in general, while 31.1% only *sometimes*. One-fifth of the participants (20.3%) did not try to modify the interface (Fig. 14).

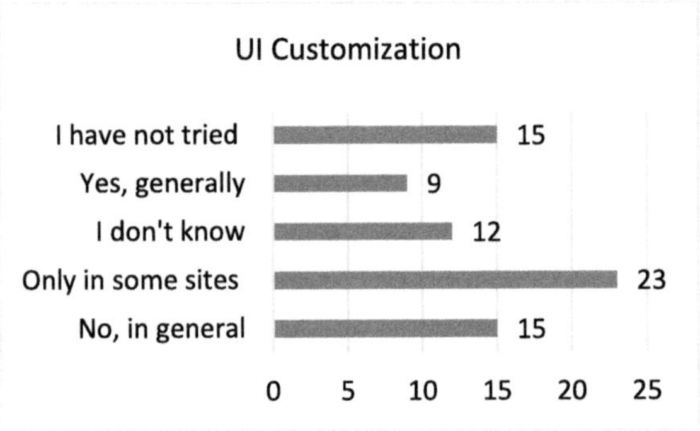

Fig. 14. Usability: interface customization (Q13).

Finally, concerning the aspects to be improved urgently (Q14), most of the users (45 out 76, i.e. 59.2%) suggested to make the UIs easier to use and simplify the interaction, while others suggested to add new online services (25%) and shortening the service response time (5.3%) as shown in Fig. 15. Some skilled users also suggested more specific improvements, e.g., better structuring of the content, use of logical templates to make it easier to find what you are looking for, addition of instant chat for immediate assistance, subtitling and possibly translation in the sign language of videos.

Authentication and Citizen Rights. Section 3 of the questionnaire contains four questions concerning authentication and citizen rights. The first question (Q15) is

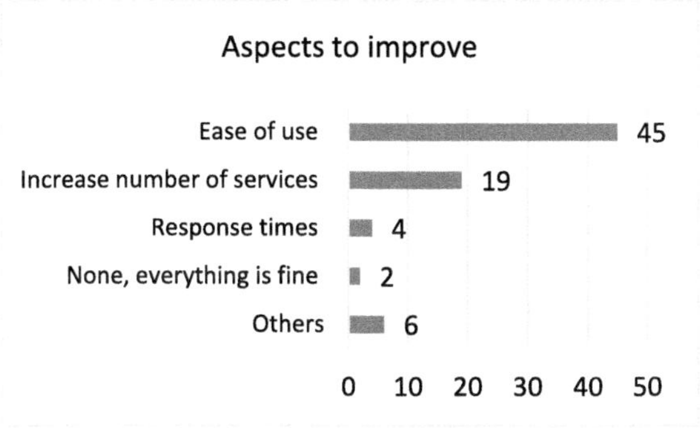

Fig. 15. Aspect to Improve (Q14).

about authentication via SPID (Public Digital Identity System). Overall, 63 participants (82.9%) use SPID for accessing PA services, 4 users (5.3%) have the authentication credentials but do not use them, 6 users (7.9%) know about the credentials but never requested them, and only 3 users (3.9%) still do not know anything about the SPID credentials, as shown in Fig. 16.

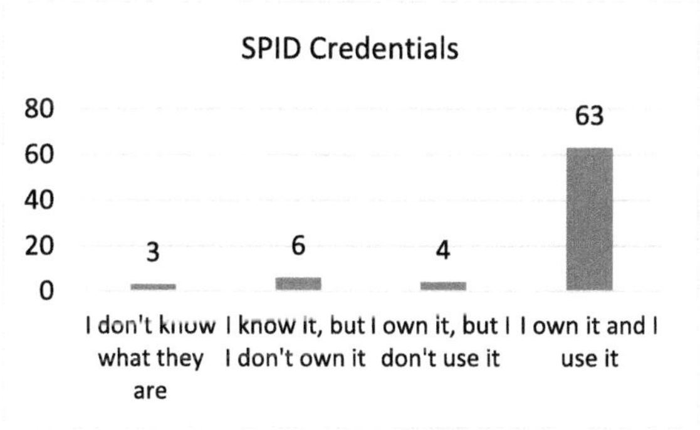

Fig. 16. Authentication: SPID credentials (Q15).

The other three questions are about citizen rights. Almost all participants (72, 94.7%) know their right to easily access all PA services (Q16), as shown in Fig. 17.

On the other hand, question Q17 investigates the awareness of users on the possibility of reporting accessibility problems to AgID (Agency for digital Italy), the authority in charge of monitoring and fueling the digitalization of the Italian PAs. Most of the participants (48.7%) are aware of this possibility but did not report any accessibility issues while 42.1% was not aware and only 9.2% sent some reports (Fig. 18).

Finally, question Q18 was about the knowledge of the role of the Public Advocate (*Ombudsman*) for the Digital Rights and the possibility to provide feedback for contributing to improve PA services. As shown in Fig. 19, most participants (47 out of 76,

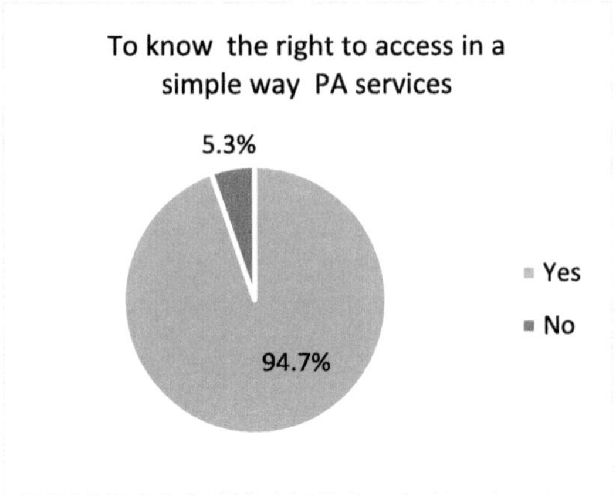

Fig. 17. Knowledge of citizen rights (Q16).

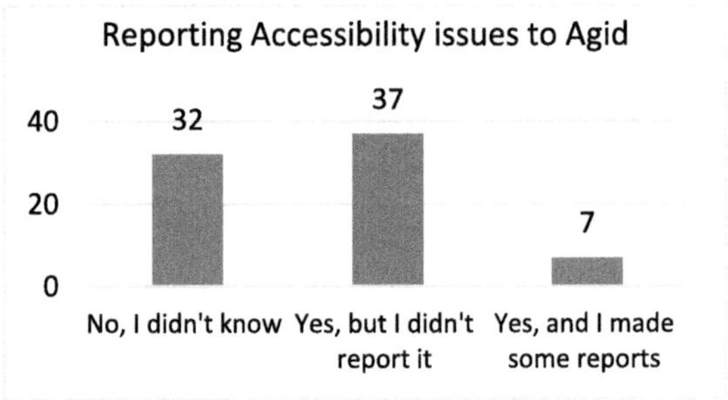

Fig. 18. Reporting Accessibility feedback to AgID (Q17).

61.8%) are interested in contributing, while 19 users (25%) do not consider themselves able to do it, one has no time, and 9 are not interested (11.8%).

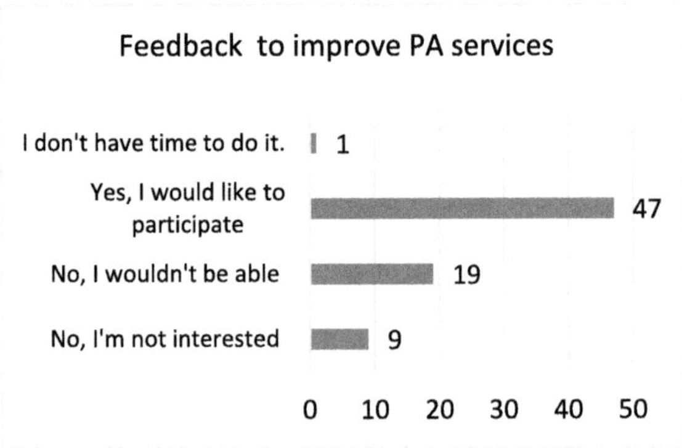

Fig. 19. Reporting Accessibility feedback to the Public Advocate (Q18).

4.5 Semi-structured Interviews Results

Table 1 summarizes the key findings from the participants' answers to the four interview questions (IQ1 – IQ4).

Table 1. Key findings from the interviews.

Issues	Affected Tasks
Lack of clarity in the contents	*Exploring, searching*
Complex structure of the websites	*Exploring, searching*
Audio of the commercial advertisements overlaps the reading of the screen reader	*Listening the vocal synthesizer*
Documents created as an image (unreadable and unstructured)	*Reading a document*
Request to change the password frequently	*Access to website areas with authentication*
Tight and insufficient time for strong authentication	*Access to website areas with authentication*

A common obstacle encountered in accessing PA websites among all the users interviewed is the difficulty in finding what they are looking for, due to the lack of clarity in the contents and the complex structure of the websites. In particular, a totally blind user, who uses PA services daily, reported two major problems: the presence of advertisements on the sites that interfere with blind users' listening to the content, and the difficulty in using downloaded documents, as they are not readable by the screen reader being not in text format (e.g., PDF documents containing scans of paper documents).

Most notably, many of the interviewed users reported difficulties in accessing PA sites due to the SPID authentication or the need to periodically change their passwords, as required by the current national security guidelines. In some cases, participants were even unable to complete the service. Such users often require the assistance of family members or friends to accomplish these tasks.

An experienced blind user declared that the SPID authentication was accessible for him only through a specific provider. Indeed, such provider grants the user with 1 min and 30 s to enter and confirm (SPID uses a two-factor authentication) the user credentials, as opposed to the 30 s granted by other providers.

5 Discussion

5.1 Current Results

The survey and interviews results show that, although there is a high awareness about PA digital services and users with disabilities are usually able to benefit from them, indicators of user satisfaction, effectiveness and ease of use are generally still very low.

Indeed, while the increasing number of services offered through the PA websites is generally a positive aspect for all users, often this implies complex user interfaces that, if not suitably structured, are more difficult to understand and navigate via screen readers.

Moreover, the increased security and authentication constraints (as the multi-factor authentication mechanisms) introduced in the last years represent new accessibility barriers.

Clearly, both the current authentication processes and the user interfaces need further design simplification efforts to make users with disabilities able to successfully access the PA services, asking for urgent improvement of usability such as ease of use of the user interfaces that should be logically organized for easier navigation, offering more help and a simplified interaction (also for authentication purposes).

5.2 Comparison with Previous Study

Having determined that the PA digital services still suffer from severe accessibility issues in 2023, it is interesting to investigate if and how much such indicators changed in the last years.

The previous study contains data collected in 2017, so in the following, we write "2017" to indicate the results of [12] and "2023" for the results of the current study. Figure 20 compares the disabilities of the survey participants.

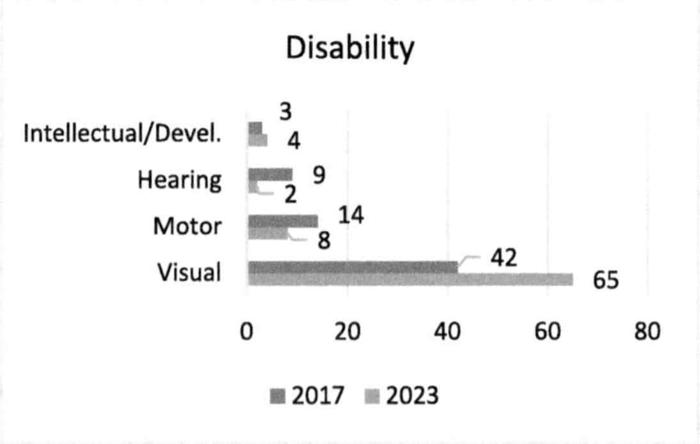

Fig. 20. Type of disability of participants: 2017 vs 2023 (bottom).

A main aspect concerns the participants' experience in accessing PA websites. Most users (89.4%) are still experiencing issues: 60.5% (vs 67.2% of 2017) of participants had problems sometimes and 28.9% (vs 19.7% of 2017) often, as shown in Fig. 21.

The situation seems to have become worse: the number of people who never experienced problems has decreased by 2.6% (10.5% in 2023 vs 13.1% in 2017) and the total number of people experiencing some issues has increased (89.4% of the sample in 2023 vs 86.9% in 2017). This could be caused by the new requirement of strong authentication for accessing the PA services, introduced by the European eIDAS Regulation on electronic Identification, Authentication and trusts services.[1] Indeed, in Italy, access to PA

[1] https://digital-strategy.ec.europa.eu/en/policies/eidas-regulation).

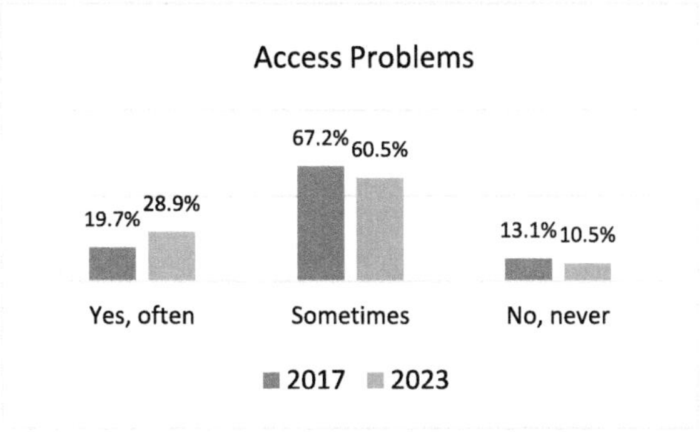

Fig. 21. Frequency of problems encountered when accessing online PA services: 2017 vs. 2023.

services exploits digital identification through an Identity Service Provider (SPID) or via an Electronic Identity Card (CIE). As discovered in the interviews, strict temporal limits can obstacle the access by blind users interacting with such authentication mechanisms via a screen reader. Also regarding effectiveness, i.e. the ability to complete a service, the situation seems slightly worse than in 2017: only 8% of the participants (in 2023), compared to 10.6% (in 2017), has always achieved their goal, while the majority, i.e., 49.3% (in 2023), compared 51.5% (in 2017), achieved it only often and 36% (in 2023) vs 30.3% (in 2017) only sometimes (Fig. 22).

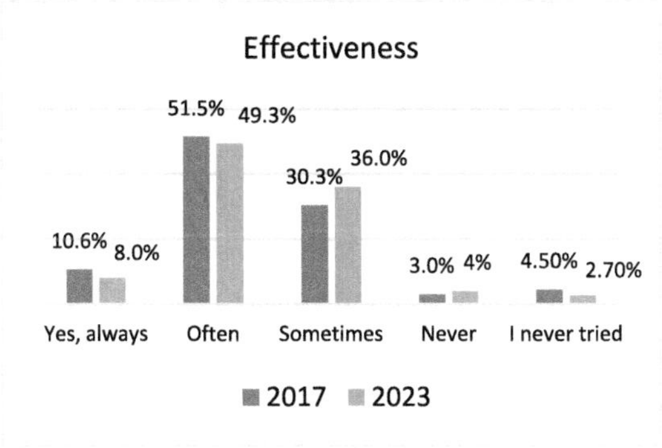

Fig. 22. Effectiveness of online PA Services: 2017 vs. 2023.

The total number of participants experiencing effectiveness issues increased to 89.3% in 2023 vs 84.8% in 2017. However, we must consider that the small samples do not allow a generalization of such results.

A slight worsening was highlighted also regarding the ease of use in the interaction with online PA websites (Fig. 23).

Participants who evaluated the interaction as simple decreased from 54.9% in 2017 (9.7% always simple + 45.2% simple) to 54.2% in 2023 (5.5% always simple + 48.7% simple). Furthermore, 46.1% of 2023 participants found the interaction quite difficult

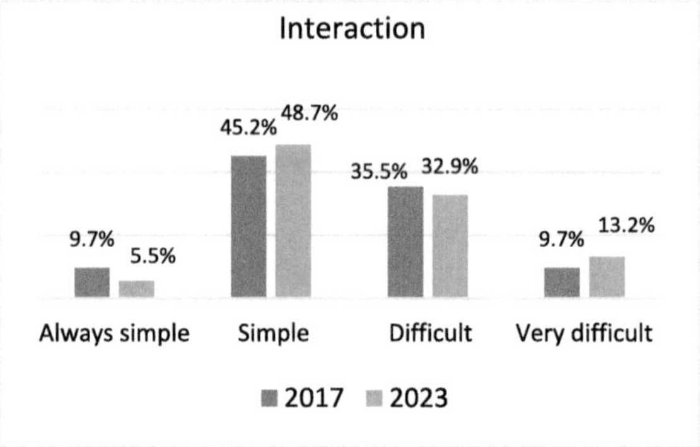

Fig. 23. Ease of interaction in accessing online PA Services: 2017 vs. 2023.

(32.9% difficult + 13.2% very difficult) compared to 45.2% in 2017 (35.5% difficult + 9.7% very difficult) of 2017.

Concerning the user satisfaction, Fig. 24 shows that most participants are still neutral (i.e., not completely satisfied), 30.3% (vs 26.5% in 2017) were satisfied and only 7.9% (vs 6.3% in 2017) very satisfied.

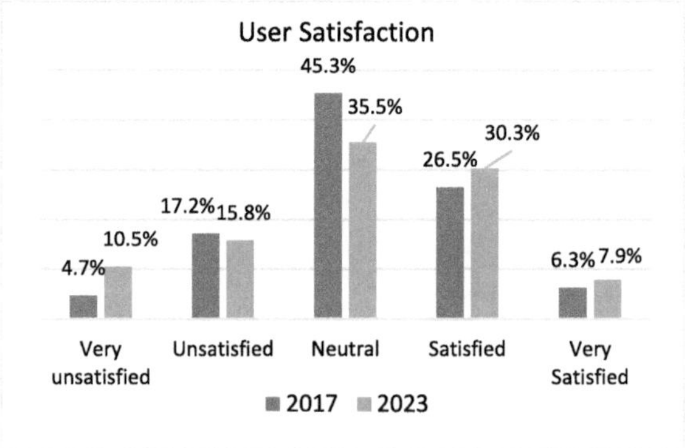

Fig. 24. User satisfaction in accessing PA website: 2017 vs. 2023.

However, since the total percentage of (vary) unsatisfied participants increased to 26.3% in 2023 vs 21.9% in 2017, the overall improvement is very little. Very often, inexperienced people having problems accessing online services believe that it is their fault, being unable to correctly interact with the interface, while the cause could be a poor usability design.

Overall, the collected users' feedback concerning the accessibility of PA websites, although on a small sample, seems to indicate a worsening, even if only slightly, compared to those of 2017, suggesting that, despite the regulation's intent, the problems seem not under resolution.

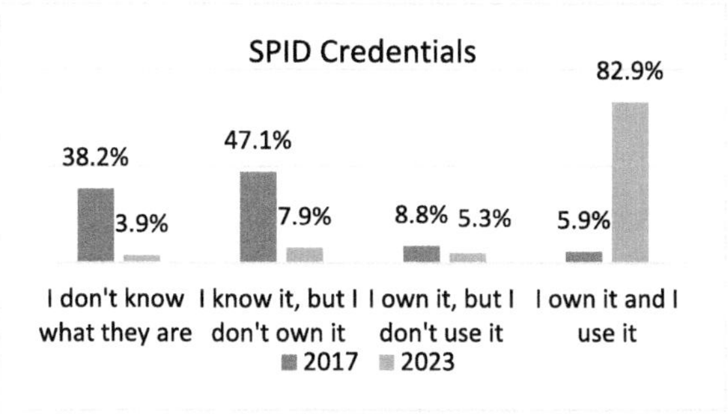

Fig. 25. SPID Credentials: 2017 vs 2023.

Concerning the Authentication & citizen rights session, with respect to the result of the 2017 questionnaire, the situation improved: 82.9% of participants use SPID for accessing PA services (Fig. 25) with respect to only 5.9% in 2017, 5.3% have the authentication credentials but do not use them vs. 8.8% in 2017, 7.9% know about the credentials are but never requested them (vs 47.1 in 2017), and only 3.9% still do not know anything about the SPID credentials (vs 37.2 in 2017).

Analogously, most of participants (94.7% vs 89.7% in 2017) know their right to easily access all PA services (Fig. 26).

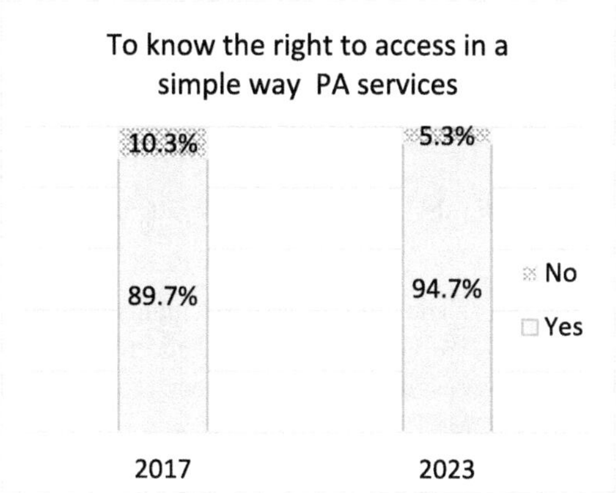

Fig. 26. Right to access PA services in a simple way: 2017 vs 2023.

Concerning reports on accessibility issues of PA services to AGID (Fig. 27), awareness increased a lot. Most of the participants (48.7%) are aware of this possibility but did not report any accessibility issues (vs 29.4% in 2017) while 42.1% were not aware in comparison to 61.8% in 2017. However, few users exploit this possibility: only 9.2% in 2023 and 8.8% in 2017.

Last, most of the participants would contribute by sending feedback to the Public Advocate (Fig. 28): 61.8% vs 47.1% in 2017, while 25% do not consider themselves able

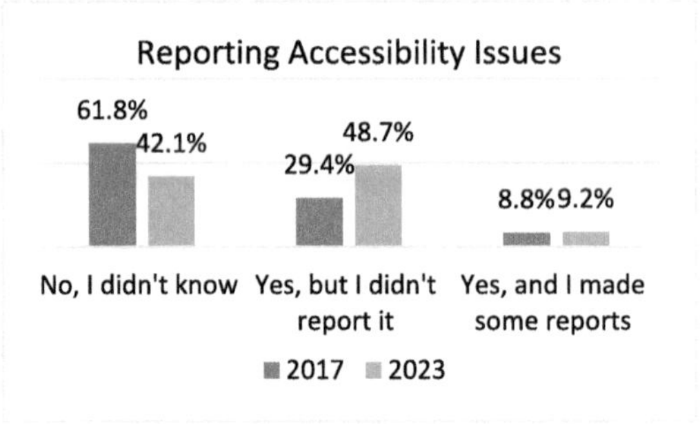

Fig. 27. Reporting accessibility issues: 2017 vs 2023.

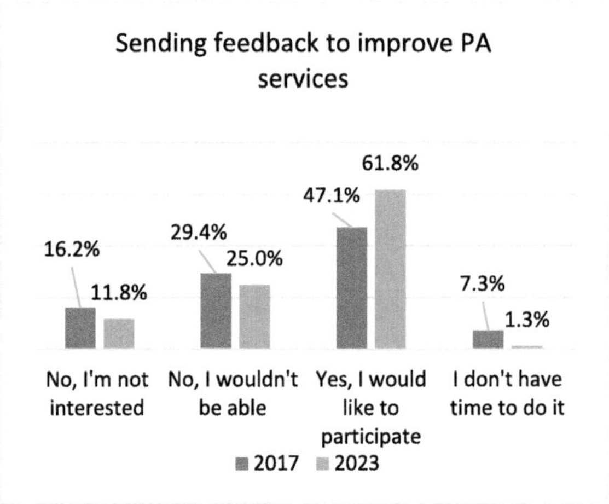

Fig. 28. Feedback: 2017 vs 2023.

to do it, respect 29.4% in 2017. The number of participants not interested or not having time slightly decreased to 11.8% vs 16.2% in 2017 the first, and to 1.3% vs 7.3% in 2017 the second one. This suggest that overall awareness and availability to participate are slightly increasing over time.

5.3 Comparison with Official Studies

To further validate the issues reported by the survey participants, and at the same time to check if the PA itself is aware of them, we compared our results with the Accessibility Monitoring reports published by the Agency for Digital Italy,[2] which is indeed responsible for implementing the project "Inclusion of Citizens: Improvement of the Accessibility of Public Digital Services", part of the National Recovery and Resilience Plan (PNRR). Such project implements a completely automated evaluation process through

[2] https://accessibilita.agid.gov.it/.

the MAUVE++[3]tool [34], to perform a "detailed evaluation of the accessibility of public administration services". The work of Iannuzzi et al. [22] describes the Mauve++ extension to allow Large Scale Automatic Web Accessibility Validation, thus providing the largest and most updated view of the accessibility situation of public websites in Italy. Results of the analysis of 2.7 million web pages highlighted that there are 14 success criteria with more than 1 million violations detected by their validation tool. Furthermore, the Accessibility Percentage metric, which provides info about the number of distinct elements successfully evaluated out of the total number of techniques for which the tool was able to make a successful or unsuccessful evaluation, reported results among 71% and 81%, highlighting a considerable margin of improvement.

As discussed in the introduction, we feel that such kind of automatic analysis cannot capture many usability problems as well as several complex accessibility issues. However, it may be a useful complement to the data reported in this paper.

Indeed, the analysis results show very high failure percentages for WGAG criteria 2.4.7 (*Visible Focus*) and 1.4.1 (*Use of Color*), 40.7% and 25.5% respectively, which may negatively affect visually impaired users, together with *Minimum Contrast* (criterion 1.4.3, 13.7%) and *Non-text Content* (1.1.1, 5.2%). Overall, 50.46% of the detected problems were about *perception* (WCAG point 1, strictly related to accessibility problems, especially for visually impaired users) and 43.67% about *usability* (WCAG point 2). Moreover, the tool also detected high (above 20%) error rates in published PDF files with respect to WCAG criteria such as "*Page Titled*" or "*Info and Relationships*", possibly connected with the "scanned PDF" problem raised by the participants to our survey.

Therefore, such results clearly support the main findings of both our surveys and interviews.

6 Conclusions

Recent years have shown an increasing attention to legislative, academic, and social levels, for accessibility of public services. However, many websites of the Public Administration are still not compliant with the minimum level of accessibility requirements [19].

In this study, the accessibility of PA services in Italy has been investigated. Although some improvements have been implemented, new difficulties have also been introduced such as the mandatory usage of digital identity for accessing services of the PAs and the time constraint in some implementation of the two-factor authentication mechanisms. These results are indeed coherent with the official studies performed by the Italian Government.

We could represent accessibility as a virtuous cycle including Steps and Tools, as showed in Fig. 29.

The Accessibility virtuous cycle includes four main steps:

Monitoring & Evaluation
- Strategic Planning (accessibility Target)

3 https://mauve.isti.cnr.it/.

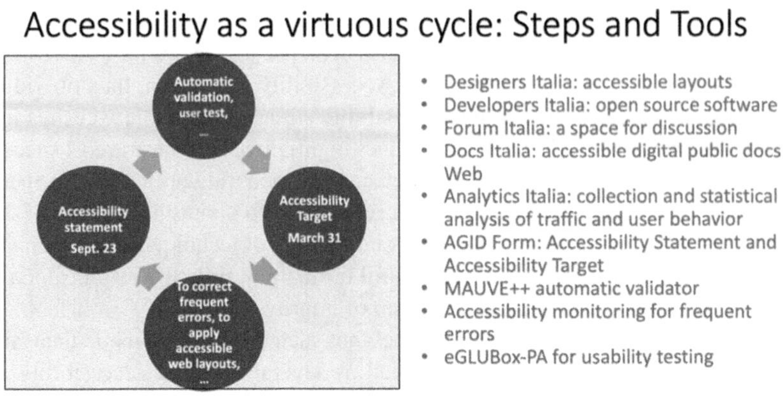

Fig. 29. Accessibility Improvements as a cyclic process.

– Implementing Accessibility Steps
– Assessing & Define (Accessibility Statement)

The path for reaching full accessibility is long, and requires times and resources, due to the peculiarity of administrations and the complexity of the bureaucracy apparatus. This process drawn by the European legislation is incremental, guarantees to improve accessibility over time, according to the resources of each Public Administration.

In Italy, Agid with the Department for the Digital Transformation, and in cooperation with the Designers and Developers Community of Italy, has provided a set of tools and resources to effectively support Public Administration in the Accessibility path, as shown in Fig. 29, and this will hopefully lead to further improvements in the coming years.

Each little accessibility enhancement is a small but important step toward the safeguard of human rights.

Acknowledgments. We thank all the study participants for their valuable contribution. We thank reviewers for their precious feedback and suggestions.

Disclosure of Interests. The authors have no competing interests to declare that are relevant to the content of this article.

References

1. Marin, A.M., Popa, L.: Web Accessibility in the Field of Social Services and Benefits in Romania. Calitatea Vietii, **35**(3), 171–195 (2024)
2. Agrawal, G., Kumar, D., Singh, M.: Assessing the usability, accessibility, and mobile readiness of e-government websites: a case study in India. Univ. Access Inf. Soc. **21**, 737–748 (2022). https://doi.org/10.1007/s10209-021-00800-8
3. Akram, M., Ali, G.A., Sulaiman, A., et al.: Accessibility evaluation of Arabic University websites for compliance with success criteria of WCAG 1.0 and WCAG 2.0. Univ. Access Inf. Soc. (2022). https://doi.org/10.1007/s10209-022-00921-8

4. Alam, T., Aftab, M., Abbas, Z., Ugli, K.M.M., Bokhari, S.A.A.: Impact of E-government initiatives to combat corruption mediating by behavioral intention: a quantitative analysis from emerging economies. Sustainability **15**, 2694 (2023). https://doi.org/10.3390/su15032694
5. Alajarmeh, N.: Evaluating the accessibility of public health websites: an exploratory cross-country study. Univ. Access Inf. Soc. **21**, 771–789 (2022). https://doi.org/10.1007/s10209-020-00788-7
6. Alcaraz-Quiles, F.J., Urquia-Grande, E., Muñoz-Colomina, C.I., Rautiainen, A.: E-Government Implementation: Transparency, Accessibility and Usability of Government Websites. In International E-Government Development, pp. 291–306. Palgrave Macmillan, Cham (2018)
7. Al-Sakran, H.O., Alsudairi, M.A.: Usability and accessibility assessment of Saudi Arabia mobile E-government websites. IEEE Access **9** (2021). https://doi.org/10.1109/ACCESS.2021.3068917
8. Julio, A.: Web accessibility and beyond in eGovernment: does web accessibility ensure accessibility to administration's websites? In: Proceedings of the 19th International Web for All Conference (W4A 2022), Article 21, pp. 1–3. Association for Computing Machinery, New York (2022). https://doi.org/10.1145/3493612.3520443
9. Babu, R., Singh, R., Ganesh, J.: Understanding blind users' Web accessibility and usability problems. AIS Trans. Hum.-Comput. Interact. **2**(3), 73–94 (2010)
10. Barricelli, B.R., Sciarelli, P., Valtolina, S., Rizzi, A.: Web accessibility legislation in Italy: a survey 10 years after the Stanca Act. Univ. Access Inf. Soc. **17**, 211–222 (2018)
11. Buzzi, M.C., Buzzi, M., Della Penna, G., Leporini, B., Ricci, F.: Accessibility of e-Government Websites in Italy: The User Experience of People with Disabilities (2023)
12. Buzzi, M.C., Buzzi, M., Ragni, F.: Accessibility of Italian E-government services: the perspective of users with disabilities. In: Electronic Governance and Open Society: Challenges in Eurasia, EGOSE 2018, Communications in Computer and Information Science, vol. 947. Springer (2019). https://doi.org/10.1007/978-3-030-13283-5_21
13. Craven, J., Brophy, P.: Non-visual access to the digital library: the use of digital library interfaces by blind and visually impaired people. Technical report, Manchester: Centre for Research in Library and Information Management (CERLIM) (2003). http://www.cerlim.ac.uk/pubs/index.php
14. DL22/2005. Normattiva. DecretoLegge 28 febbraio 2005, n. 22, sez. II Guida dei diritti di cittadinanza digitale. https://www.normattiva.it/uri-res/N2Ls?urn:nir:stato:decreto.legge:2005-02-28;22
15. Duran, M.: What we found when we tested tools on the world's least-accessible webpage (2017). https://accessibility.blog.gov.uk/2017/02/24/what-we-found-when-we-tested-tools-on-the-worlds-least-accessible-webpage/. Accessed 9 Apr 2024
16. European Union. European Commission. Union of equality: Strategy for the rights of persons with disabilities 2021–2030 (2021). https://ec.europa.eu/social/main.jsp?catId=1484
17. European Union. European Digital Rights and Principles (2024). https://digital-strategy.ec.europa.eu/en/policies/digital-principles. Accessed Apr 2024
18. Ferri, D., Favalli, S.: Web accessibility for people with disabilities in the European Union: Paving the road to social inclusion. Societies **8**(2), 40 (2018)
19. Gaggi, O., Perinello, L.: Improving accessibility of web accessibility rules. In: Proceedings of the 2022 ACM Conference on Information Technology for Social Good, pp. 167–174 (2022)
20. Galvez, R.A., Youngblood, N.E.: E-Government in Rhode Island: what effects do templates have on usability, accessibility, and mobile readiness? UAIS **15**(2), 281–296 (2016)
21. Huber, W., Vitouch, P.: Usability and accessibility on the internet: effects of accessible web design on usability. In: Computers Helping People with Special Needs: 11th International

Conference, ICCHP 2008, Linz, Austria, 9–11 July 2008. Proceedings 11, pp. 482–489. Springer, Heidelberg (2008)
22. Iannuzzi, N., Manca, M., Paternò, F., Santoro, C.: Large scale automatic web accessibility validation. In: Proceedings of the 2023 ACM Conference on Information Technology for Social Good, pp. 307–314 (2023)
23. Inal, Y., Mishra, D., Torkildsby, A.B.: An analysis of web content accessibility of municipality websites for people with disabilities in Norway: Web accessibility of Norwegian municipality websites. In: Nordic Human-Computer Interaction Conference, pp. 1–12 (2022)
24. Ivory, M.Y., Yu, S., Gronemyer, K.: Search result exploration: a preliminary study of blind and sighted users' decision making and performance. In: Extended Abstracts of CHI 2004, pp. 453–1456 (2023)
25. Jaeger, P.T.: Disability and the Internet. In: Disability and the Internet. Lynne Rienner Publishers (2022)
26. Kous, K., Kuhar, S., Pavlinek, M., Heričko, M., Pušnik, M.: Web accessibility investigation of Slovenian municipalities' websites before and after the adoption of European Standard EN 301 549. Univ. Access Inf. Soc. **20**, 595–615 (2021)
27. Lazar, J., Allen, A., Kleinman, J., Malarkey, K.: What frustrates screen reader users on the web: a study of 100 blind users. Int. J. Hum.-Comput. Interact. **22**(3), 247–269 (2007). https://doi.org/10.1080/10447310709336964
28. Leporini, B., Buzzi, M.: Home automation for an independent living: investigating the needs of visually impaired people. In: Proceedings of the 15th International Web for All Conference, pp. 1–9 (2018)
29. Leporini, B., Paternò, F.: Increasing usability when interacting through screen readers. Univ. Access Inf. Soc. **3**, 57–70 (2004)
30. Leporini, B., Paternò, F.: Applying web usability criteria for vision-impaired users: does it really improve task performance? Int. J. Hum.-Comput. Interact. **24**(1), 17–47 (2008)
31. Lynn, T., et al.: Web accessibility of Irish local government websites. In: ICDS 2023 (Seventeenth International Conference on Digital Society) (2023)
32. Macakoglu, S.S., Peker, S.: Web accessibility performance analysis using web content accessibility guidelines and automated tools: a systematic literature review. In: 2022 International Congress on Human-Computer Interaction, Optimization and Robotic Applications (HORA), Ankara, Turkey, pp. 1–8 (2022). https://doi.org/10.1109/HORA55278.2022.9799981
33. Moreno, L., Martínez, P., Ruiz-Mezcua, B.: A bridge to web accessibility from the usability heuristics. In: HCI and Usability for e-Inclusion: 5th Symposium of the Workgroup Human-Computer Interaction and Usability Engineering of the Austrian Computer Society, USAB 2009, Linz, Austria, 9–10 November 2009 Proceedings 5, pp. 290–300. Springer, Heidelberg (2009)
34. Parvin, P., Palumbo, V., Manca, M., Paternò, F.: The transparency of automatic accessibility evaluation tools. In: Proceedings of the 18th International Web for All Conference, pp. 1–5 (2021)
35. Paul, S.: Accessibility analysis using WCAG 2.1: evidence from Indian e-government websites. Univ. Access Inf. Soc. (2022). https://doi.org/10.1007/s10209-021-00861-9
36. Petrie, H., Hamilton, F., King, N.: Tension, what tension?: website accessibility and visual design. In: Proceedings of 2004 International Cross-disciplinary Workshop on Web Accessibility (W4A), pp. 13–18 (2004)
37. Petrie, H., Kheir, O.: The relationship between accessibility and usability of websites. In: Proceedings of the SIGCHI Conference on Human Factors in Computing Systems, pp. 397–406 (2007)
38. Seljan, S., Miloloža, I., Pejić Bach, M.: e-Government in European countries: gender and ageing digital divide. Interdisc. Manag. Res. **16**, 1563–1584 (2020)

39. Silcock, R.: What is e-government. Parliamentary Aff. **54**(1), 88–101 (2001)
40. Siqueira, M.S.S., Dias, F.S., Rigatto, S.H., Carvalho, M.C.N., Marques, T.A.M., Freire, A.P.: Who watches the watchers? Accessibility of the public prosecutor's office websites in Brazil and implications for e-government accessibility surveillance policies. Electron. Gov. J. **19**(1), 72–94 (2022). https://doi.org/10.1504/EG.2023.127576
41. Valtolina, S., Fratus, D.: Local government websites accessibility: evaluation and finding from Italy. Digit. Gov. Res. Pract. **3**(3), Article 17, 16 p (2022). https://doi.org/10.1145/3528380
42. Vigo, M., Brown, J., Conway, V.: Benchmarking web accessibility evaluation tools. In: Proceedings of the 10th International Cross-Disciplinary Conference on Web Accessibility, pp. 1–10 (2013)
43. W3C. Introduction to Web Accessibility (2022). https://www.w3.org/WAI/fundamentals/accessibility-intro/#evaluate
44. W3C. Web Content Accessibility Guidelines (WCAG) 2.2 (2023). https://www.w3.org/TR/WCAG22/
45. Yesilada, Y., Harper, S.: Web Accessibility. Springer, London (2019)

Assessing the Validity of the Trust Factor Within User Experience Questionnaire Plus (UEQ+): A Comprehensive Validation Study

Andreas Hinderks[1(✉)], Martin Schrepp[2], Maria Rauschenberger[3], and Jörg Thomaschewski[3]

[1] University of Applied Sciences and Arts, Hannover, Germany
andreas@hinderks.org
[2] SAP SE, Walldorf, Germany
schrepp@sap.de
[3] Faculty of Technology, University of Applied Science Emden/Leer, Emden, Germany
{maria.rauschenberger,joerg.thomaschewski}@hs-emden-leer.de

Abstract. As digital technologies continue to evolve, the significance of user experience (UX) in determining the success of software and services has become increasingly apparent. The User Experience Questionnaire Plus (UEQ+) serves as a versatile tool designed to assess UX through questionnaires that are specifically tailored to address unique challenges. Within this context, a critical yet frequently overlooked factor is Trust. Trust encompasses a user's confidence in a software's ability to perform reliably, maintain security, and respect user data privacy, and is particularly crucial in sectors such as financial services, health informatics, and e-commerce platforms. This paper details the development and validation of *Trust* as a novel factor within the UEQ+. The initial phase of this work involved assembling a preliminary set of potential items to define the trust factor. A rigorous study with 405 participants then enabled the refinement of these items to four essential elements through a structured factor analysis approach. Subsequent stages encompassed three distinct phases of validation, engaging a total of 1497 participants, during which these items underwent thorough validation. This extensive process culminated in the establishment of a robustly validated factor, *Trust*, comprised of the following key items: *insecure-secure*, *untrustworthy-trustworthy*, *unreliable-reliable*, and *non-transparent-transparent*.

Keywords: User experience questionnaire · Trust · Measurement · Questionnaires · UX · Validation · Factor analysis

1 Introduction

In the rapidly evolving domain of digital technology, software and services are not only becoming more complex but also increasingly functional and influential. This evolution has made user experience (UX) a critical determinant in fostering user engagement and satisfaction, and ultimately, in the success of these applications. A common method employed by numerous corporations to measure and evaluate the user experience of

M. Marchiori and F. García Peñalvo (Eds.): WEBIST 2023, LNBIP 543, pp. 208–229, 2025.
https://doi.org/10.1007/978-3-031-89621-7_11

their products and services is through the deployment of questionnaires. UX questionnaires provide a structured quantitative measure of user experience and have gained widespread acceptance across various industries [13]. The literature currently includes a variety of UX questionnaires, such as the Visual Aesthetics of Websites Inventory (VisAWI) [16], Standardized User Experience Percentile Rank Questionnaire (SUPR-Q) [18], the User Experience Questionnaire (UEQ) [10], and the User Experience Questionnaire Plus (UEQ+) [21].

The ISO 9241-210 standard provides a well-established definition of user experience, describing it as ‘a person’s perceptions and responses that result from the use or anticipated use of a product, system or service’ [5]. As such, user experience embodies a comprehensive concept that incorporates an array of emotional, cognitive, and physical reactions to the specific, or even potential, use of a product before, during, and after its operation. However, this standard does not provide a detailed list of factors or methods for assessing user experience comprehensively.

The User Experience Questionnaire Plus (UEQ+) [21] represents an adaptable framework designed to construct a UX questionnaire specifically tailored to address particular research questions. It integrates 20 UX factors that can be selected and combined to create a customized questionnaire, allowing researchers and practitioner to pinpoint the factors most relevant to the product under investigation and to evaluate it accordingly. While these factors offer a detailed insight into a user’s interaction with software or a service, one critical yet often overlooked factor remains: *Trust*.

In the context of software or services, trust can be understood as a user’s confidence in the software’s ability to consistently, securely, and respectfully perform its functions while safeguarding user data privacy. This factor is crucial in shaping the relationship between the user and the product or service, especially in today’s landscape marked by frequent data breaches and growing privacy concerns. The importance of trust becomes even more significant in sectors such as financial services, health informatics, e-commerce platforms, and others, where users must entrust sensitive personal or financial information to the system.

The focus of this paper is on the construction and validation of the factor *Trust* within the User Experience Questionnaire Plus. Building upon our publication in Hinderks et al. [4], we have conducted an additional study to validate this factor, the results of which are now being presented in this paper.

Section 2 explores the background and relevant work. Section 3 delineates our methodology for developing the UX factor *Trust*. Section 4 presents the findings of our studies. Section 5 provides a discourse on the results.

2 Background and Related Work

User experience (UX) is defined as a holistic and multidimensional concept that encapsulates users’ perceptions of a product (testing object) before, during, or after its use [5]. Influential factors on this perception may include emotions, prior experiences, or deeply held beliefs. UX is commonly divided into several factors such as *Attractiveness*, *Efficiency*, *Value*, *Perspicuity*, and *Trust*, among others. These factors, for example, are utilized by the User Experience Questionnaire Plus (UEQ+) [21], which currently offers twenty factors to measure UX tailored to the specific needs of the testing object. The name UEQ+ builds on its predecessor, the UEQ, whose core factors (*Attractiveness, Efficiency, Perspicuity, Dependability, Stimulation, Novelty*) form the foundation. Additional factors such as *Haptics* and *Acoustics* [1], as well as *Aesthetics, Adaptability, Usefulness, Intuitive Use, Value, Trustworthiness of Content*, and *Quality of Content* [21], have since been integrated.

Each factor can be chosen and combined in any configuration to craft a product-specific questionnaire suited to a particular research inquiry. This is facilitated using a special modular factor format, where each selected factor is introduced with a brief contextual statement, followed by four items presented as a semantic differential combined with a seven-point *Likert-scale*. Immediately following, the significance of the UX aspect represented by this factor is explored (example provided in Table 1).

Table 1. Example of the *Efficiency* factor [21].

To achieve my goals, I consider the product as...		
slow	o o o o o o o	fast
inefficient	o o o o o o o	efficient
impractical	o o o o o o o	practical
cluttered	o o o o o o o	organized
The product property described by these terms is for me...		
completely irrelevant	o o o o o o o	highly relevant

For practical applications, the selected UEQ+ factors are organized sequentially to ensure that the effort and time required to complete the questionnaire remain manageable, ideally covering no more than 6 to 7 distinct factors. Detailed guidance on factor selection is provided in the UEQ+ handbook [22], which, along with the tool itself, is freely accessible at http://ueqplus.ueq-research.org/. This resource also houses various linguistic versions, including German [11], English [10], and Spanish [17].

Unlike traditional UX questionnaires such as the original UEQ or other similar tools, the UEQ+ does not offer a ready-to-use UX measurement tool but instead emphasizes the modularity of its factors, enabling the selection of components that best align with the objectives of the testing object, thus ensuring relevant UX aspects are measured.

Trust, as a UX factor, has been emphasized increasingly due to the growing prevalence of sensitive data exchanges over the Internet in activities like e-commerce, online banking, and cloud services [20]. Factors like Attractiveness, Trust, and Value are considered overarching and pertain to a user's holistic assessment of a product. The factor of trust, in particular, accentuates the user's confidence in the safe handling and processing of data within platforms such as online banking. Additionally, the importance of trust in voice user interfaces has been empirically validated in recent studies by Klein et al. [7]. Despite cultural differences, users share concerns about privacy and distrust of VUI companies, which influence their adoption behavior [6,8].

While questionnaire-based UX measurement has traditionally been a staple [12, 15, 18–20], only a select few tools specifically measure trust, and these are typically tailored for particular contexts or purposes. Alongside the UEQ+, instruments like *Trust* are measured and specific guidelines are offered for contexts such as automation [9], mobile applications (SUPR-Qm) [19], management information systems (MIS) [15], or website design [14].

The importance of trust in facilitating social interactions under conditions of uncertainty, particularly in the realms of online services and e-commerce where anonymity and a lack of physical control prevail, is further explored in the development of the TrustDiff scale, a semantic differential that measures user trust across three dimensions: Benevolence, Integrity, and Competence of an online vendor [2]

Although existing guidelines aim to enhance trust in MIS by addressing uncertainties related to trustworthiness, user experience, intentions to use, and actual usage, they stop short of providing a measurement tool [15].

Given that the current UX questionnaires including the *Trust* factor are highly context-specific and resist easy adaptation to new requirements, it becomes crucial to enhance the construct validity of these tools, particularly for practical applications with the UEQ. By presenting a detailed validation of the *Trust* factor for the UEQ+, this paper aims to facilitate its easy adaptation across diverse contexts and requirements, obviating the need for developing new questionnaires for each unique scenario.

3 Research Methodology

Within the framework of the User Experience Questionnaire Plus (UEQ+), the element of trust was initially integrated through preliminary validation procedures [21]. This integration meant that the UEQ+ already provided a Trust factor ready for selection. However, this factor had not undergone thorough validation. In light of this, we found it necessary to reconstruct and validate this *Trust* factor by employing a mix of existing and new items, thereby approaching the construct from a fresh perspective.

Following the reconstruction, we proceeded to validate this evolved *Trust* factor across seven distinct studies, thus ensuring a substantial base of data to support our findings. Our research methodology for constructing the *Trust* factor was methodically segmented into four distinct phases. The initial phase (Table 2) was dedicated solely to the development of the *Trust* factor. This foundational phase was crucial for setting the parameters for further validation.

The first group of validation (Table 3) focused on the practical validation of the Trust factor by applying it to two different products. During this phase, the same set of UX factors, including *Trust*, was used for both products to gather preliminary insights into their effectiveness and reliability.

Progressing to the second group of validation (Table 4), we expanded the validation process to encompass a wider array of UX factors alongside *Trust*. This phase was designed to evaluate the stability and consistency of the newly developed Trust factor when applied across various products and UX dimensions, thereby offering a more comprehensive assessment of its universal applicability.

The final phase, detailed in Table 5), involved an in-depth examination of two products selected from the third phase, integrating the valence dimension of attractiveness. This holistic approach across all four phases ensured that various combinations of UX factors, products, and valence dimensions were systematically explored and validated. This thorough methodology reinforced the robustness and reliability of the *Trust* factor, confirming its effectiveness across a diverse range of applications.

This revised introduction expands slightly on each phase of the study and clarifies the structured approach taken in validating the *Trust* factor within the UEQ+ framework, thereby providing a clearer understanding of the methodological rigor and the comprehensive nature of the validation process.

Table 2. Overview of the Studies for Reconstruction of the Factor *Trust* [4].

	Reconstruction			
	1. Validation	2. Validation	3. Validation	4. Validation
1. Conducted in	2022	2022	2022	2022
2. Testing object	Airbnb	BookingCom	TikTok	TradingApp
3. Participants	103	100	104	98
4. Item set Trust	insecure - secure untrustworthy - trustworthy *unreliable - reliable* *non-transparent - transparent* indiscreet - discreet unserious - serious non trustful - trustful non responsibly - responsibly unhonest - honest			
5. Goal	Construction	Construction	Construction	Construction
6. Method	EFA	EFA	EFA	EFA
7. Factor	Trust Efficiency Dependability Intuitive of Use			

Table 3. Overview of the Studies for the Frist Group of Validation of the Factor *Trust* [4].

	Frist Group of Validation	
	5. Validation	6. Validation
1. Conducted in	2019	2019
2. Testing object	Facebook	Youtube
3. Participants	248	195
4. Item set Trust	insecure/secure untrustworthy/trustworthy unreliable/reliable non-transparent/transparent	
5. Goal	Validation	Validation
6. Method	CFA	CFA
7. Factor	Trust Intuitive of Use Quality of Content Stimulation	

Table 4. Overview of the Studies for the Second Group of Validation of the Factor *Trust* [4].

	Second Group of Validation				
	7. Validation	8. Validation	9. Validation	10. Validation	11. Validation
1. Conducted in	2022	2022	2022	2022	2022
2. Testing object	AirBnB	Amazon	TikTok	Skype	Booking.com
3. Participants	91	206	51	57	49
4. Item set Trust	insecure/secure untrustworthy/trustworthy unreliable/reliable non-transparent/transparent				
5. Goal	Validation	Validation	Validation	Validation	Validation
6. Method	CFA	CFA	CFA	CFA	CFA
7. Factor	Efficiency Dependability Trust Quality of Content Clarity	Dependability Trust Intuitive use Quality of Content Clarity	Dependability Stimulation Trust Intuitive use Quality of Content	Efficiency Dependability Trust Usefulness Intuitive use	Efficiency Dependability Trust Quality of Content Clarity

Table 5. Overview of the Studies for the Third Group of Validation of the Factor *Trust*.

	Third Group of Validation	
	12. Validation	13. Validation
1. Conducted in	2024	2024
2. Testing object	Booking.com	Tiktok
3. Participants	300	300
4. Item set Trust	insecure/secure untrustworthy/trustworthy unreliable/reliable non-transparent/transparent	
5. Goal	Validation	Validation
6. Method	CFA	CFA
7. Factor	Efficiency Dependability Trust Quality of Content Attractiveness	

In our research methodology, we executed all investigations using English as the primary language. The elements that were methodically chosen post-construction and rigorously validated throughout the course of these studies are positioned to serve as foundational benchmarks for subsequent translations into a multitude of other languages.

3.1 Reconstruction of the Factor Trust

In the initial phase, a panel of three subject matter specialists (the authors of the current study) meticulously sifted through a variety of resources, including digital platforms and lexical databases, to gather items that semantically represented the construct of *Trust*. These items were systematically aggregated and refined to suit different validation studies.

A iterative internal feedback mechanism was established among these experts. The purpose of this mechanism was to guarantee the validity and quality of the *Trust* factor's items for different contexts. This collaborative critique and validation process ensured that each item accurately depicted *Trust*. The items selected for the initial study are as follows:

- insecure - secure
- untrustworthy - trustworthy
- unreliable - reliable
- non-transparent - transparent
- indiscreet - discreet
- unserious - serious
- non trustful - trustful
- non responsibly - responsibly
- dishonest - honest

In our initial research investigation, we conduceted a first study with this particular set of items.

Object of Study. The primary goal of the first study was to streamline the item set down to four distinctive items, which are expected to exhibit a strong correlation to the *Trust* factor. The item selected for study within the scope of this investigation include AirBnb, Booking.com, TikTok, and various trading applications.

In order to facilitate this analysis, we constructed a User Experience Questionnaire Plus (UEQ+) that incorporates the initial items. The UEQ+ evaluated the *Trust* factor along with several additional factors-efficiency, dependability, and intuitive usability.

It's important to note that the choice to examine the same suite of factors across all four test objects was intentional. This strategic decision was driven by our desire to conduct a factor analysis upon the conclusion of our study. It is paramount that the data extracted from each test item retains a consistent structure, to ensure the validity and reliability of our eventual analytical output.

Context. The research was conducted in April 2022, originating from England and facilitated via an online survey. For the dataset, we have choosen a social panel (Prolific Academic https://www.prolific.co/) to collect the data. The survey's initial stage prompted the participants to select one of the four options: AirBnb, Booking.com, TikTok, or a specified Trading App. Subsequently, the participants were required to fulfill the User Experience Questionnaire Plus (UEQ+) with the selected items for the factor

Trust, supplemented by items related to three additional factors: Efficiency, Dependability, and Intuitive of Use. Along with these, participants were also asked to provide demographic details such as age and gender. Each participant restricted their evaluation to a singular product. The study drew the participation of 405 individuals, as outlined in Table 6.

The research outcome was the identification of four items linked to the *Trust* factor. These items are scheduled for further validation using additional products in the subsequent studies.

Table 6. Participant Count for each Test Object in the Reconstruction Group [4] (M = male, F = Female, N/A = no answer).

Test object	Total	F	M	N/A
AirBnb	103	70	33	0
Booking.com	100	76	23	2
TikTok	104	80	23	2
Trading app	98	65	33	3
Total	405			

3.2 First Group of Validation

We conducted an evaluation of two prominent social media platforms, YouTube and Facebook, using the User Experience Questionnaire Plus (UEQ+). For this assessment, we selected items representing the factor *Trust* from a preceding study. In addition, we chose to include the factors Intuitive Use, Quality of Content, and Stimulation, which differ from those used in our initial construction study. This strategic selection was made to determine the robustness and applicability of the *Trust* item set across different contexts.

Object of Study. This study focused on products with a high level of user familiarity to ensure accurate assessments by the participants. The primary aim was to validate the item set related to the *Trust* factor, specifically to see if the newly integrated items would yield meaningful and interpretable results.

A confirmatory factor analysis (CFA) was conducted with the objective of establishing whether the *Trust* items could independently and distinctly represent the factor. The aim was to confirm that these items primarily loaded on the *Trust* factor, thereby minimizing their association with other factors. Successful demonstration of this independence would substantiate the *Trust* items as reliable indicators of the factor.

The overarching goal of this initial validation group was to establish a solid foundation for the *Trust* factor's validity within the context of user experience evaluation.

Context. The validation study was conducted in Germany for YouTube and in England for Facebook, utilizing an online version of the questionnaire. The German dataset was collected in collaboration with the University of Applied Sciences Emden/Leer, while the English dataset was gathered using a social panel, Prolific Academic. A total of 443 participants were involved in the study, with demographic details including age and gender also collected to enrich the dataset analysis. The participant responses were distributed with 195 evaluations for YouTube and 248 for Facebook (Table 7).

In pursuit of further validating the *Trust* factor, a subsequent study was meticulously designed and executed. The results of this initial research are intended to be strengthened and confirmed through a secondary, confirmatory study.

This revised introduction enhances the professional tone and elaborates slightly on the methodological approach, aims, and context of the study to provide a clearer understanding of the research objectives and validation process.

Table 7. Participant Count for Each Test Object in the First Validation Group [4].

Test object	Total	F	M	N/A
YouTube	195	65	123	7
Facebook	248	132	112	4
Total	443			

To ensure the validity of the *Trust* factor, a subsequent study was designed and executed. The findings of this research are intended to be corroborated through a second, confirmatory study.

3.3 Second Group of Validation

To enhance the robustness of our previous findings and demonstrate their repeatability, we conducted an additional study, this time incorporating a broader array of products along with an expanded set of factors. While the initial study provided promising results, the purpose of this subsequent research was to further confirm the repeatability and thus, the reliability of those initial findings.

For this secondary validation group, our selection of test objects was strategically chosen based on their diverse applicability across various User Experience Questionnaire Plus (UEQ+) factors. The selected entities for this phase included AirBnb, Amazon, TikTok, Skype, and Booking.com. Notably, some of these test objects had been part of the initial development phase, though they were assessed under different factors.

Object of Study. The selection of a varied range of factors for these test objects was driven by the goal to validate the independence of the Trust factor's items from other UX factors. Therefore, unique factors were specifically assigned to each test object to ensure a comprehensive evaluation. Additionally, a diverse array of factors was used for identical test items relative to their construction, aiming to test the robustness of these factors under varied conditions. An overview of the factors chosen for each test object is detailed in Table 4.

A Confirmatory Factor Analysis (CFA) was subsequently performed to examine whether the Trust factor's items exhibited minimal to low loading on other factors, consistent with our expectations from the initial analysis. Confirmation of this hypothesis would provide strong support for the successful validation of the *Trust* factor, reinforcing its independence and reliability (Table 8).

Context. This study was conducted in Germany at the University of Applied Sciences Emden/Leer, utilizing an online version of the questionnaire. The study involved a total of 454 participants. In addition to responses collected via the UEQ+, demographic information such as age and gender was also gathered to provide further depth to the analysis. The data was segmented with 195 responses pertaining to YouTube and 248 to Facebook, as detailed in Table 7.

Table 8. Participant Count for Each Test Object in the Second Validation Group [4].

Test object	Total	F	M	N/A
AirBnb	91	49	39	3
Amazon	206	92	110	4
TikTok	51	29	21	1
Skype	57	24	26	7
Booking.com	49	26	20	3
Total	454			

3.4 Third Group of Validation

In the third phase of validation, we aimed to select two previously utilized products and validate the UX factor of Trust alongside different UX factors than those used in the prior study. This approach was intended to ensure that the Trust factor could be effectively applied across various UX factors. Booking.com and TikTok were chosen as the test subjects.

Object of Study. The same UX factors were selected for the two test subjects, Booking.com and TikTok: Efficiency, Dependability, Trust, Quality of Content, and Attractiveness. By utilizing the same UX factors, results can be compared across both platforms, allowing conclusions to be drawn about their validity. The objective of this study is to further validate the UX factor of *Trust* in combination with other UX factors. An overview of the study is presented in Table 9. Subsequently, a confirmatory factor analysis (CFA) was conducted.

Context. The study was conducted in March 2024 in England using an online questionnaire. For data collection, we utilized a social panel from Prolific Academic (https://www.prolific.co/). A total of 600 participants were involved in the study (Table 9). In addition to the UEQ+, we also asked for their age and gender.

Table 9. Participant Count for Each Test Object in the Validation Group.

Test object	Total	F	M	N/A
Booking.com	299	230	68	1
TikTok	301	193	107	1
Total	600	423	175	2

4 Results

Before initiating each phase of our study, we conducted thorough data screening to ensure the integrity of our analysis. Any questionnaires that were incomplete or exhibited potential anomalies, such as a respondent providing identical responses across all items, were meticulously identified and subsequently excluded. It is critical to note that all figures cited in this paper represent the count of valid questionnaires that were ultimately utilized after this stringent filtration process.

Factor analysis was integral to all three groups involved in our study. During the initial reconstruction phase, our primary aim was to refine and reduce the number of items representing the *trust* factor. This effort led to the identification and selection of four key items. In the subsequent validation phases-both the first and the second groups-the focus was on performing confirmatory factor analysis (CFA). The objective here was to validate and substantiate the items designated for the *trust* factor identified during the earlier reconstruction phase.

In the following sections, we delve into the specific results of the factor analysis for each phase.

4.1 Results from the Reconstruction

During the reconstruction phase, factor analysis was performed on a dataset that encompassed evaluations of four distinct products. We began with a pool of nine potential items for the *trust* factor. The goal of this factor analysis was to identify the most representative four items from this pool. As a preliminary step in this process, a scree plot was constructed to aid in the decision-making (Fig. 1).

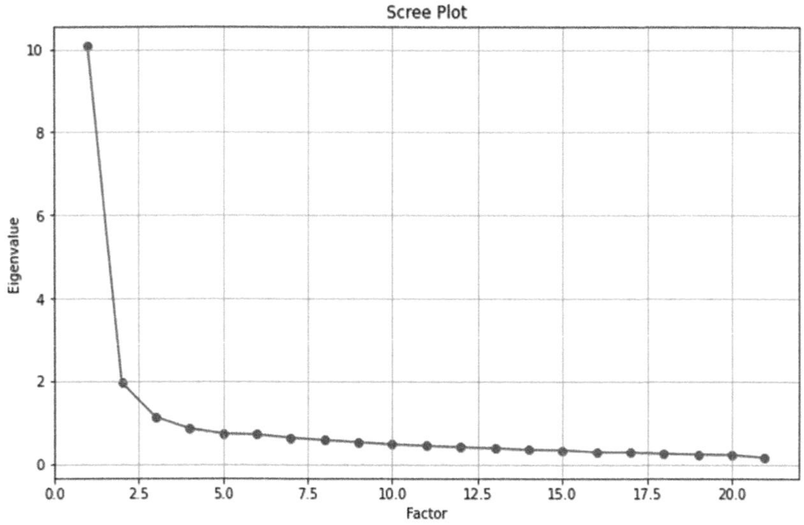

Fig. 1. Screeplot for the Reconstruction Group [4].

The scree plot (Fig. 1) clearly delineates the number of factors, suggesting that both three and four factors were plausible choices. This observation was in alignment with our initial selection of four factors.

Following this, an exploratory factor analysis (EFA) using Principal Component Analysis (PCA) was carried out, targeting these four factors specifically. To improve the clarity and interpretability of the factor loadings, Varimax rotation was applied. The results of this analysis, which detail the factor loadings for each item, are displayed in Table 10.

The item loadings across the factors did not yield entirely uniform outcomes as initially projected. A detailed examination was specifically directed toward the item loadings for the *trust* factor. To facilitate easier visualization and evaluation, loadings exceeding 0.4 were highlighted in red, signaling their potential candidacy for selection. Following the guidelines suggested by Comrey et al. [3], loadings greater than 0.4 are generally considered acceptable. The items such as 'insecure-secure' (0.739), 'untrustworthy-trustworthy' (0.839), 'unreliable-reliable' (0.748), 'non-transparent-transparent' (0.707), 'non-trustful-trustful' (0.733), 'non-responsible-responsible' (0.678), and 'dishonest-honest' (0.733) all demonstrated loadings significantly above this threshold on Factor 0 (Table 10). These high loadings qualified them as strong candidates for the *Trust* factor.

However, the items 'indiscreet-discreet' (0.301) and 'unserious-serious' (0.389) did not achieve sufficient loadings and were thus excluded from further consideration.

Table 10. The outcomes of the EFA (Principal Component Analysis (PCA) with Varimax Rotation) conducted on the Reconstruction Group. Loadings exceeding the value of 0.4 are distinctively highlighted [4].

Factor	Number	Item	Factor 0	Factor 1	Factor 2	Factor 3
Trust	0	insecure - secure	0.739	0.244	0.196	0.072
	1	untrustworthy - trustworthy	0.839	0.241	0.202	0.056
	2	unreliable - reliable	0.748	0.304	0.082	0.118
	3	non-transparent - transparent	0.707	0.214	0.191	0.140
	4	indiscreet - discreet	0.301	0.007	0.712	−0.020
	5	unserious - serious	0.389	0.170	0.696	−0.113
	6	non trustful - trustful	0.733	0.308	0.385	0.099
	7	non responsibly - responsibly	0.678	0.237	0.451	0.057
	8	dishonest - honest	0.733	0.215	0.334	0.050
Efficiency	9	slow - fast	0.082	0.158	0.031	0.923
	10	inefficient - efficient	0.292	0.520	0.155	0.562
	11	impractical - practical	0.316	0.575	0.309	0.276
	12	cluttered - organized	0.328	0.532	0.124	0.066
Dependability	13	unpredictable - predictable	0.191	0.229	0.632	0.216
	14	obstructive - supportive	0.525	0.291	0.447	0.200
	15	not secure - secure	0.674	0.333	0.308	0.183
	16	does not meet expectations - meets expectations	0.406	0.606	−0.089	0.348
Intuitive of Use	17	difficult - easy	0.188	0.683	−0.229	0.281
	18	illogical - logical	0.234	0.730	0.384	0.060
	19	not plausible - plausible	0.213	0.668	0.441	0.042
	20	inconclusive - conclusive	0.277	0.712	0.267	0.011

In addition to these, items such as 'obstructive-supportive' (0.525), 'not secure-secure' (0.674), and 'does not meet expectations-meets expectations' (0.406) also exhibited loadings above 0.4 on Factor 0 (Table 10). Despite surpassing the minimum threshold, their loadings were relatively lower compared to those items explicitly designated for the *Trust* factor. Consequently, these items were not selected for further detailed analysis.

After careful deliberation and review of the loading values, the decision was made to select the four items 'insecure-secure', 'untrustworthy-trustworthy', 'unreliable-reliable', and 'non-transparent-transparent' for inclusion in the *Trust* factor. Detailed analyses with these selected items were carried out to further explore their applicability and validity, which will be discussed in subsequent sections.

4.2 Results from the First Validation Group

In the phase involving the first validation group, we chose two well-established digital platforms, Facebook and YouTube, as the subjects for evaluation. This part of the study was pivotal for assessing the stability and suitability of the items selected during the earlier reconstruction phase.

The primary methodological tool employed in this validation phase was confirmatory factor analysis (CFA), a statistical technique renowned for its effectiveness in confirming the factor structure of a set of observed variables. In this phase, the CFA was specifically used to verify whether the data collected for the four chosen items were consistent with the theoretical expectations established from the reconstruction group. As an initial step in this process, a scree plot was constructed to visually assess the factor structure (Fig. 2).

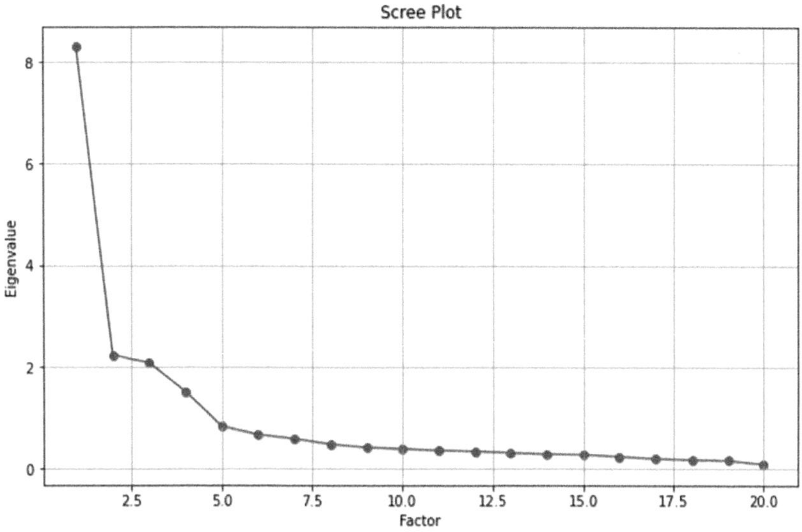

Fig. 2. Screeplot for the First Validation Group [4].

The scree plot displayed a distinct inflection at the point indicating five factors. This visual indication supports the assumption of a five-factor structure, which was intentionally targeted in this study. The alignment of the scree plot results with our predefined research design reinforced the appropriateness of the chosen factor model.

Following the visual assessment, a detailed CFA was conducted on the dataset using Principal Component Analysis (PCA). To enhance the clarity and interpretability of the factor loadings, Varimax rotation-a widely used method for orthogonal rotation-was applied. The resulting factor loadings, which provide a quantitative measure of how well the observed variables represent the hypothesized factors, are illustrated in Table 11.

The four items specifically selected to represent the *Trust* factor demonstrated remarkably high factor loadings, all exceeding 0.8 on Factor 3. This notable result indicates a strong and exclusive correlation with this particular factor, with no other items showing loadings above 0.4 on Factor 3. Such a distinct pattern strongly suggests that these *Trust* factor items are capable of being evaluated independently of the other factors, thereby affirming their specificity and relevance to the *Trust* construct.

Table 11. The outcomes of the CFA (Principal Component Analysis (PCA) with Varimax Rotation) conducted on the First Validation Group. Loadings exceeding the value of 0.4 are distinctively highlighted [4].

UX Factor	Number	Item	Factor 0	Factor 1	Factor 2	Factor 3	Factor 4
Intuitive Use	0	difficult - easy	0.020	0.726	0.105	0.074	0.029
	1	illogical - logical	0.167	0.835	0.073	0.073	0.140
	2	not plausible - plausible	0.083	0.858	0.055	0.080	0.168
	3	inconclusive - conclusive	0.167	0.802	0.121	0.146	0.179
Quality of Content	4	obsolete - up-to-date	0.303	0.154	0.115	0.090	0.725
	5	not interesting - interesting	0.433	0.104	0.235	0.188	0.636
	6	poorly prepared - well prepared	0.322	0.135	0.190	0.239	0.700
	7	incomprehensible comprehensible	0.124	0.218	0.132	0.196	0.777
Trustworthiness of Content	8	useless - useful	0.720	0.154	0.179	0.239	0.313
	9	implausible - plausible	0.745	0.202	0.137	0.246	0.274
	10	untrustworthy - trustworthy	0.773	0.095	0.164	0.368	0.238
	11	inaccurate - accurate	0.796	0.102	0.133	0.287	0.239
Trust	12	insecure - secure	0.224	0.074	0.116	0.883	0.123
	13	untrustworthy - trustworthy	0.289	0.137	0.133	0.862	0.186
	14	unreliable - reliable	0.264	0.155	0.132	0.843	0.205
	15	non-transparent - transparent	0.185	0.083	0.121	0.804	0.127
Stimulation	16	not interesting - interesting	0.177	0.138	0.810	0.064	0.107
	17	boring - exiting	0.127	0.140	0.857	0.051	0.157
	18	inferior - valuable	0.009	0.033	0.815	0.212	0.092
	19	demotivating - motivating	0.178	0.065	0.772	0.115	0.143

An exception was noted with the item "not interesting-interesting", which displayed loadings above 0.4 on both Factor 0 and Factor 4 (Table 11). This outcome diverges from the clear and exclusive factor assignments observed with the other items, indicating a potential overlap in factor associations that may require further scrutiny.

Given the high loadings and the specific alignment of the chosen items with the *Trust* factor, these results provide strong preliminary evidence supporting their validity. Nevertheless, to ensure the robustness and reliability of these findings, an additional study was initiated. This subsequent investigation aimed to further scrutinize and confirm the independence and validity of the *Trust* factor items. Details of this further validation effort are presented in the following section.

4.3 Results from the Second Validation Group

During the analysis of the second validation group, an evaluation was undertaken involving five distinct test objects. Besides the four items previously identified for the *Trust* factor, additional factors were also incorporated, as detailed in Table 4.

Each dataset was subjected to a confirmatory factor analysis, specifically employing PCA with Varimax Rotation. Owing to spatial constraints and in the interest of maintaining clarity, only the loadings of the items pertaining to the *Trust* factor have been included in Table 12.

Table 12. The outcomes of the CFA (PCA with Varimax Rotation) for only the Factor *Trust* conducted on the Second Validation Group. Loadings exceeding the value of 0.4 are distinctively highlighted [4].

UX Factor	Number	Item	Factor 0	Factor 1	Factor 2	Factor 3	Factor 4
AirBnb	0	insecure - secure	0.021	0.917	0.048	0.068	0.128
	1	untrustworthy - trustworthy	0.149	0.899	0.213	0.089	0.083
	2	unreliable - reliable	0.216	0.871	0.255	0.098	-0.046
	3	non-transparent - transparent	0.027	0.871	0.036	0.156	0.101
Amazon	0	insecure - secure	0.101	0.835	0.039	0.185	0.131
	1	untrustworthy - trustworthy	0.078	0.855	0.189	0.191	0.057
	2	unreliable - reliable	0.082	0.840	0.120	0.135	0.110
	3	non-transparent - transparent	0.148	0.747	0.049	-0.032	0.235
TikTok	0	insecure - secure	0.062	0.904	0.016	0.061	0.036
	1	untrustworthy - trustworthy	0.169	0.914	0.063	0.070	0.058
	2	unreliable - reliable	0.258	0.915	0.031	0.090	0.094
	3	non-transparent - transparent	-0.009	0.865	0.081	0.095	0.110
Skype	0	insecure - secure	0.154	0.880	0.137	0.174	0.000
	1	untrustworthy - trustworthy	0.046	0.856	0.230	0.232	0.143
	2	unreliable - reliable	0.057	0.876	0.216	0.211	0.187
	3	non-transparent - transparent	0.398	0.753	0.090	0.144	-0.108
Booking.com	0	insecure - secure	0.150	0.919	0.147	0.037	0.029
	1	untrustworthy - trustworthy	0.192	0.919	0.176	0.001	0.106
	2	unreliable - reliable	0.051	0.905	0.212	0.132	-0.029
	3	non-transparent - transparent	0.095	0.745	0.246	0.063	0.374

The empirical findings provide strong evidence that all items associated with the *Trust* factor distinctly load on the first factor. Importantly, no significant loadings of these items on any alternate factors were observed during the analysis. This indicates a clear demarcation and specificity of these components towards the *Trust* factor. Thus, it is reasonable to infer that these four items stand as independent representatives of the *Trust* factor, devoid of substantial interference from or dependencies on other factors.

4.4 Results from the Third Validation Group

Confirmatory Factor Analysis (CFA) with Varimax Rotation was individually conducted for each test subject to assess the validity of the *Trust* factor across different platforms while using the same UX factors. The results of the CFA for Booking.com are depicted in Table 13, and for TikTok in Table 14.

For both Booking.com and TikTok, the Trust factor was distinctly separated from the remaining factors, with one additional item included. For both test subjects, the item 'not secure - secure' demonstrated factor loadings of 0.549 for Booking.com and 0.534 for TikTok, indicating a moderate association with the Trust factor. The other items were predominantly clearly aligned with their respective UX factors.

5 Discussion

Throughout the phases of reconstruction and validation, we employed both exploratory and confirmatory factor analyses. This combination is a well-established methodology

Table 13. The outcomes of the EFA (Principal Component Analysis (PCA) with Varimax Rotation) conducted for the Booking.com dataset. Loadings exceeding the value of 0.4 are distinctively highlighted.

UX Factor	Number	Item	Factor 0	Factor 1	Factor 2	Factor 3	Factor 4
Efficiency	0	slow - fast	0.214	0.181	0.215	0.715	−0.063
	1	inefficient - efficient	0.215	0.146	0.183	0.846	−0.081
	2	impractical - practical	0.240	0.159	0.198	0.743	0.100
	3	cluttered - organized	0.397	0.202	0.088	0.617	0.056
Quality of Content	4	obsolete - up-to-date	0.446	0.188	0.150	0.545	0.317
	5	not interesting - interesting	0.725	0.188	0.132	0.215	0.108
	6	poorly prepared - well prepared	0.773	0.215	0.206	0.290	0.146
	7	incomprehensible - comprehensible	0.740	0.065	0.249	0.284	0.216
Dependability	8	unpredictable - predictable	0.155	0.117	0.840	0.184	−0.051
	9	obstructive - supportive	0.364	0.159	0.710	0.238	0.190
	10	not secure - secure	0.215	0.549	0.385	0.059	0.583
	11	does not meet expectations - meets expectations	0.286	0.288	0.622	0.392	0.133
Attractiveness	12	annoying - enjoyable	0.497	0.349	0.331	0.385	−0.345
	13	bad - good	0.604	0.395	0.340	0.403	−0.108
	14	unpleasant - pleasant	0.708	0.343	0.256	0.288	−0.069
	15	unfriendly - friendly	0.725	0.281	0.118	0.204	−0.159
Trust	16	insecure - secure	0.123	0.896	0.073	0.155	0.150
	17	untrustworthy - trustworthy	0.189	0.887	0.122	0.233	0.040
	18	unreliable - reliable	0.288	0.821	0.157	0.230	0.090
	19	non-transparent - transparent	0.328	0.746	0.184	0.130	−0.159

Table 14. The outcomes of the EFA (Principal Component Analysis (PCA) with Varimax Rotation) conducted for the TikTok dataset. Loadings exceeding the value of 0.4 are distinctively highlighted.

UX Factor	Number	Item	Factor 0	Factor 1	Factor 2	Factor 3	Factor 4
Efficiency	0	slow - fast	0.040	0.130	0.847	0.162	0.186
	1	inefficient - efficient	0.118	0.197	0.868	0.145	0.190
	2	impractical - practical	0.267	0.181	0.813	0.051	0.197
	3	cluttered - organized	0.225	0.351	0.530	0.152	0.264
Quality of Content	4	obsolete - up-to-date	0.210	0.097	0.241	0.100	0.753
	5	not interesting - interesting	0.452	0.154	0.240	0.127	0.593
	6	poorly prepared - well prepared	0.358	0.263	0.221	0.106	0.687
	7	incomprehensible - comprehensible	0.293	0.125	0.192	0.231	0.733
Dependability	8	unpredictable - predictable	0.143	0.067	0.065	0.853	0.108
	9	obstructive - supportive	0.310	0.237	0.396	0.601	0.219
	10	not secure - secure	0.159	0.534	0.233	0.525	0.199
	11	does not meet expectations - meets expectations	0.470	0.199	0.289	0.501	0.309
Attractiveness	12	annoying - enjoyable	0.821	0.216	0.106	0.205	0.226
	13	bad - good	0.776	0.234	0.141	0.189	0.293
	14	unpleasant - pleasant	0.839	0.158	0.121	0.178	0.263
	15	unfriendly - friendly	0.731	0.245	0.184	0.040	0.224
Trust	16	insecure - secure	0.194	0.871	0.158	0.171	0.127
	17	untrustworthy - trustworthy	0.216	0.881	0.127	0.113	0.157
	18	unreliable - reliable	0.205	0.856	0.170	0.069	0.112
	19	non-transparent - transparent	0.154	0.854	0.197	0.054	0.099

in the field of questionnaire construction, adept at elucidating the interrelationships among items and their corresponding factors.

In our analytical process, we selected Varimax rotation, an orthogonal method, for factor rotation. We opted for this method due to the clarity it provides in the loading of items on the factors. Varimax rotation enhances interpretability by maximizing the variance of factor loadings, thereby making it easier to discern and understand the distinct contributions of each item to its factor.

Conversely, we decided against using Promax rotation, which involves an oblique transformation. Our rationale was that this method might introduce distortions in the loading of items on the factors, potentially complicating the interpretation of these loadings. Therefore, aiming for precision and clarity in our results, Varimax rotation was deemed the most suitable choice.

5.1 Reconstruction

The interpretability of the scree plot (Fig. 1) presents a degree of ambiguity, indicating the potential for either a three- or four-factor structure. While this ambiguity does not significantly impact our item selection process, it is notable that all items associated with the potential *Trust* factor load distinctly on Factor 0 (Table 10). Should we opt for a three-factor structure, Factor 3 (Table 10) would be eliminated, although only the items 'slow-fast' and 'inefficient-efficient' load on this factor. Given the high loading of 0.923 for 'slow-fast', this item nearly constitutes a separate factor in itself (Table 10). Thus, a four-factor structure appears to be the most rational configuration for our dataset.

The loadings for the potential items (0-8, as shown in Table 10) generally meet the high standard required, with the exceptions of 'indiscreet-discreet' and 'unserious-serious'. Therefore, all but these two items were incorporated into the *Trust* factor. The final selection included items 0-3 ('insecure-secure', 'untrustworthy-trustworthy', 'unreliable-reliable', and 'non-transparent- transparent'), which are reflective of trust dimensions used in prior research, thereby supporting the validation of these items by preconstruction.

5.2 First Group of Validation

The factor analysis results for the initial validation group demonstrated a strong performance. Each item within the *Trust* factor displayed a loading value exceeding 0.8 on Factor 3 (Table 11), with no other items showing comparable loadings on this specific factor. This indicates a high validity for the trust items, clearly differentiating them from other factors.

Items associated with other factors, except for the item 'not interesting-interesting', clearly loaded onto their respective factors, further confirming their validity. This outcome reinforces the reliability of our factor analysis beyond the scope of the *Trust* factor, affirming the overall structural integrity of our questionnaire.

5.3 Second Group of Validation

Involving a diverse set of factors and test objects, the second validation group aimed to ascertain whether the items tied to the *Trust* factor maintained robust and adequate factor loadings across different contexts. The positive results from this group's analysis affirm the validity of the *Trust* factor, demonstrating its stability and appropriateness across varied applications.

5.4 Third Group of Validation

The third validation study was designed to verify the consistency of the Trust factor across different test subjects while maintaining identical UX factors. This goal was successfully met, with the exception of the item 'not secure - secure'. The Trust factor proved to be largely independent, as confirmed by the results of the Confirmatory Factor Analysis (CFA). The item 'not secure - secure' registered lower loadings and thus was considered less critical to the factor's integrity.

The factor loadings for the remaining items were consistent with our expectations, typical for CFAs, which seldom produce perfectly unambiguous results. Consequently, the findings from this study are deemed satisfactory, with only the item 'not secure - secure' requiring further consideration due to its ambiguous factor alignment.

5.5 Limitations

This paper details the process of reconstructing and validating the *Trust* factor within the User Experience Questionnaire Plus (UEQ+). The results lend credibility to the selected items for the *Trust* factor, affirming their validity in the assessed context. However, it is crucial to verify this validity continuously in future applications of the UEQ+ that incorporate the *Trust* factor. Confirmatory factor analysis remains a critical methodology for ensuring this ongoing validity.

Moreover, the outcomes from the factor analyses reveal certain nuances. Identically phrased items originating from different factors may not consistently align with a single factor. This observation underscores the need for heightened scrutiny in future applications, particularly when handling items with similar wording but different factor origins. Further research could elucidate these findings more clearly, refining methodologies for a more precise item assignment in diverse contexts.

6 Conclusions and Future Work

This paper details the comprehensive construction and validation process for the *Trust* factor within the User Experience Questionnaire Plus (UEQ+). The initial stage of this endeavor, referred to as 'preconstruction', involved the meticulous collation and preliminary assessment of potential items that might effectively capture the essence of trust. This foundational phase yielded a selection of items, which were subsequently evaluated through a rigorous study involving four distinct test objects and a total of 405 participants. The exploratory factor analysis performed during this stage successfully distilled the *Trust* factor into the following four key items:

- insecure-secure
- untrustworthy-trustworthy
- unreliable-reliable
- non-transparent-transparent

In the subsequent 'First Group of Validation', these four items underwent further scrutiny in a study involving 443 participants, who assessed platforms such as Facebook and YouTube. The confirmatory factor analysis (CFA) conducted in this phase substantiated the structural integrity and distinctiveness of these items as components of the *Trust* factor.

An additional 'Second Group of Validation' study was carried out involving five test objects and 454 participants. The CFA executed during this phase reaffirmed the validity of the four *Trust* items, underscoring their consistent performance across different contexts.

The final phase, known as the 'Third Group of Validation', was executed with two test objects and 600 participants. The CFA in this phase once again validated the four *Trust* items, demonstrating their robustness and stability as components of the *Trust* factor across various conditions.

Throughout these three phases of validation, we were able to successfully validate the *Trust* factor under diverse conditions, confirming that the four selected items could be stably validated as a distinct factor within the UEQ+. Consequently, the primary objective of this manuscript-to construct and robustly validate a new *Trust* factor for the UEQ+-has been effectively achieved.

Considering the wide applicability of the UEQ+, it is pertinent to acknowledge that not all product categories and environments were covered within the scope of our studies. Future research endeavors or practical applications utilizing the UEQ+ and the *Trust* factor should aim to confirm its validity across a broader spectrum of contexts and products. This ongoing validation is essential to ensure the *Trust* factor's relevance and effectiveness in capturing user perceptions reliably across varied user experience studies.

References

1. Boos, B., Brau, H.: Erweiterung des UEQ um die Dimensionen Akustik und Haptik [Extension of the UEQ by the dimensions acoustics and haptics]. In: Mensch & Computer im Strukturwandel (Humans Humans & computers computers in times of change in structure), pp. 321–327. Mensch & Computer (Humans & computers) 2017, Usability Professionals, Regensburg: Gesellschaft für Informatik e.V., (2017)
2. Brühlmann, F., Petralito, S., Rieser, D.C., Aeschbach, L.F., Opwis, K.: TrustDiff: development an dValidation ofa semantic differential for user trust on the web. J. Usability Stud. **16**(1), 29–48 (2020)
3. Comrey, A.L., Lee, H.B.: A First Course in Factor Analysis. Taylor and Francis, Hoboken, 2nd edn. (2013). http://gbv.eblib.com/patron/FullRecord.aspx?p=1562106
4. Hinderks, A., Schrepp, M., Rauschenberger, M., Thomaschewski, J.: Reconstruction and validation of the UX factor trust for the user experience questionnaire plus (UEQ+). In: Proceedings of the 19th International Conference on Web Information Systems and Technologies - Volume 1: WEBIST, pp. 319–329. INSTICC, SciTePress (2023). https://doi.org/10.5220/0012186700003584

5. ISO/TC 159/SC 4 Ergonomics of human-system interaction: Part 210: Human- centred design for interactive systems. In: Ergonomics of human-system interaction, vol. 1, p. 32. International Organization for Standardization (ISO), Brussels (2010)
6. Klein, A.M., Deutschländer, J., Kölln, K., Rauschenberger, M., Escalona, M.J.: Exploring the context of use for voice user interfaces: toward context dependent user experience quality testing. J. Softw. Evol. Process (2023). https://doi.org/10.1002/smr.2618
7. Klein, A.M., Hinderks, A., Schrepp, M., Thomaschewski, J.: Measuring user experience quality of voice assistants. In: 2020 15th Iberian Conference on Information Systems and Technologies (CISTI), pp. 1–4. IEEE (2020). https://doi.org/10.23919/CISTI49556.2020.9140966
8. Klein, A.M., Rauschenberger, M., Thomaschewski, J., Escalona, M.J.: Comparing voice assistant risks and potential with technology-based users: a study from Germany and Spain. J. Web Eng. **20**, 1991–2016 (2021). https://doi.org/10.13052/jwe1540-9589.2071
9. Körber, M.: Theoretical considerations and development of a questionnaire to measure trust in automation. In: Bagnara, S., Tartaglia, R., Albolino, S., Alexander, T., Fujita, Y. (eds.) IEA 2018. AISC, vol. 823, pp. 13–30. Springer, Cham (2019). https://doi.org/10.1007/978-3-319-96074-6_2
10. Laugwitz, B., Held, T., Schrepp, M.: Construction and evaluation of a user experience questionnaire. In: Holzinger, A. (ed.) USAB 2008. LNCS, vol. 5298, pp. 63–76. Springer, Heidelberg (2008). https://doi.org/10.1007/978-3-540-89350-9_6
11. Laugwitz, B., Schrepp, M., Held, T.: Konstruktion eines fragebogens zur messung der user experience von softwareprodukten. In: Heinecke, A.M., Paul, H. (eds.) Mensch & Computer 2006, pp. 125–134. Oldenbourg, München (2006)
12. Laugwitz, B., Schrepp, M., Held, T.: Konstruktion eines Fragebogens zur Messung der User Experience von Softwareprodukten [Construction of a questionnaire for the measurement of user experience of software products]. In: Mensch & Computer im Strukturwandel (Humans & computers in times of change in structure), pp. 125–134. Mensch & Computer (Humans & computers) 2006 (2006)
13. Lazar, J., Feng, J.H., Hochheiser, H.: Research methods in human-computer interaction. Wiley, Chichester and West Sussex and U.K (2010)
14. Loiacono, E.T., Watson, R.T., Goodhue, D.L.: WebQual: an instrument for consumer evaluation of web sites. Int. J. Electron. Commer. **11**(3), 51–87 (2007). https://doi.org/10.2753/JEC1086-4415110302
15. Meeßen, S.M., Thielsch, M.T., Hertel, G.: Trust in management information systems (MIS): a theoretical model. Zeitschrift fur Arbeits- und Organisationspsychologie **64**(1), 6–16 (2020). https://doi.org/10.1026/0932-4089/a000306
16. Moshagen, M., Thielsch, M.T.: Facets of visual aesthetics. Int. J. Hum. Comput. Stud. **68**(10), 689–709 (2010). https://doi.org/10.1016/j.ijhcs.2010.05.006
17. Rauschenberger, M., Schrepp, M., Perez-Cota, M., Olschner, S., Thomaschewski, J.: Efficient measurement of the user experience of interactive products. how to use the user experience questionnaire (ueq).example: Spanish language version. Int. J. Interact. Multimedia Artif. Intell. **2**(1), 39 (2013). https://doi.org/10.9781/ijimai.2013.215
18. Sauro, J.: SUPR-Q: a comprehensive measure of the quality of the website user experience. J. Usability Stud. **2015**(10), 68–86 (2015)
19. Sauro, J., Zarolia, P.: SUPR-Qm: a questionnaire to measure the mobile app user experience. J. Usability Stud. **13**(1), 17–37 (2017)
20. Schrepp, M.: User Experience mit Fragebögen messen (Deutsch) Taschenbuch. CreateSpace Independent Publishing Platform (7. April 2018), 1 edn. (2018)

21. Schrepp, M., Thomaschewski, J.: Design and validation of a framework for the creation of user experience questionnaires. Int. J. Interact. Multimedia Artif. Intell. (2019). https://doi.org/10.9781/ijimai.2019.06.006
22. Schrepp, M., Thomaschewski, J.: Handbook for the modular extension of the User Experience Questionnaire. - All you need to know to apply the UEQ+ to create your own UX questionnaire. Technical report, University of Applied Sciences Emden/Leer (2019). https://doi.org/10.13140/RG.2.2.15485.20966

Digital Transition to a Paperless Checklist Integrated into the Industrial Information System

José Cosme[1(✉)], Anabela Ribeiro[2], Vítor Filipe[1,3], Eurico Vasco Amorim[1,3], and Rui Pinto[4]

[1] School of Science and Technology, University of Trás-os-Montes e Alto Douro (UTAD), 5000-801 Vila Real, Portugal
al74139@alunos.utad.pt, {vfilipe,eamorim}@utad.pt

[2] Continental Advanced Antenna, Sociedade Unipessoal, Lda, Vila Real, Portugal
anabela.ribeiro@continental-corporation.com

[3] INESC Technology and Science, 4200-465 Porto, Portugal

[4] SYSTEC-ARISE, Faculty of Engineering of the University of Porto, Porto, Portugal
rpinto@fe.up.pt

Abstract. The Digital Twin concept involves the transition to digital representations of factory floor equipment, the computerized simulation of processes and the visualization of data in real time. This type of digital transformations can be considered radical, encountering barriers in its implementation either due to resistance to change by the different elements that make up the industry or due to the disruption it can cause in the production process. The start of production on an assembly line is usually preceded by a checking procedure of parameters/conditions of the equipment present on the assembly line, using a sheet of paper containing the list of items to check and validate. In this article we describe the adoption of a paperless checklist to verify the configuration of assembly line equipment at production bootstrapping. A training program to coach the employees for a successful digital transition is also presented and discussed. Both the digital checklist and the training program are validated in a real-world industrial scenario. The results highlight the advantages of the digital approach given to the checklist with a multi-access viewing and maintenance of data for later analysis, with the training plan demonstrating effectiveness in breaking down barriers and resistance to the adoption of a new working method.

Keywords: Industrial information system · Manufacturing process control · Validation checklist · Digital transition

1 Introduction

An information system brings together a set of interconnected components that, through their interoperability, collect, process, store and distribute data and information within an organization or within a specific context. In the systems we can consider people,

M. Marchiori and F. García Peñalvo (Eds.): WEBIST 2023, LNBIP 543, pp. 230–244, 2025.
https://doi.org/10.1007/978-3-031-89621-7_12

hardware, software, data and associate procedures as integral elements. The main purpose attributed to an information system is the ability to support management, processes, production process operations and decision making at an organization level. As mentioned, the ability to collect, store, process and disseminate data and information efficiently, quickly and effectively is important for the various interested parties.

If we approach in comparative terms the traditional approaches used to facilitate communication, collaboration and data management, we can characterize an information system by: I) Refinement in decision making supported with data-driven insights; II) Increased Efficiency and Productivity (automation of repetitive tasks and improvement of the production flow); III) Improvement of communication and collaboration channels between elements; IV) Enhancement in data management, with improved storage, organization and recovery of significant amounts of data, ensuring its integrity, security and clarity.

Information systems are crucial in the maintenance and management of industrial equipment and assets. As added value associated with these, we can consider the assistance provided in tracking equipment maintenance schedules, managing production orders and monitoring the performance of the machines that make up the factory floor. The inclusion of predictive maintenance techniques must be considered, with the appropriate treatment and subsequent analysis of the data transmitted by the sensors and by the machines to which they may be associated, and as a consequence the identification of possible failures and systematic imperfections, being able to program adequate proactive maintenance, reducing downtime and optimizing asset productivity. They provide a centralized repository for maintenance documentation, from equipment manuals, standard production processes, maintenance checklist and troubleshooting guides. This ensures that the personnel responsible for maintenance have adequate access to the necessary information, improving the procedures associated with maintenance activities.

In addition to the various functionalities previously described, information systems have the ability to collect and store maintenance knowledge and empirically obtained information, allowing knowledge sharing and continuous improvement. Furthermore, the ability to access these systems through mobile devices allows human operators to receive work orders, update maintenance records and access relevant documentation while on the factory floor. Mobile access facilitates not only collaboration but also effective and real-time communication and data capture, improving efficiency and reducing paperwork.

Despite all the advantages of information systems, the transition to a digital paradigm presents challenges and obstacles that go beyond the simple implementation. The resistance to change that occurs encompasses both operators on the factory floor and hierarchical superiors, supported by an empirically successful process and a production flow that is already well-rooted. In addition to this barrier, which can only be overcome through a clear explanation and training of the technologies and new methods to be implemented, some of the main challenges that companies face when trying to embrace Industry 4.0 are related to the fact that there is no qualified talent with the necessary capabilities to idealize, develop and manage high technology systems, the need to develop an internal training program to prepare the employees with the necessary skills and tools, the existence of a reduce budget for investment and integration of technologies in existing processes and systems, and the economic risk associated with a possible stoppage or decrease in production capacity.

The training of all elements associated with the production process is essential to overcome all the possible obstacles, ensuring that the digital transition is carried out gradually in order to avoid losses of revenue and production capacity as much as possible. For this purpose it must also take into account the iterative nature that must be associated with the training process, listening to employees ideas and questions, promoting proactivity in relation to continuous learning and innovation.

This study proposes an information system approach to enable a digital checklist, used to manage equipment maintenance and validate process status at the beginning of production shifts. The development of this paper follows a previous one, which addresses the beginning of the development of this system (Cosme et al., 2023), so in this study we add the component associated with the transition process to this digital tool, also ending up identifying and characterizing the changes that were made to the initial solution. The solution implemented in a real industrial scenario, also takes into account a training program directed for the workers involved in the production process with the necessary requirements for an adequate transition to this system. In the current context, the work methodology is based on the use of a paper checklist to manage the necessary information, and with its digitalization, paperwork is eliminated and information is easily accessible between employees, teams and departments. Therefore, in this study, we intend to address the following research questions:

RQ 1. How are the information processing and the respective ramp-up checklist management affected by the implementation of paperless practices in terms of process performance with emphasis on the efficiency, flexibility and quality aspects?

RQ 2. What are the specific challenges encountered in the implementation of a paperless checklist system, and how can the system be further developed to enhance its integration, usability, and efficiency in collaboration with operators?

After the introductory chapter, the paper consists of five more sections. The following section, Sect. 2, frames the technological context with an analysis of the state of the art. Section 3 details the case study covered and the proposed approach and solution for it. The validation methodology and results obtained are described in Sect. 4, while Sect. 5 presents discussions regarding the approach presented. Last but not least, Sect. 6 concludes the paper with final considerations and observations in relation to the study presented.

2 Related Work

The adoption of paperless strategies in an industrial context has been a hot topic among researchers (Palade and Møller, 2022; Müller, 2017; Djassemi and Sena, 2006). With all the associated premises, digitalization is increasingly important in companies that want to eliminate paper-based processes. An example of the approach to this topic is the requirement for documentation of factory production and how it is possible to produce without having documentation on paper (Bulut et al., 2020; Mleczko et al., 2013). What the literature demonstrates is that an organization based on digital documentation can present potential improvements in processes, improvements that have an impact on both efficiency, quality and flexibility. Ramp-up checklists (Christensen and Rymaszewska, 2016; Biffl et al., 2022) are documents commonly used at the start of shifts or during production changes to ensure proper transition and performance. These checklists help operators or maintenance personnel carry out essential verification and tasks to prepare equipment, methodologies and systems for the production process. They are essential to ensure that equipment, systems and processes are properly prepared, minimizing potential risks and ensuring a fluid transition to production. The items on an acceleration checklist may vary depending on the specificity of the sector in which it operates and operational requirements (Cosme et al., 2023).

Information systems play a crucial role in streamlining asset management at the factory floor level, providing mechanisms that contribute to increasing the efficiency of maintenance processes, reducing costs and improving equipment reliability. With the digitalization of checklists, not only does the use of paperwork reduce, but it also allows organizations to adapt them to their specific maintenance or production needs, also ensuring the automation of the distribution of checklists with timely access and action by the qualified operators (Doltsinis et al., 2020).

Real-time updating and tracking are inherent benefits of these digital checklists. Operators can update tasks, record measurements and report problems as they occur, providing accurate and transparent tracking of progress. Mobile access further facilitates real-time collaboration, allowing supervisors and managers to remotely monitor progress and provide timely guidance or resolve concerns.

In this context, the ability to integrate digital checklists with flow management tools plays an important role, allowing effective coordination between checklist completion tasks and generating, for example, work orders or notifications to initiate maintenance actions or forward the problem for relevant personnel. The permanence of the list data guarantees access to historical data and the record of maintenance activities, resulting in the possibility of analyzing performance levels and compliance requirements (Doltsinis et al., 2020).

Despite the benefits demonstrated by the literature in relation to the implementation of paperless strategies, digitizing checklists and their completion, this transition not only faces challenges in terms of the resistance of all elements associated with the production process, but is also necessary to develop and implement methodologies and strategies to recognize the pre-existing skills of professionals in relation to the demands of this transformation, training programs capable of providing the operators with the necessary skills at the level of Industry 4.0 and evaluating the best way to carry out training and the implementation of the system without compromising the company's workflow, productivity and revenue.

The challenges associated with implementing a system that facilitates the digital transition at the factory floor level arise not only from the resistance of top hierarchical levels, reluctant to change empirically successful processes, but also from line operations employees accustomed to the methods of existing production (Ikegami and Iijima, 2020). The transformative impact of Digital Transformation can be considered radical and profound, therefore requiring complex coordination efforts, as these strategies can intersect with other business strategies already implemented. In a comprehensive study involving eight Japanese companies and surveys among their employees, it was observed that more than two-thirds of companies fall into the reactive category towards digital transformation. This suggests that a significant part of these companies are at an early stage in relation to implementing actions related to this transition, with the leaders and managers who supervise it indicating some barriers that make this process difficult, giving rise to the need to reinforce internal systems of training, the inability to integrate old and new systems and the limited budgetary capacity that companies allocate to digital activities are just beginning to take initial actions towards this transition. The lack of consensus between the company's different departments regarding the need for digital transition appears as a widespread issue across the business landscape, exacerbated by the shortage of quality workers and leaders capable of devising the strategies necessary for this transformation.

The transition to paperless checklists represents a significant step forward, simplifying operations and reducing dependence on traditional paper-based methods, however, this shift entails challenges and structural changes, such as the need for a skills-matched workforce inherent to the use of a digital system like this, and the integration of the proposed technologies into existing systems (Hernandez-de-Menendez et al., 2020). Therefore, there is a need to develop educational proposals to promote the learning of the necessary skills to use, verify and manage digital checklists in an effective, quick and fluid manner with the premise of adapting to a constantly evolving technological context.

Assessment methods are essential for evaluating and improving employee skills in line with Industry 4.0 requirements. The use of learning management systems allows the performance and progress of operators to be monitored during the learning, training and transition phases, associating practical tests and simulations that evaluate the success of applying theoretical knowledge in practice (Shevyakova et al., 2021).

Finally, the change to paperless strategies and the digitization of checklists in industrial environments present opportunities and challenges and although the literature extols the advantages related to increased efficiency, flexibility and fluidity of production, the transition and implementation of these digital solutions isn't free of obstacles. Resistance to change, transversal to all the elements that make up and integrate the production process, budget restrictions and the lack of adequate qualifications, can prevent the intended changes from being implemented. Covering all the premises mentioned above, it is also necessary to take into account the need to design and develop adequate and comprehensive training programs, as well as homogeneous integration into existing systems.

3 Proposed Solution

An information system is proposed to enable a paperless acceleration checklist, with a training program to support its implementation. While the information system aims to promote the fluid flow of information and facilitate communication between operators and supervisors when managing acceleration checklists, the training program aims to provide these elements with the necessary skills to make good use of this solution. The proposed checklist in digital format as well as the outlined training program considers a specific use case at the Portuguese industrial company Continental Advance Antenna (CAA) (Continental, 2024).

3.1 Industrial Use Case

CAA has extensive experience in R&D activities related to new connectivity features for antenna systems in automotive vehicles. CAA is one of the major players in the automotive sector and one of the most important industrial companies in Portugal, producing around 20 million antennas per year and typically, each type has a dedicated assembly line, which can work in multiple work shifts. The following figure, Fig. 1, represents a paper acceleration checklist used in CAA.

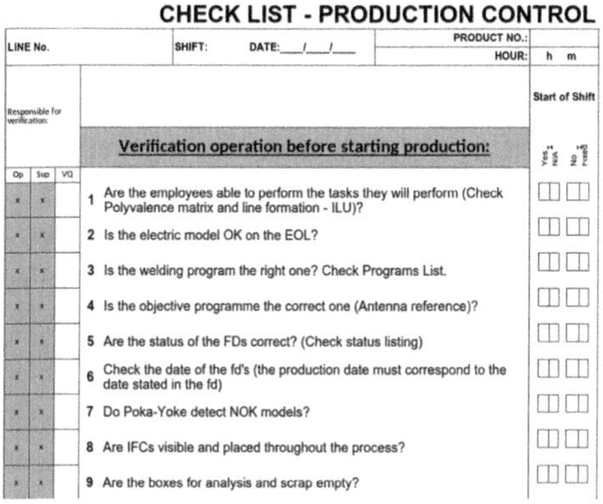

CHECK LIST - PRODUCTION CONTROL

LINE No. | SHIFT: | DATE:___/___/___ | PRODUCT NO.: | HOUR: h m

Responsible for verification:

Verification operation before starting production: — Start of Shift (Yes / N/A; No / Fixed)

Op	Sup	VQ	No.	Verification	Yes / N/A	No / Fixed
x	x		1	Are the employees able to perform the tasks they will perform (Check Polyvalence matrix and line formation - ILU)?		
x	x		2	Is the electric model OK on the EOL?		
x	x		3	Is the welding program the right one? Check Programs List.		
x	x		4	Is the objective programme the correct one (Antenna reference)?		
x	x		5	Are the status of the FDs correct? (Check status listing)		
x	x		6	Check the date of the fd's (the production date must correspond to the date stated in the fd)		
x	x		7	Do Poka-Yoke detect NOK models?		
x	x		8	Are IFCs visible and placed throughout the process?		
x	x		9	Are the boxes for analysis and scrap empty?		

Fig. 1. Example of a Paper-based Ramp-up Checklist in CAA (Adapted from Cosme et al. 2023).

The CAA ramp-up checklist is a document that must be completed mandatorily and carefully at the beginning of each shift, at each reference change and after prolonged maintenance/intervention in the process or quality problem. The importance of this document and its completion comes from the need to verify that all production requirements are in accordance with the required criteria. At this moment, the checklist is presented in paper format to users and is therefore completed manually, making the process time-consuming and susceptible to errors. The main checkpoints of this checklist are:

- ILUO: It consists of a methodology based on 4 levels in order to formalize the qualitative capacity of production operators (I - Operator with a basic qualification level, and will normally have associated supervision of their work, with the need to follow pre-conditions and instructions pre-defined; L - Operator already with considerable knowledge of the production process, but still subject to some guidance and supervision; U - Operator capable of producing with quality and at the pace defined for the assembly line in which he is evaluated, therefore qualified to work autonomously;

O - Operator with a high level of theoretical and practical knowledge of the processes, capable of supervising and training operators cataloged at other levels of the system).
- Electric models: A binary test of the electrical model at the end of the assembly line assessed. The test passes if the result is OK.
- Welding program: In the welding robots, the operator must check if the welding program loaded in the robot is the correct one for the reference antenna being assembled.
- Preventive maintenance: For preventive maintenance purposes, there is an established number of assembled pieces as a threshold. This test assesses if the threshold number of assembled pieces for preventive maintenance was reached.
- Hotmelt: The operator must know the injection process and its adjacent equipment in order to check the operating mode of the machine and the state of the injection material.
- Welding: To validate this point, the operator must know the documentation and associated procedure for replacing soldering nozzles, check if nozzle temperature is within limits, check if smoke extractors are working, nitrogen levels, and reservoir of water.
- Screwing: To validate this point, the operator must know the exchange history of the screwdriver and check the torque of semi-automatic screwdrivers.
- Ultrasounds: To validate this operation, the supervisor must check the pressure in the ultrasound nanometer.

3.2 Paperless Checklist

The developed information system is supported in Python programming language and the Flask framework (Lokhande et al., 2015; Grinberg, 2018), which allows CAA's operators and supervisors to manage the ramp-up checklist in a completely digital way. They can visualize and edit the available information in an efficient manner, while being managed and persistently stored in a database. In order to allow access to the platform, the system provides a centralized login system which, using the username entered, creates an association in the database with the role corresponding to that information, thus presenting the same interface for all elements involved in the production process but with different permissions on data manipulation and visualization.

One of the new features of the system in relation to the prototype version discussed in the previous study is the indication in the interface of a new section, the backoffice section. While the visualization of the checklist is now the same, regardless of the role, presenting only differences in relation to the permissions to interact with it, hierarchical superiors have a tab that is restricted to them that allows them to manage lines, shifts, users, functionalities and permissions.

Regarding the management of the digital start-up checklist, operators on the factory floor are only authorized to consult and complete it, while supervisors have a section adjacent to the backoffice section that allows them to carry out various actions, from the validation and parametrization of the checklist, up to the management of parameters and verification groups.

In relation to the main page, which translates the checklist and the actions related to it, significant changes are noted in relation to the previously developed interface model. This interface was designed and built to provide a unified graphics model capable of meeting the needs of both operators and supervisors. Therefore, it is important to note the continuous presence, during the completion of the checklist, of the components that refer to the shift and production line related to the checklist in question. At the heart of the page we have the presentation of the parameters, which in this case refer to the checks that need to be carried out before and after the start of production, as indicated in Fig. 2.

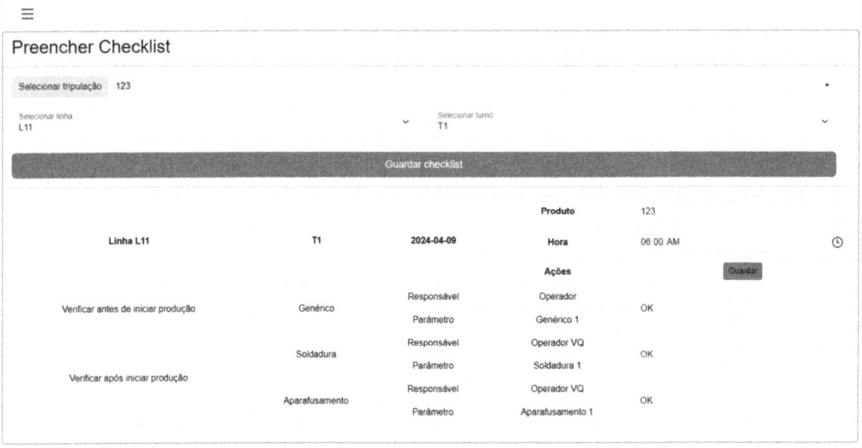

Fig. 2. Checklist Completion View.

These parameters are associated with the field that indicates which checkpoint they belong to, those responsible for checking them, the time indication of when they were checked, and the verification status, that is, whether the state of the parameter complies with the conditions and rules anticipated. In addition, the functionality of saving the current state is available, giving the user the opportunity to save changes made whenever he wants, instead of being in a continuous autosave state.

3.3 Training Program

The transition and implementation of this digital checklist solution, given the industrial context in which it is inserted, will have to follow a structured training and transition plan, a decision that is in line with what the literature shows us and with the opinion of hierarchical superiors of the company, who consider essential to implement a rigorous and properly planned methodology so that this structural change is successful, at least in the medium-long term.

The training program described in this section aims to support the transition from already rooted paper checklists to the digital system, addressing not only technical but also human issues.

Before starting a training methodology and a transition plan, it was necessary to listen to those responsible for administration and other elements linked to the production process. It was necessary to understand some questions:

- Operational Requirements: Understand and analyze the operational requirements and challenges in relation to the processes where the information system will be an integral part.
- Workflow and Feedback: Obtain a perspective of the production flow, identifying possible weaknesses that could jeopardize the implementation of the system, seeking in this regard to obtain feedback from operators, contemplating their experience with the current approach and what are their expectations and concerns about the future solution.
- Physical and training resources: Identify possible constraints in terms of logistics, hardware and software integration, and what types of most prominent skills are seen as necessary to promote in the training program.

Currently the company operates on a three-shift basis. Their schedules are:

- Shift 1 (S1) - 06:00 AM - 03:00 PM.
- Shift 2 (S2) - 03:00 PM - 12:00 AM.
- Shift 3 (S3) - 12:00 AM - 06:00 AM.

Taking these working hours into account, and the need to ensure continuous production, the training program is designed to cover both theoretical and practical aspects of the digital checklist system implementation, without compromising the minimum production services.

The proposed training model is represented in Fig. 3, and comprises two days of training. The first day will feature two sessions, and will be dedicated to the theoretical component, addressing the reasons and benefits of this transition, the importance that derives from the new solution in relation to improving the efficiency and quality of the production process, comprehensive user interface concepts and lastly, a space for questions, clarification of doubts and obtaining feedback. The second day will also consist of two sessions, and complementing the previous day, it will address the practical perspective, starting with system experimentation, with guidance on data insertion and validation, flow demonstrations and system integration, ending with practical exercises to simulate scenarios on the factory floor.

With the schedule staggered in this way, all shifts are able to comply with the training program, never having these extra hours coincide with their working hours, ensuring one of the requirements initially defined, which concerned not affecting the production rhythm.

The period after this training program, which concerns the testing phase on production lines, is a topic that was also addressed when carrying out this solution and this study. The company, recognizing the complexity of the production environment, and

	Day One		Day Two	
12:00 PM			S2 & S3	A Pratical Start
01:00 PM	S2 & S3	Why to Implement a Paperless Checklist?	S2 & S3	Data Validation and Process Fluidity
02:00	S2 & S3	Concepts and Q&A	S2 & S3	Scenarios simulation
03:00	S1	Why to Implement a Paperless Checklist?	S1	A Pratical Start
04:00	S1	Concepts and Q&A	S1	Data Validation and Process Fluidity
05:00			S1	Scenarios simulation

Fig. 3. Training Program Schedule.

the disruptive impact that a minimal change may imply, intends to carry out a gradual transition, initially focusing on individual production lines. This strategy allows not only a more effective and rapid response to any unforeseen events, but also an increased focus and attention on the system implementation, with a phased approach favoring the smoothness of the transition and adequate monitoring of the impact of the solution on each production line.

Iterative process characterization is essential to the success of this transition. As the new digital checklist system is implemented across different production lines, it is essential to continually collect feedback from operators and other stakeholders. This iterative process allows the company to identify areas for improvement, make necessary adjustments, and ensure that the system meets the specific needs and requirements of each production line. At an initial stage, it is also important, and the company's desire, to maintain digital and paper systems in parallel, providing security and protection in the event of failures in the information system, or even an inability of operators to the digital system. This dual work system, despite having disadvantages in terms of speed of production, guarantees a more competent transition and provides the necessary support to operators during the change phase.

4 Solution Validation

The validation of the proposed solution adopts a descriptive approach, specifically adapted to the CAA operational context. To effectively answer our research questions, we focused on understanding the complexities of current shop floor processes for managing the ramp-up checklists and evaluating how these processes are affected by the adoption of paperless practices.

Most of the data to validate the proposed approach was qualitative in nature, and it was collected through interviews, field observations and informal field conversations with different employees from CAA. The data was collected mainly through semi-structured interviews, where a set of questions and topics were prepared in advance. The interviews were conducted with the main stakeholders, i.e., CAA employees who were either involved in planning the implementation of paperless practices or whose current tasks were affected by the proposed approach. The set of questions that were used in the interviews is enclosed next.

Question 1. What is your role in CAA and how are you involved in the paperless checklist?
Question 2. Have you already had experiences regarding a shift towards paperless practices before?
Question 3. What does the collaboration between the different stakeholders look like when managing the ramp-up checklist?
Question 4. What impact does the implementation of a paperless checklist has in terms of information flow and the respective workflow?
Question 5. What are the main challenges for a transition from a paper-based towards a paperless checklist?
Question 6. In what ways does the implementation of a paperless checklist change your working tasks?
Question 7. How do you believe training can help overcome possible challenges or obstacles in this transition?
Question 8. What improvements or features do you think would make the paperless checklist more user-friendly for operators?
Question 9. How can the system be optimized to improve efficiency and collaboration during the checklist management process?

4.1 Analysis of Results

Most of the employees interviewed are supervisors of assembly lines, responsible for the continuous improvement of manufacturing processes and project managers. These employees are mainly associated with production, checklist monitoring and editing. They didn't have experience in the past with process digitalization and paperless practices before.

Through the information transmitted by these employees, the paper checklist is provided on the assembly line by the supervisor to the operators working on that shift, who will check all parameters. Unverified parameters are immediately reported to the supervisor, who must go to the assembly line and, together with the operators, resolve the problem. Eventually, depending on the problem identified, other employees may be requested. After validating all checklist parameters, both the operator and supervisor sign the checklist for an archive.

The feedback received regarding the developed information system was predominantly positive. Several positive aspects were observed, such as a notable improvement in the storage and organization of checklists, enabling a more efficient retrieval of different checklists and parameters, essentially resulting in an improvement in the quality of the process. The system also allows the integration of new functionalities, such as parameter filtering and real-time monitoring of digital checklists, ultimately promoting the simplification of the validation process, as it is no longer necessary to deliver the current paper checklists to the different employees responsible for final approval and archiving.

The main impacts of a digital checklist are:

- Elimination of paper;
- Supervisor avoids moving around to be on-site;
- Supervisor can consult updated information in real-time;
- Supervisor has easy access to the history of checklists.

In addition to the positive aspects mentioned, the developed information system also offers several other advantages. A significant benefit is the reduction of errors and inconsistencies when completing the checklist. With the digital system, mandatory fields can be implemented, ensuring that all necessary information is entered correctly. Automated validation checks can be performed to identify any missing or incorrect data, reducing the likelihood of human errors and improving overall data accuracy. The change to this digital system, ensuring the implementation of adequate protection, encryption, and backup measures, also offers greater security and confidentiality of data, with the added value of being more difficult for data to be lost or damaged compared to checklists in physical format.

Finally, before analyzing the results related to the training program, another notable advantage, which is an area where the company intends to invest, is the ability to analyze data and prepare reports. A digital checklist system can generate comprehensive reports and provide valuable insights into productivity, efficiency and compliance, as well as analyzing data trends and patterns when looking for areas of process improvement and optimization. The availability of historical data also allows for better monitoring of performance over time and facilitates auditing and regulatory compliance.

Regarding the training program, it aims to promote a successful transition to the new digital process. Given that the majority of employees interviewed, namely supervisors and project managers, had no previous experience with digitizing processes, the need for training that addressed all cross-cutting points in the proper use of the paperless checklist became evident. Through this training, the aim is to develop the technical skills of employees and familiarize them with the system's functionalities, in an attempt to maximize the potential of the solution.

The feedback obtained highlighted the need for ongoing training and support to keep employees up to date as the implementation phases of the verification system progress. Consideration of sporadic clarification and training sessions are also planned, addressing specific issues and functions. In this way, adaptive training guarantees not only a more fluid transition, but also the promotion of a critical spirit and learning autonomy related to the skills required in the new industry.

5 Discussion

Considering the qualitative validation and the results found, we are able to answer the research questions. In relation to RQ 1., the use of the digital checklist allowed the measurement and analysis of several parameters to assess the extent of their effects. The results of the study suggest that the implementation of the digital checklist positively influenced information processing and improved checklist management. The digital system simplified the data collection process, enabling real-time analysis and monitoring of data and metrics connected to the verification process. Associated with this improvement in efficiency and simplification of procedures, we have the reduction of errors in data entry and the evolution in the speed of decision-making, in the case of errors in the production line or in the general optimization of the system.

Also, the flexibility of the digital checklist facilitated immediate adjustments to production processes in response to changing requirements. In relation to quality aspects, the digital checklist presented benefits by providing more accurate, consistent and practical data input, resulting in improved product quality, improved monitoring parameters and detection of anomalies at an early stage. As a result, the transition to a paperless approach has improved the efficiency, flexibility and quality of the production process.

In response to RQ 2., the implementation of the paperless checklist system revealed significant resistance to change, especially among older employees and therefore more accustomed to traditional paper-based methods. This resistance, although expected, highlights the importance of designing effective strategies to support operators and train them. Openly communicating about the benefits of the digital system and providing ongoing support can help ease this transition, making employees more receptive to the new approach.

The training plan stood out as an essential component, and according to the feedback obtained, its personalization and individualization is essential in order to respond not only to general needs across all types of operators, but also to the difficulties and distinct characteristics of each worker. Finally, the need to ensure adequate integration with existing organizational processes became clear, with an employee-oriented approach, collecting their opinions and contributions, always with the aim of providing a better user experience that leads to management optimization of the ramp-up checklists in this industrial context.

6 Conclusions

Based on a previous study and solution, this article presents an information system representative of the checklist used on the CAA factory floor, from the perspective of this paperless approach, helping operators to manage equipment information and more effectively monitor the production process. Validation of the proposed solution was important to discern the advantages of the system, the limitations and the implications of its implementation on the flow of information and employees tasks. The conclusions highlight the potential of the information system in increasing productivity, collaboration and data integrity and clarity.

The positive feedback received indicates that, despite the challenges that may underlie the idealized solution, with appropriate support and training strategies proposed and described in this study, it is possible to make a successful transition to the digital checklist system.

The addition of a training program in this version results from the recognition of the importance that is providing to employees the necessary technical and interpersonal skills, with training and transition methodologies designed from the perspective of a smooth adaptation to the paperless checklist. This holistic and iterative approach, promoting the evolution of the system and seeking to offer the best conditions for production, highlights the commitment to digital transformation within the CAA organization.

Acknowledgments. This work was supported under the Mobilizing Agenda "A-MOVER - Development of Products & Systems towards an Intelligent and Green Mobility", supported by the PRR-Recovery and Resilience Plan and the European NextGeneration EU Funds.

References

Cosme, J., Pinto, T., Ribeiro, A., Filipe, V., Amorim, E. V., Pinto, R.: Paperless checklist for process validation and production readiness: an industrial use case. In: Proceedings of the 19th International Conference on Web Information Systems and Technologies, pp. 95–103. SCITEPRESS (2023)

Biffl, S., Kropatschek, S., Kiesling, E., Meixner, K., Lüder, A.: Risk-driven derivation of operation checklists from multi-disciplinary engineering knowledge. In: 2022 IEEE 20th International Conference on Industrial Informatics (INDIN), pp. 7–14. IEEE (2022)

Bockelmann, M., Nickel, P., Nachreiner, F.: Development of an Online Checklist for the Assessment of Alarm Systems and Alarm Management in Process Control. In: Nah, F.F.-H., Tan, C.-H. (eds.) HCIBGO 2017. LNCS, vol. 10294, pp. 325–332. Springer, Cham (2017). https://doi.org/10.1007/978-3-319-58484-3_25

Bulut, S., Giese, T. G., Wende, M., Anderl, R.: The paperless factory: from document driven factories to model driven factories. In: 2020 IEEE International Conference on Teaching, Assessment, and Learning for Engineering (TALE), pp. 725–730 (2020)

Christensen, I., Rymaszewska, A.: Lean application to manufacturing ramp-up: a conceptual approach. Qual. Manag. J. **23**(1), 45–54 (2016)

Continental: Advanced Antenna Solutions. https://www.continental-automotive.com/en-gl/Passenger-Cars/Architecture-and-Networking/Vehicle-Telematics-Solutions. Accessed June 2023

Djassemi, M., Sena, J.A.: The paperless factory: a review of issues and technologies. Int. J. Comput. Sci. Netw. Secur. **6**(12), 185 (2006)

Doltsinis, S., Ferreira, P., Mabkhot, M. M., Lohse, N.: A decision support system for rapid ramp-up of industry 4.0 enabled production systems. Comput. Ind. **116**, 103190 (2020)

Gonçalves, G., Reis, J., Pinto, R., Alves, M., Correia, J.: A step forward on intelligent factories: a smart sensor-oriented approach. In: Proceedings of the 2014 IEEE Emerging Technology and Factory Automation (ETFA), pp. 1–8. IEEE Press (2014)

Graf, H., Soubra, S., Picinbono, G., Keough, I., Tessier, A., Khan, A.: Lifecycle building card: toward paperless and visual lifecycle management tools. In: SpringSim (SimAUD), pp. 5–12 (2011)

Grinberg, M.: Flask web development: developing web applications with python. Inc, O'Reilly Media (2018)

Gunasekaran, A., Ngai, E.: Information systems in supply chain integration and management. Eur. J. Oper. Res. **159**(2), 269–295 (2004)

Lokhande, P., Aslam, F., Hawa, N., Munir, J., Gulamgaus, M.: Efficient way of web development using python and flask (2015)

Love, P.E., Zhou, J., Matthews, J., Luo, H.: Systems information modelling: enabling digital asset management. Adv. Eng. Softw. **102**, 155–165 (2016)

Mleczko, J., Podhora, P., Patka, J.: Distribution of information and documentation for the manufacturing processes of the high-variety products-case study. Appl. Comput. Sci. **9**(2) (2013)

Müller, R.: Impacts of implementing paperless practices at shop floor level on the information flow and the working processes (2017)

Nagendra, A., Deshpande, M.: Human resource information systems (HRIS) in HR planning and development in mid to large sized organizations. Procedia. Soc. Behav. Sci. **133**, 61–67 (2014)

Palade, D., Møller, C.: Paperless production demonstrator: jumpstarting digital transformation. In: The Future of Smart Production for SMEs: A Methodological and Practical Approach Towards Digitalization in SMEs, pp. 225–238. Springer, Cham (2022)

Palade, D., Møller, C.: Paperless production demonstrator, pp. 225–238. Springer, Cham (2023)

Plocher, T., Yin, S., Laberge, J., Thompson, B., Telner, J.: Effective shift handover. In: Harris, D. (ed.) Engineering Psychology and Cognitive Ergonomics, pp. 332–341. Springer, Heidelberg (2011)

Ikegami, H., Iijima, J.: Unwrapping Efforts and Difficulties of Enterprises for Digital Transformation. In: Agrifoglio, R., Lamboglia, R., Mancini, D., Ricciardi, F. (eds.) Digital Business Transformation. LNISO, vol. 38, pp. 237–250. Springer, Cham (2020). https://doi.org/10.1007/978-3-030-47355-6_16

Hernandez-de-Menendez, M., Morales-Menendez, R., Escobar, C.A., McGovern, M.: Competencies for industry 4.0. Int. J. Interact. Des. Manufact. (IJIDeM) **14**, 1511–1524 (2020)

Shevyakova, A., Munsh, E., Arystan, M., Petrenko, Y.: Competence development for Industry 4.0: qualification requirements and solutions. Insights Reg. Dev. **3**(1), 124–135 (2021)

Personalized User Experience and Engagement in Interactive Mobile Augmented Reality Software Through Fuzzy Weights

Christos Papakostas(✉), Christos Troussas, Akrivi Krouska, Phivos Mylonas, and Cleo Sgouropoulou

Department of Informatics and Computer Engineering, University of West Attica, Egaleo, Greece
{cpapakostas,ctrouss,akrouska,mylonasf,csgouro}@uniwa.gr

Abstract. The integration of digital content into the real world through mobile augmented reality (AR) systems presents captivating prospects, yet user engagement and experience within such environments are largely contingent upon individual computer skills, which can vary significantly among users, thereby influencing their proficiency in utilizing this technology. This paper aims to bridge this gap by examining the impact of users' computer skills on interactions within mobile AR settings. To this end, a novel fuzzy logic-based model is introduced to evaluate and refine users' computer skills within the realm of mobile AR systems. By modeling users' computer aptitude, the system is capable of delivering tailored assistive prompts and feedback. These prompts are strategically aligned with established fuzzy weights, thereby augmenting users' interactions within the mobile AR sphere. The proposed methodology is seamlessly integrated into a personalized mobile AR platform designed for spatial ability training. Evaluation findings reveal promising outcomes, underscoring the substantial potential of fuzzy logic modeling to elevate user experiences and propel advancements in mobile augmented reality technology.

Keywords: Fuzzy logic · Mobile augmented reality · Spatial ability training · Intelligent tutoring system · Computer skills · Feedback · Assistive messages · Personalization · Education

1 Introduction

Recent advancements in augmented reality (AR) technology, particularly in the realm of mobile AR, have uncovered significant opportunities across various applications, spanning entertainment such as gaming, as well as educational and professional training contexts [1]. Mobile AR provides users with an immersive experience that seamlessly blends digital and physical worlds, enabling virtual elements to be integrated into the real environment through mobile devices [2]. Despite the potential advantages offered by mobile AR, users often encounter challenges when interacting with these systems. These challenges may stem from a lack of familiarity with the technology, limited proficiency

M. Marchiori and F. García Peñalvo (Eds.): WEBIST 2023, LNBIP 543, pp. 245–266, 2025.
https://doi.org/10.1007/978-3-031-89621-7_13

in computer skills, or an insufficient understanding of the capabilities and functionalities of such systems [3].

In addition to these challenges, users may also face issues related to spatial orientation, cognitive load, and user interface complexities, which can hinder their overall experience and limit the effectiveness of mobile AR applications [4]. Moreover, the diverse range of user backgrounds, preferences, and skill levels further complicates the design and implementation of user-friendly mobile AR interfaces [5].

To address these challenges and enhance user engagement in mobile AR environments, there is a growing interest in leveraging advanced technologies such as fuzzy logic-based modeling [6]. By incorporating fuzzy logic techniques, mobile AR systems can dynamically adapt to users' individual characteristics and preferences, providing personalized assistance and guidance tailored to their specific needs and skill levels [7, 8]. This approach not only improves user interactions within mobile AR settings but also contributes to a more inclusive and accessible user experience for individuals with varying levels of technological proficiency.

As mobile AR continues to evolve and diversify across different domains, further research efforts are needed to explore innovative strategies for enhancing user engagement, mitigating usability challenges, and maximizing the potential benefits of this transformative technology. Through interdisciplinary collaborations and a user-centered design approach, researchers and practitioners can work together to unlock new opportunities and drive advancements in the field of mobile AR.

The proficiency level and technological acumen of users constitute pivotal factors influencing their capacity to engage effectively with mobile augmented reality (AR) systems. Individuals possessing advanced computer skills and knowledge tend to exhibit greater adeptness in navigating and harnessing the features and capabilities of mobile AR applications. Conversely, those with limited computer literacy may encounter challenges in comprehending and leveraging the diverse interaction modalities and functionalities inherent in mobile AR platforms. Hence, there arises a pressing need to model and assess users' computer skills and knowledge [9–11] as a means of optimizing interaction paradigms and augmenting the overall user experience within mobile AR environments.

Moreover, the effective modeling of users' computer skills and knowledge not only facilitates tailored guidance and support but also fosters inclusivity by accommodating individuals across varying proficiency levels. By incorporating insights from cognitive science and human-computer interaction, researchers can develop sophisticated models that dynamically adapt to users' cognitive abilities and learning preferences, thereby promoting a more intuitive and user-centric interaction framework in mobile AR systems. Additionally, the integration of machine learning algorithms and data-driven approaches holds promise for enhancing the accuracy and granularity of user skill assessments, enabling personalized recommendations and interventions to optimize user engagement and satisfaction in mobile AR contexts.

Numerous methodologies exist for modeling users' computer skills and knowledge within the realm of mobile augmented reality (AR) systems. One prominent approach is fuzzy logic, a technique acknowledged for its capacity to handle imprecise and uncertain data [12–14]. Fuzzy logic offers a versatile framework for representing and analyzing the ambiguity and unpredictability inherent in users' computer competencies. By employing fuzzy logic-based models, it becomes feasible to evaluate users' computer skills along a spectrum rather than relying on a binary classification. This nuanced approach facilitates a deeper understanding of users' proficiency levels and enables the customization of interaction experiences within mobile AR systems.

In addition to fuzzy logic, enhancing user interaction in mobile AR systems can be accomplished through the integration of various techniques, such as feedback mechanisms and informative messaging [15, 16]. Feedback mechanisms serve to guide users, furnish assistance, and offer real-time recommendations to augment their comprehension and utilization of mobile AR functionalities. Concurrently, informative messages can be deployed to apprise users of specific features, tips, or updates pertinent to the mobile AR application they are utilizing. These enhancements collectively aim to elevate the user experience, foster heightened engagement, and facilitate more efficient interactions within mobile AR environments.

Furthermore, the implementation of these techniques underscores a broader effort to develop user-centered design principles tailored to the unique attributes and requirements of mobile AR users. Through iterative refinement and user feedback, designers and developers can iteratively optimize mobile AR applications to better align with users' needs, preferences, and skill levels. By prioritizing user engagement and interaction fluidity, these advancements contribute to the ongoing evolution and maturation of mobile AR technology, fostering its broader adoption across diverse domains and user demographics.

This study focuses on the examination of users' computer proficiencies and their consequential impacts on interactions within mobile augmented reality (AR) systems. In pursuit of this objective, a novel fuzzy logic-based model is introduced to evaluate and refine users' computer skills within the context of mobile AR environments. The evaluation of users' computer aptitude concerning their interactions with mobile AR environments was conducted through a meticulously crafted questionnaire, developed by domain experts to target various facets of mobile AR interaction. Furthermore, this paper illustrates the integration of the proposed approach into a personalized mobile AR system designed specifically for spatial ability training. Within this framework, users receive tailored assistive messages and feedback based on the established fuzzy weights derived from the fuzzy logic-based model. Through this personalized approach, users are guided and supported in navigating the mobile AR environment, thus enhancing their overall experience and efficacy in spatial ability training. The evaluation findings reveal a notably positive outlook, underscoring the efficacy and promise of fuzzy logic modeling in bolstering user experiences within mobile AR contexts. This positive reception underscores the potential of leveraging fuzzy logic techniques not only to optimize user interactions but also to propel advancements in mobile AR technology.

The present paper is an extended and updated version of our preliminary conference paper that was presented in [6]. This paper significantly expands the evaluation of the proposed mobile AR system through fuzzy logic.

2 Related Work

Analyzing the body of literature pertaining to mobile augmented reality (AR) systems reveals a notable surge in growth over recent years. Research endeavors within this domain predominantly concentrate on enhancing the user experience, refining interface design for greater user-friendliness, and introducing novel forms of interaction [17–35]. Additionally, there is a growing emphasis on infusing intelligence into these systems to facilitate user modeling and knowledge representation [7, 8, 36–39]. The incorporation of intelligence into mobile AR systems encompasses a spectrum of techniques, with fuzzy logic emerging as a noteworthy approach among them.

Fuzzy logic offers a robust methodology for handling uncertainty and imprecision, making it well-suited for modeling user behavior, preferences, and knowledge within the context of mobile AR applications. By leveraging fuzzy logic-based models, researchers can effectively capture the nuances of user interactions and tailor AR experiences to align with individual user profiles.

In addition to fuzzy logic, advancements in artificial intelligence (AI) and machine learning (ML) have also contributed to the intelligence augmentation of mobile AR systems. Techniques such as neural networks, deep learning, and reinforcement learning are being explored to enable more sophisticated user modeling, context awareness, and adaptive system behavior. These intelligent capabilities pave the way for personalized AR experiences that are finely tuned to meet the unique needs and preferences of individual users.

Considering the discussed developments, it can be inferred that there exists a notable gap in the literature concerning the personalization of such systems. While considerable attention has been devoted to improving user interface and friendliness, there remains a dearth of research addressing the customization of mobile AR experiences to individual user characteristics and preferences. This gap underscores the need for further exploration into personalized approaches that can tailor mobile AR systems to the unique needs and preferences of users, thereby maximizing engagement and satisfaction.

3 Fuzzy Weighting Framework

Supporting users of mobile AR systems necessitates a nuanced understanding of their computer skills, a multifaceted undertaking fraught with uncertainties. In our investigation, we adopted a rigorous methodology leveraging a questionnaire devised by a consortium of 15 faculty members of informatics from public universities. The questionnaire comprised 10 questions, each carefully designed to gauge proficiency levels across a spectrum. Participants were tasked with selecting from a range of options for each query, with each option corresponding to a distinct proficiency grade. For instance, option A corresponded to a single grade, while option B warranted two grades, and so forth, culminating in a maximum grading potential of 40 for each participant.

The questionnaire's design was intricately tailored to probe participants' knowledge base and hands-on experience with mobile augmented reality systems, thereby offering valuable insights into their aptitude and familiarity with the technology. Table 1 presents a comprehensive overview of the questionnaire items, delineating the specific inquiries devised to elicit nuanced responses from participants.

Table 1. Questionnaire.

Questions	Answers
How would you rate your knowledge of AR technology?	a) Limited or no knowledge b) Basic understanding c) Moderate understanding d) Advanced knowledge
Have you previously used AR applications?	a) No b) Yes, but only a few times c) Yes, occasionally d) Yes, frequently
How comfortable are you with using mobile devices?	a) Not comfortable at all b) Somewhat comfortable c) Moderately comfortable d) Very comfortable
How familiar are you with the concept of virtual objects and their interaction in AR?	a) Not familiar at all b) Somewhat familiar c) Moderately familiar d) Very familiar
How proficient are you in installing and setting up AR applications on a mobile device?	a) Not proficient at all b) Basic proficiency c) Moderate proficiency d) Advanced proficiency
How comfortable are you with the process of calibrating an AR system to align virtual objects with the real world?	a) Not comfortable at all b) Somewhat comfortable c) Moderately comfortable d) Very comfortable
How experienced are you in creating or designing augmented reality experiences?	a) No experience b) Limited experience c) Moderate experience d) Extensive experience
How knowledgeable are you about the different types of AR technologies, such as marker-based, marker-less, or projection-based AR?	a) Limited knowledge b) Basic knowledge c) Moderate knowledge d) Advanced knowledge
How confident are you in troubleshooting common issues that may arise while using AR applications or devices?	a) Not confident at all b) Somewhat confident c) Moderately confident d) Very confident
How comfortable are you with integrating AR technology with other digital tools or platforms, such as social media or e-learning systems?	a) Not comfortable at all b) Somewhat comfortable c) Moderately comfortable d) Very comfortable

For example, categorizing a user with a computer knowledge test score of 7.5 out of 10 as either "good" or "very good" presents a challenge due to the inherent ambiguity in such classifications, as both designations hold some degree of validity. To address this complexity, fuzzy logic emerges as a viable solution. In this paradigm, learners' computer skills are delineated by four fuzzy weights: Novice (N), Basic (B), Advanced (A), and Proficient (P), each characterized by trapezoidal membership functions (refer to Table 2 and Fig. 1). These functions are defined by four boundary values (a_1, a_2, a_3, a_4), wherein the degree of membership gradually ascends from 0 to 1 between a_1 and a_2, remains constant at 1 between a2 and a3, and then descends from 1 to 0 between a_3 and a_4. Trapezoidal membership functions were specifically selected for their ability to accurately delineate the interval within which students' scores unequivocally belong to a specific knowledge category.

Expanding on this framework, the incorporation of fuzzy logic not only addresses the challenge of categorizing users with nuanced skill levels but also enables a more detailed and granular representation of learners' proficiencies. By employing trapezoidal membership functions (Table 2), the fuzzy logic model accommodates the inherent uncertainty and imprecision associated with human cognition (Fig. 1), thereby facilitating a more flexible and adaptable approach to assessing and refining users' computer skills within the context of mobile augmented reality systems. This sophisticated framework lays the groundwork for personalized and targeted interventions aimed at enhancing user engagement and interaction efficacy in mobile AR environments.

As reiterated earlier, the current methodology employs membership functions to represent a user's current level of computer knowledge in AR interactions. These membership functions serve as the basis for defining the values of the fuzzy weights, ranging from 0 to 1. A value of 1 assigned to the knowledge level indicates that the user has attained mastery in the domain and possesses comprehensive understanding. Consequently, the cumulative value of each segmented fuzzy set signifies the knowledge level of a domain learning unit and collectively sums up to 1, as denoted by the equation $\mu_N(x) + \mu_B(x) + \mu_A(x) + \mu_P(x) = 1$.

The establishment of the fuzzy weights and the determination of thresholds for their membership functions were entrusted to a panel of 15 faculty members from esteemed public universities specializing in informatics. These seasoned faculty members were tasked with furnishing descriptive definitions of learners' computer knowledge levels throughout the learning continuum, along with delineating the ranges of achievement corresponding to each knowledge level. Leveraging their extensive collective experience spanning over 12 years in university settings, these faculty members possess the requisite expertise to accurately depict users' computer knowledge skills.

Furthermore, the engagement of domain experts in this process ensures the robustness and validity of the fuzzy weighting framework, instilling confidence in the accuracy and reliability of the fuzzy logic model utilized for assessing and refining users' computer skills within the AR domain. Their contributions not only enrich the methodology but also enhance the credibility and applicability of the research findings within both academic and practical contexts.

Table 2. Membership functions [6].

Computer Knowledge Level
Membership Function
Novice (N)
$\mu_N(x) = \begin{cases} 1 & x \leq 10 \\ 1 - \frac{x-10}{5} & 10 < x < 15 \\ 0 & x \geq 15 \end{cases}$
Basic (B)
$\mu_B(x) = \begin{cases} \frac{x-10}{5} & 10 < x < 15 \\ 1 & 15 \leq x \leq 20 \\ 1 - \frac{x-20}{5} & 20 < x < 25 \\ 0 & x \leq 10 or x \geq 25 \end{cases}$
Advanced (A)
$\mu_A(x) = \begin{cases} \frac{x-20}{5} & 20 < x < 25 \\ 1 & 25 \leq x \leq 35 \\ 1 - \frac{x-35}{2} & 35 < x < 37 \\ 0 & x \leq 20 or x \geq 37 \end{cases}$
Proficient (P)
$\mu_P(x) = \begin{cases} \frac{x-35}{2} & 35 < x < 37 \\ 1 & 37 \leq x \leq 40 \\ 0 & x > 40 \end{cases}$

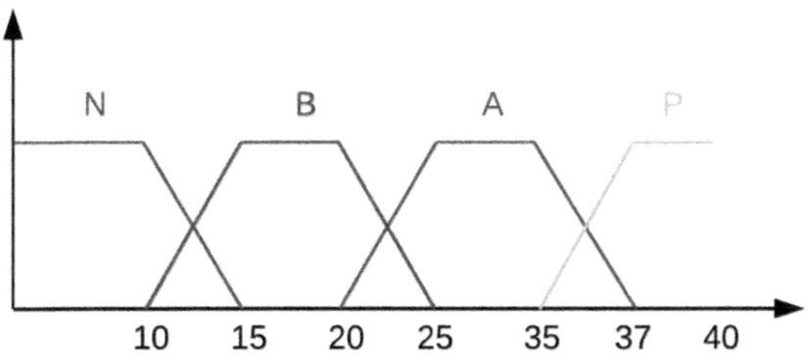

Fig. 1. Schemes of Fuzzy Weights [6].

4 Facilitating Assistive Messages Delivery

This section offers a comprehensive overview of the personalized assistive messages and feedback meticulously tailored to correspond with the fuzzy weights previously established. These bespoke messages are strategically disseminated to users at various junctures throughout their interaction with the mobile AR environment, with the overarching objective of enriching their overall experience and fostering deeper engagement with the technology.

Each fuzzy weight encompasses several categories of assistive messages. These are provided to the user randomly, offering diverse avenues for support. For instance, within the novice user category, specific message categories include the "Initiation Phase", "Discovery and Movement", and "Support and Advice". After determining the user's fuzzy weight, the system selects the most relevant message category for each interaction. To achieve this, the system employs an if-then algorithmic approach taking into consideration the user's interaction history as well as the prevailing task or context.

Moreover, within each designated category, the system employs a random selection process to choose a message tailored to the individual user, ensuring a personalized and customized experience. This dynamic approach enables users to receive targeted assistance and guidance that aligns with their proficiency level and immediate needs, thereby enhancing user satisfaction and efficacy in navigating the mobile AR environment.

4.1 Feedback to Novice Users

Novice users, characterized by their limited or nonexistent prior experience with AR systems, necessitate comprehensive, step-by-step instructions and guidance to navigate the system, utilize basic features, and grasp the interface. The personalized feedback messages tailored for novices encompass:

Initiation Phase. This section introduces the preliminary messages aimed at guiding and assisting novice users during their initial engagements with the AR system. The messages are as follows:

- "Welcome to the world of augmented reality! Let's start by guiding you through the basic features. Follow the step-by-step instructions to navigate the system and explore virtual objects."
- "To interact with virtual objects, simply tap on them. Try tapping on the floating cube to see it respond!"
- "Remember to maintain a comfortable distance from virtual objects for better interaction. Experiment with different gestures to explore their functionalities."

Additionally, these introductory messages serve as the cornerstone for novice users, offering clear and concise guidance to facilitate their acclimatization to the AR environment and empower them to explore its potential with confidence.

Discovery and Movement. In this segment, we introduce informative messages tailored to aid users in navigation and exploration tasks within the mobile AR system. The messages are outlined as follows:

- "To navigate through the AR environment, swipe left, right, up, or down to move around and discover more virtual content."
- "Do not be afraid to explore! Move your device around to find hidden objects and discover exciting AR experiences."

- "If you're feeling lost, refer to the on-screen guide or consult the user manual for assistance in navigating the AR environment."

Moreover, these informative messages serve as valuable navigational aids, empowering users to seamlessly navigate the AR landscape while fostering a sense of curiosity and discovery.

Support and Advice. This section introduces contextual messages meticulously crafted to furnish users of the mobile AR system with assistance and guidance. The messages are outlined as follows:

- "Do not worry if you are unsure about something. Remember, you can always refer to the user guide for detailed explanations and helpful tips."
- "If you need assistance at any point, tap on the help icon in the menu to access the FAQ section or get in-app support from our team."
- "Take your time to familiarize yourself with the interface and features. Learning at your own pace will help you build confidence in using augmented reality."

Furthermore, these personalized feedback messages tailored for novice users aim to offer lucid instructions, foster a spirit of exploration, and extend support when necessary. By receiving these messages, novices can gradually cultivate confidence and competence in navigating the augmented reality system.

4.2 Feedback to Basic Users

Basic users have acquired a degree of proficiency with AR systems and mobile devices. They are capable of executing routine tasks such as initiating the app, engaging with virtual objects, and configuring fundamental settings. Nonetheless, they may encounter occasional hurdles or seek guidance from user manuals to explore more intricate functionalities. The personalized feedback messages aimed at assisting basic users from the system are:

Interactive Skill Enhancement. This section introduces messages tailored to enrich the interaction experience for basic users within the mobile AR system. The messages are outlined as follows:

- "Congratulations on becoming familiar with augmented reality! Now, let's enhance your interaction skills. Experiment with different gestures to explore the functionalities of virtual objects."
- "To interact with virtual objects, tap on them, and observe how they respond. You can also try using long presses or double taps for additional actions."
- "Remember to maintain a comfortable distance from virtual objects for better interaction. Experiment with different gestures and distances to find what works best for you."

Moreover, these tailored messages aim to empower basic users by providing clear instructions, encouraging exploration, and promoting effective interaction practices.

Through these messages, basic users can further enhance their proficiency and enjoyment of the augmented reality experience.

Advanced Functionality Exploration. This section introduces messages tailored to facilitate the exploration of advanced features and functionalities for basic users within the mobile AR system. The messages are outlined as follows:

- "You're making great progress! It's time to dive into more advanced features. Explore the settings menu to customize preferences such as lighting, object placement, and interactions."
- "Adjust the sensitivity of your device's motion tracking to ensure a smoother and more immersive augmented reality experience."
- "Feel free to experiment with additional features like voice commands, object scaling, or integrating external data sources. Unlock the full potential of augmented reality!"

Furthermore, these customized messages aim to empower basic users by encouraging them to explore beyond the fundamentals, unlock advanced functionalities, and fully embrace the immersive possibilities of augmented reality. Through these messages, basic users can expand their horizons and discover the rich array of features available within the mobile AR environment.

Support and Learning Tools. In this segment, we present messages offering assistance and supplementary resources to aid basic users in their mobile AR interactions. The messages are outlined as follows:

- "If you encounter any challenges or have questions about specific features, refer to the user guide for detailed explanations and troubleshooting tips."
- "Don't hesitate to explore the help section in the app for quick access to FAQs and video tutorials. It can provide valuable insights and guidance."
- "As an intermediate user, your growing familiarity with the system enables you to explore new possibilities. Challenge yourself to discover hidden features or collaborate with other users to exchange tips and tricks."

These personalized feedback messages for basic users are designed to inspire further exploration, highlight advanced features, and offer assistance when necessary. By receiving these messages, basic users can continue to enhance their AR skills and deepen their comprehension of the system's capabilities.

4.3 Feedback to Advanced Users

Advanced users possess an extensive comprehension of the AR system and demonstrate proficiency in utilizing its functionalities. They exhibit adeptness in navigating intricate interfaces, tailoring settings to their preferences, seamlessly interacting with virtual objects, and resolving potential issues that may surface. The system delivers personalized feedback messages to advanced users, comprising:

Enhancing Complex Interactions. This section presents messages designed to assist advanced users in mastering intricate interactions within the mobile AR system. The messages are outlined as follows:

- "Impressive skills! You've mastered the basics and are ready for more advanced features. Navigate through the complex interface confidently and make the most of the additional tools at your disposal."
- "Experiment with advanced interactions such as multi-touch gestures, pinch-to-zoom, or two-finger rotations to manipulate and interact with virtual objects more precisely."
- "Consider exploring hand tracking or body tracking options for a more immersive and natural interaction with the augmented reality environment."

Furthermore, these tailored messages aim to empower advanced users by guiding them towards advanced features, offering suggestions for refining interaction techniques, and promoting exploration of innovative functionalities. Through these messages, advanced users can further elevate their mastery of the AR system and maximize their engagement with the augmented reality environment.

Personalization and Performance Tuning. In this section, we present messages crafted to aid advanced users in customizing and refining their experience within the mobile AR system. The messages are outlined as follows:

- "As an advanced user, customization is key. Dive into the settings menu to fine-tune preferences and tailor the augmented reality experience to your liking."
- "Optimize system performance by adjusting parameters like rendering quality, frame rate, or lighting effects. Maximize the quality and fluidity of your AR experiences."
- "Explore the advanced settings to enable features like occlusion, physics simulations, or real-time reflections. Push the boundaries of realism and create captivating AR scenes."

Furthermore, these tailored messages are designed to empower advanced users by guiding them through the customization process, optimizing system performance, and unlocking advanced features to elevate the quality of their AR experiences. Through these messages, advanced users can further refine their AR encounters and unleash their creativity to create immersive and captivating scenes.

Expert Guidance and Problem Solving. In this section, we present messages tailored to troubleshooting and offering support to advanced users within the mobile AR system. The messages are outlined as follows:

- "With your expertise, troubleshooting common issues becomes second nature. Share your knowledge with others in the community and offer assistance when needed."
- "Stay updated with the latest software updates and firmware releases to ensure compatibility, stability, and access to new features and improvements."
- "If you encounter more complex problems or need in-depth technical support, don't hesitate to reach out to our dedicated support team. They are available to assist you in resolving any challenges you may face."

These personalized feedback messages for advanced users recognize their proficiency and aim to empower them to explore advanced interactions, customize settings, and address issues effectively. By receiving these messages, advanced users can further hone their AR skills, enhance their experiences, and contribute to the AR community by sharing their knowledge and offering support.

4.4 Feedback to Proficient Users

Proficient users demonstrate high skills and extensive experience in navigating mobile AR systems. They boast a deep understanding of intricate features, excel in optimizing system performance, and may even possess the capability to craft or tailor their own augmented reality encounters. The system delivers personalized feedback messages to expert users, comprising:

Expanding Creativity. In this section, we introduce messages tailored to harnessing the creative potential of proficient users within the mobile AR system. The messages are outlined as follows:

- "Welcome, AR expert! Your knowledge and skills are highly valuable. Use your expertise to push the boundaries of augmented reality and unleash your creative potential."
- "Consider developing your own custom AR experiences using scripting or coding. Let your imagination soar and create unique interactive narratives or games."
- "Explore advanced techniques like object recognition, markerless tracking, or machine learning integration to create immersive and intelligent AR applications."

Furthermore, these messages are designed to inspire proficient users to venture into innovative avenues, experiment with advanced techniques, and pioneer groundbreaking AR experiences. Through these messages, proficient users can expand their horizons, push the boundaries of creativity, and leave their mark on the mobile AR landscape.

Enhancing System Efficiency and Personalization. This section presents messages concerning performance enhancement and personalization within the mobile AR system, as follows:

- "As an expert user, you have the power to optimize system performance. Fine-tune advanced settings like shaders, level of detail, or occlusion culling to achieve the best possible results."
- "Experiment with advanced lighting and shadow techniques to add depth and realism to your AR scenes. Leverage the full potential of the rendering engine to create visually stunning experiences."
- "Consider implementing performance optimization strategies such as asset bundling, texture compression, or dynamic resource loading to ensure smooth performance on various devices."

Moreover, these messages aim to empower expert users to maximize system efficiency, elevate visual quality, and ensure smooth performance across a range of devices.

Through these recommendations, expert users can further enhance their AR endeavors and deliver immersive experiences that captivate and engage audiences.

Engagement and Knowledge Exchange. In this section, we introduce messages centered on fostering collaboration and sharing within the mobile AR system, as follows:

- "Share your expertise with the AR community! Engage in forums, online communities, or developer conferences to exchange knowledge, collaborate on projects, and inspire others with your creations."
- "Consider publishing your AR experiences or apps on dedicated platforms to reach a wider audience and make a lasting impact in the world of augmented reality."
- "Remember to document your learnings, techniques, and best practices. By sharing your insights and tutorials, you can contribute to the growth and development of the AR ecosystem."

Furthermore, these messages aim to encourage users to engage with the AR community, share their creations, and contribute to the collective growth of the field. Through collaboration and knowledge-sharing, users can collectively propel the evolution and adoption of augmented reality technologies.

Pioneering Technological Advancements. This section introduces messages aimed at fostering cutting-edge research and innovation within the mobile AR system, as follows:

- "You are at the forefront of AR innovation! Stay informed about the latest advancements, research papers, and emerging technologies to continue expanding your expertise."
- "Consider exploring cutting-edge areas such as spatial mapping, real-time object recognition, or collaborative AR experiences. Push the boundaries of what's possible in augmented reality."
- "Share your findings and contribute to the advancement of the field through publishing research papers, attending conferences, or even mentoring aspiring AR developers."

These personalized feedback messages for expert users acknowledge their remarkable proficiency and encourage them to continue pushing the boundaries of augmented reality. By receiving these messages, expert users are inspired to explore advanced techniques, optimize performance, engage with the community, and drive ongoing innovation in the realm of augmented reality.

5 Evaluation Results and Discussion

In this section, we provide an assessment of the system's performance and efficacy.

5.1 Descriptive Analysis

The evaluation process extended over an entire academic semester and focused on spatial ability training tutoring within the framework of the undergraduate course "Educational

Technologies and IT Didactics" at a public university in the capital city. 80 undergraduate students participated in the evaluation. It is noteworthy that gender and age metrics were collected from a randomly selected sample and did not influence the research outcomes. The questionnaire (Table 3) was designed to gauge participants' familiarity, experience, and comfort levels with augmented reality (AR) technology.

Comprising six questions, the questionnaire explored various facets of AR, encompassing knowledge, usage, proficiency, familiarity with virtual objects, comfort with mobile devices, experience in crafting AR experiences, acquaintance with diverse AR technologies, troubleshooting skills, and integration of AR with digital tools. The responses aimed to offer valuable insights into participants' comprehension and competence in AR, aiding in pinpointing areas for enhancement and tailoring future AR-related initiatives to their preferences.

Table 3. Evaluation questionnaire [6].

Constructs	Question
Personalization	Rate the effectiveness of the feedback messages you received regarding your AR usage and proficiency
	How well did the personalized feedback messages address your specific areas of improvement and help you enhance your understanding and skills in AR?
Design	Rate the user interface of the AR application you used
	How intuitive and user-friendly was the user interface of the AR application in terms of navigation, interactions, and accessing features?
User experience	How would you rate your overall user experience with the AR application?
	How likely are you to recommend it to others who are interested in exploring AR technology?

To perform a descriptive analysis of the questionnaire responses, the authors aggregated the data by computing the frequency for each question (Fig. 2).

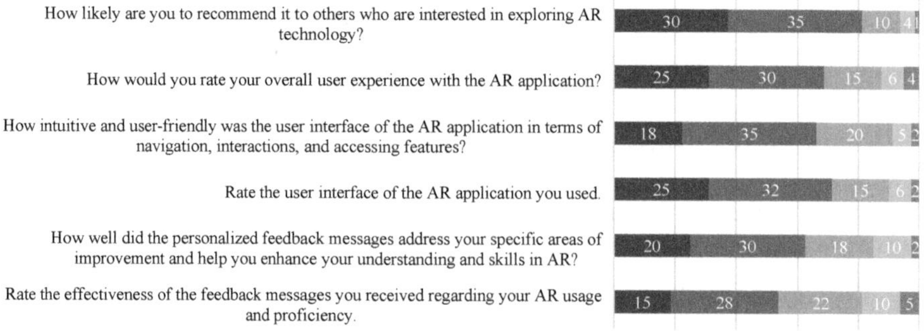

Fig. 2. Frequency of the answers in stacked-bar mode [6].

Based on participants' responses, the descriptive analysis offers intriguing insights into their familiarity, experience, and comfort level with AR technology.

The analysis of feedback message efficacy concerning AR usage and proficiency indicates that a significant portion of participants found the feedback helpful. Combining responses from the "extremely helpful" and "very helpful" categories, 53.75% of participants rated the feedback as highly advantageous. This suggests that the feedback messages provided valuable insights and guidance for enhancing participants' understanding and skills in AR. However, there is room for improvement, as 18.75% of participants expressed lower levels of effectiveness. This underscores the importance of tailoring feedback to address individual improvement areas and accommodating diverse proficiency levels.

Examining the effectiveness of personalized feedback messages in targeting specific improvement areas, results reveal that a notable percentage of participants (62.5%) felt the feedback adequately addressed their needs, rating it as either "very well" or "well." This signifies that personalized feedback messages effectively targeted specific improvement areas and enhanced participants' understanding and skills in AR. Nonetheless, addressing concerns from participants who found the feedback less effective (15%) is crucial to ensure a comprehensive and tailored approach.

Regarding the AR application's user interface, the analysis indicates that a majority of participants (71.25%) rated it as either "excellent" or "good." This suggests positive evaluation of the application's design and usability, indicating a satisfactory experience for most participants. However, not all ratings for the user interface were uniformly positive, with 10% expressing lower satisfaction levels. This underscores the importance of continuous refinement to cater to diverse user preferences and needs.

Overall, participants' ratings for their overall user experience with the AR application are predominantly positive. The majority (68.75%) rated their experience as either "excellent" or "good," suggesting the application provided a positive user experience, meeting expectations in performance, usability, and enjoyment. However, a small percentage of participants (12.5%) rated their experience as average or poor, indicating areas for improvement to enhance user satisfaction.

Concerning the likelihood of recommending the AR application to others interested in AR technology exploration, analysis reveals a positive inclination among participants. A significant majority (81.25%) expressed likelihood of recommending the application, either "very likely" or "likely." This high recommendation potential suggests participants perceived the AR application as valuable and beneficial, indicating an overall positive impression. Nonetheless, addressing concerns from participants expressing reluctance or neutrality (6.25%) is essential to maximize the application's reach and impact.

In conclusion, the descriptive analysis provides valuable insights into participants' perceptions and experiences with AR technology. It underscores the effectiveness of feedback messages, usability of the application's user interface, overall user experience, and likelihood of recommendation. These insights can inform further enhancements and refinements to the feedback process, user interface design, and overall AR experience, ultimately boosting engagement, satisfaction, and proficiency in AR technology.

5.2 Statistical Analysis

The instructors divided the student population into two cohorts, each comprising 40 students. The first group, termed the experimental group, was instructed to utilize an AR application featuring feedback messaging, while the second group, referred to as the control group, did not have access to the feedback generation module.

The incorporation of adaptive feedback messaging is pivotal in assessing the efficacy of an intelligent tutoring system integrating personalized feedback via fuzzy logic. To gauge users' perceptions of the system's ability to tailor feedback, a questionnaire was administered.

To determine if a statistically significant difference in personalized messages exists between students using the AR application with personalized feedback and those without, a *t*-test analysis was conducted. This analysis encompassed several steps:

- Formulation of the null hypothesis (H0) and alternative hypothesis (H1) based on the research inquiry and objectives.
- Selection of the significance level (alpha) to establish the threshold for accepting or rejecting the null hypothesis, with a value of 0.05 utilized for alpha.
- Collection of data from both the experimental and control groups.
- Computation of mean scores for each group separately, reflecting perceptions of personalized messages.
- Calculation of the variance for each group, indicating data spread or variability.
- Conducting the *t*-test to compare group means and evaluate statistical significance.
- Determination of the p-value associated with the calculated t-statistic, contrasting it with the significance level (alpha) to ascertain null hypothesis acceptance or rejection.
- Interpretation of findings: Rejection of the null hypothesis if the p-value is below the significance level, signifying a statistically significant difference in personalized messages between groups. Conversely, failure to reject the null hypothesis suggests no significant difference between the groups.

At the semester's end, both cohorts completed a questionnaire employing a 5-point Likert scale to express their agreement or disagreement, ranging from 1 (strongly disagree) to 5 (strongly agree).

To conduct the *t*-test analysis on the effectiveness of feedback messages regarding AR usage and proficiency between the experimental and control groups, ratings from both cohorts were compared. The *t*-test aimed to determine if there was a notable difference in group means.

Initially, the null hypothesis (H0) and alternative hypothesis (H1) were established as follows:

H0: There is no significant disparity in feedback message effectiveness between the experimental and control groups.

H1: A significant difference exists in feedback message effectiveness between the experimental and control groups.

Subsequently, a *t*-test analysis was performed using the ratings provided by both groups, resulting in the computation of t-values and p-values (Table 4).

Based on the t-value and degrees of freedom, we can ascertain the statistical significance of the findings. The p-value falls below the predetermined significance threshold

Table 4. t-test results [6].

	Group A	Group B
Mean	4.129	3.083
Variance	0.558	0.742
Observations	40	40
Hypothesized Mean Difference	0	
df	186	
t Stat	11.400	
P (T ≤ t) one-tail	<.001	
t Critical one-tail	1.582	
P (T ≤ t) two-tail	<.001	
t Critical two-tail	1.861	

of 0.05, leading us to reject the null hypothesis and infer a notable discrepancy in the efficacy of feedback messages between the experimental and control cohorts.

The *t*-test analysis unveils a t-statistic of 11.400, with 186 degrees of freedom, yielding a p-value below 0.001 for a one-tailed test. This indicates a statistically significant variance in the effectiveness of feedback messages between the two groups.

Under the null hypothesis (H0), no substantial difference is presumed between the two cohorts, whereas the alternative hypothesis (H1) posits a significant disparity. As the p-value is lower than the selected significance level, we dismiss the null hypothesis and confirm a significant contrast in the effectiveness of feedback messages between the experimental and control sets.

Furthermore, the *t*-test outcomes provide evidence that the mean effectiveness rating for the experimental group (M = 4.129) significantly surpasses that of the control group (M = 3.083). This implies that the feedback messages received by participants in the experimental cohort were more impactful in enhancing their comprehension and proficiencies in AR compared to their counterparts in the control group.

Impact of AR on Spatial Abilities. Given the potential of augmented reality to enhance spatial ability, an additional *t*-test was conducted to evaluate whether the use of AR technology significantly improves the spatial skills of students [40, 41]. This analysis aims to ascertain the effectiveness of AR in fostering spatial reasoning, a key skill in many educational and professional domains [42, 43].

The formulation of hypotheses is as follows:

H0: There is no significant difference in the improvement of spatial abilities between students who used the AR application and those who did not.

H1: Students who used the AR application show a significant improvement in spatial abilities compared to those who did not.

Before the experiment, all participants were tested to establish a baseline for their spatial abilities using a standard spatial reasoning test. At the end of the semester, the same test was administered again. The test scores from the beginning and the end of the semester were then used to assess improvements in spatial abilities.

The same cohorts of 40 students each from the experimental and control groups were involved. Pre-test and post-test scores on spatial reasoning were collected, while the mean improvement scores for each group were calculated by subtracting the pre-test scores from the post-test scores. The variances of these improvement scores were also computed. The *t*-test execution considered the alpha level at 0.05 and the results are presented in Table 5.

Table 5. t-test results of pre-test and post-test.

	Group A	Group B
Pre-test Mean	2.950	2.700
Post-test Mean	4.750	3.650
Difference	1.800	0.950
Standard Deviation	0.613	1.138
Pearson Correlation	0.112	−0.462
t Stat	−13.340	−3.579
p-Value	<0.001	<0.001

The results presented in Table 5 provide a comprehensive overview of the impact of AR technology on enhancing spatial abilities among students. The primary focus of this analysis lies in interpreting the *t*-test results, particularly examining the significance of the differences between the pre-test and post-test scores of the two groups. Group A, which utilized AR technology, showed a substantial improvement in their post-test scores with a mean increase of 1.800, compared to Group B, which did not use the technology and showed a smaller increase of 0.950. This difference is statistically significant, as evidenced by the very low p-values (<0.001) for both groups, suggesting that the improvements in spatial abilities are not due to chance.

The magnitude of the mean differences further underscores the effectiveness of AR in fostering spatial reasoning skills. The greater improvement observed in Group A can be linked to the interactive and immersive nature of AR, which may enhance the learning experience by providing visual and spatial cues that aid in understanding complex spatial relationships. The standard deviation in the improvements (0.613 for Group A) indicates a relatively consistent effect across participants using AR. This consistency suggests that AR could be a reliable tool in educational settings for enhancing spatial abilities. Additionally, the negative t-values, particularly −13.340 for Group A, are indicative of a significant improvement from the pre-test to the post-test, further supporting the efficacy of AR in educational enhancements.

Lastly, the Pearson correlation coefficients present interesting insights into the relationship between the initial and improved performances of the students. For Group A, a positive but weak correlation (0.112) suggests little to no relationship between students' initial abilities and the extent of their improvement. In contrast, Group B exhibits a moderate negative correlation (−0.462), which could indicate that students with lower initial scores might have shown relatively more improvement than those with higher initial scores, albeit without the aid of AR technology. This contrasting dynamic between the groups could highlight the role of AR in providing a more uniform improvement in spatial skills, irrespective of initial ability levels, potentially leveling the educational playing field for all students. These findings advocate for the integration of AR applications in educational programs that aim to develop and enhance spatial reasoning skills, contributing significantly to students' learning outcomes in fields requiring strong spatial capabilities.

6 Conclusions

This paper presents a pioneering method for augmenting user interactions within mobile AR systems. It employs fuzzy logic-based modeling of users' computer skills to dynamically tailor personalized assistive messages and feedback. By employing fuzzy logic models, the system dynamically tailors personalized assistive messages and feedback to each user's unique level of computer proficiency, thus facilitating more effective navigation and utilization of mobile AR environments.

The evaluation outcomes highlight the efficacy of this approach, demonstrating its potential to refine user interactions and catalyze future advancements in mobile AR technology. However, it is essential to acknowledge certain limitations of the current methodology. For instance, the effectiveness of the fuzzy logic model may vary depending on the complexity of AR applications and the diversity of user skill levels. Additionally, the subjective nature of user feedback could introduce biases that may influence the accuracy of the fuzzy logic model.

Future endeavors will concentrate on the development of a hybrid algorithmic framework that integrates fuzzy logic with machine learning techniques to enhance the precision and adaptability of personalized message delivery. Furthermore, extensive user studies will be conducted to validate the effectiveness and robustness of the proposed approach across diverse user demographics and AR application scenarios.

Disclosure of Interests. The authors have no competing interests to declare that are relevant to the content of this article.

References

1. Papakostas, C., Troussas, C., Krouska, A., Sgouropoulou, C.: Exploration of augmented reality in spatial abilities training: a systematic literature review for the last decade. Inform. Educ. **20**(1), 107–130 (2021). https://doi.org/10.15388/infedu.2021.06
2. White, G., Cabrera, C., Palade, A., Clarke, S.: Augmented reality in IoT. Lecture Notes in Computer Science (including subseries Lecture Notes in Artificial Intelligence and Lecture Notes in Bioinformatics). LNCS, vol. 11434, pp. 149–160 (2019). https://doi.org/10.1007/978-3-030-17642-6_13
3. Verma, A., Purohit, P., Thornton, T., Lamsal, K.: An examination of skill requirements for augmented reality and virtual reality job advertisements. Ind. High. Educ. **37**(1), 46–57 (2022). https://doi.org/10.1177/09504222221109104

4. Müller, T.: Challenges in representing information with augmented reality to support manual procedural tasks. AIMS Electron. Electr. Eng. **3**(1), 71–97 (2019)
5. Kourouthanassis, P.E., Boletsis, C., Lekakos, G.: Demystifying the design of mobile augmented reality applications. Multimed Tools Appl **74**(3), 1045–1066 (2015). https://doi.org/10.1007/s11042-013-1710-7
6. Troussas, C., Papakostas, C., Sgouropoulou, C.: Enhancing users' interactions in mobile augmented reality systems through fuzzy logic-based modelling of computer skills. In: Proceedings of the 19th International Conference on Web Information Systems and Technologies - WEBIST, SciTePress, pp. 381–390 (2023). https://doi.org/10.5220/0012204300003584
7. Pena-Rios, A., Hagras, H., Gardner, M., Owusu, G.: A fuzzy logic based system for mixed reality assistance of remote workforce. In: 2016 IEEE International Conference on Fuzzy Systems (FUZZ-IEEE). IEEE (2016). https://doi.org/10.1109/fuzz-ieee.2016.7737716
8. Papakostas, C., Troussas, C., Krouska, A., Sgouropoulou, C.: Personalization of the learning path within an augmented reality spatial ability training application based on fuzzy weights. Sensors **22**(18) (2022). https://doi.org/10.3390/s22187059
9. Irshad, S., Rambli, D.R.A., Sulaiman, S.: Design and implementation of user experience model for augmented reality systems. In: Proceedings of the 18th International Conference on Advances in Mobile Computing & Multimedia, in MoMM 2020, pp. 48–57. Association for Computing Machinery, New York (2021). https://doi.org/10.1145/3428690.3429169
10. Virvou, M., Troussas, C.: Web-based student modeling for learning multiple languages. In: International Conference on Information Society (i-Society 2011), pp. 423–428 (2011). https://doi.org/10.1109/i-Society18435.2011.5978484
11. Virvou, M., Troussas, C., Caro, J., Espinosa, K.J.: User modeling for language learning in Facebook. In: Sojka, P., Horák, A., Kopeček, I., Pala, K. (eds.) Text, Speech and Dialogue, pp. 345–352. Springer, Heidelberg (2012). https://doi.org/10.1007/978-3-642-32790-2_42
12. Campanella, P.: Neuro-fuzzy learning in context educative. In: 2021 19th International Conference on Emerging eLearning Technologies and Applications (ICETA), pp. 58–69 (2021). https://doi.org/10.1109/ICETA54173.2021.9726657
13. Krouska, A., Troussas, C., Sgouropoulou, C.: Fuzzy logic for refining the evaluation of learners' performance in online engineering education. Eur. J. Eng. Technol. Res. **4**(6), 50–56 (2019). https://doi.org/10.24018/ejeng.2019.4.6.1369
14. Troussas, C., Krouska, A., Sgouropoulou, C.: Dynamic detection of learning modalities using fuzzy logic in students' interaction activities. In: Kumar, V., Troussas, C. (eds.) Intelligent Tutoring Systems, pp. 205–213. Springer, Cham (2020)
15. Kassim, R., Johari, J., Rahim, M., Buniyamin, N.: Lecturers' perspective of student online feedback system: a case study (2017). https://doi.org/10.1109/ICEED.2017.8251186
16. Troussas, C., Krouska, A., Sgouropoulou, C.: Impact of social networking for advancing learners' knowledge in E-learning environments. Educ. Inf. Technol. (Dordr) **26**(4), 4285–4305 (2021). https://doi.org/10.1007/s10639-021-10483-6
17. Han, Y., Chen, Y., Wang, R., Wu, J., Gorlatova, M.: Intelli-AR preloading: a learning approach to proactive hologram transmissions in mobile AR. IEEE Internet Things J. **9**(18), 17714–17727 (2022). https://doi.org/10.1109/JIOT.2022.3159554
18. Ito, G., Nakajima, T.: Expanding the interaction space using hand-held-AR with hearing support. In: 2021 Thirteenth International Conference on Mobile Computing and Ubiquitous Network (ICMU), pp. 1–6 (2021). https://doi.org/10.23919/ICMU50196.2021.9638926
19. Kim, S., et al.: Bring store in my room: AR store authoring system for spatial experience in mobile shopping. In: 2022 IEEE International Symposium on Mixed and Augmented Reality Adjunct (ISMAR-Adjunct), pp. 654–656 (2022). https://doi.org/10.1109/ISMAR-Adjunct57072.2022.00135

20. Lee, B., Kim, J., Jung, S.-U.: Light-weighted network based human pose estimation for mobile AR service. In: 2020 International Conference on Information and Communication Technology Convergence (ICTC), pp. 1609–1612 (2020). https://doi.org/10.1109/ICTC49870.2020.9289085
21. Linowes, J.: Augmented Reality with Unity AR Foundation: A practical guide to cross-platform AR development with Unity 2020 and later versions. Packt Publishing (2021)
22. Lo, J.-H., Lai, Y.-F.: Effects of incorporating AR-based mobile learning system on elementary school students' perceived usefulness of M-learning. In: 2021 IEEE 3rd Eurasia Conference on Biomedical Engineering, Healthcare and Sustainability (ECBIOS), pp. 169–172 (2021). https://doi.org/10.1109/ECBIOS51820.2021.9510571
23. Petrovic, N., Roblek, V., Khokhobaia, M., Gagnidze, I.: AR-Enabled Mobile Apps to Support Post COVID-19 Tourism (2021). https://doi.org/10.1109/TELSIKS52058.2021.9606335
24. Qiao, X., Ren, P., Dustdar, S., Liu, L., Ma, H., Chen, J.: Web AR: a promising future for mobile augmented reality—state of the art, challenges, and insights. Proc. IEEE **107**(4), 651–666 (2019). https://doi.org/10.1109/JPROC.2019.2895105
25. Vardhan, H., Saxena, A., Dixit, A., Chaudhary, S., Sagar, A.: AR museum: a virtual museum using marker less augmented reality system for mobile devices. In: 2022 3rd International Conference on Issues and Challenges in Intelligent Computing Techniques (ICICT), pp. 1–6 (2022). https://doi.org/10.1109/ICICT55121.2022.10064611
26. Wang, C., Zhao, Y., Guo, J., Pei, L., Wang, Y., Liu, H.: NEAR: the NetEase AR oriented visual inertial dataset. In: 2019 IEEE International Symposium on Mixed and Augmented Reality Adjunct (ISMAR-Adjunct), pp. 366–371 (2019). https://doi.org/10.1109/ISMAR-Adjunct.2019.00-10
27. Wang, Y., Wang, Y., Fan, Z.: Current status and prospects of mobile AR applications. In: 2021 International Conference on Culture-oriented Science & Technology (ICCST), pp. 34–37 (2021). https://doi.org/10.1109/ICCST53801.2021.00018
28. Xu, J., Yang, L., Guo, M.: AR mobile video calling system based on WebRTC API. In: 2022 IEEE 5th International Conference on Computer and Communication Engineering Technology (CCET), pp. 110–114 (2022). https://doi.org/10.1109/CCET55412.2022.9906395
29. Zhao, Y., Guo, T.: FUSEDAR: adaptive environment lighting reconstruction for visually coherent mobile AR rendering. In: 2022 IEEE Conference on Virtual Reality and 3D User Interfaces Abstracts and Workshops (VRW), pp. 572–573 (2022). https://doi.org/10.1109/VRW55335.2022.00137
30. Zhou, J., Xu, Z., Yan, H., Gao, B., Yang, O., Zhao, Z.: AR creator: a mobile application of logic education based on AR. In: 2020 International Conference on Virtual Reality and Visualization (ICVRV), pp. 379–380 (2020). https://doi.org/10.1109/ICVRV51359.2020.00109
31. Ping, J., Liu, Y., Weng, D.: Study on Mobile AR Guide System to Enhance User Experience in Cultural Heritage Sites (2020). https://doi.org/10.1109/ICVRV51359.2020.00027
32. Medrano, K., Tejada, R., González, B., Juarez, S.: Development of a mobile application with VR and AR to improve the experience of visitors in cultural settings. In: 2021 IEEE CHILEAN Conference on Electrical, Electronics Engineering, Information and Communication Technologies (CHILECON), pp. 1–5 (2021). https://doi.org/10.1109/CHILECON54041.2021.9703083
33. Nguyen, P., Ha, H.-D., Vu, T.-H., Nguyen, Q.-H., Truong, H.-P., Le, H.-S.: Applying VR/AR technology in product advertising to improve user experience. In: 2021 15th International Conference on Advanced Computing and Applications (ACOMP), pp. 201–205 (2021). https://doi.org/10.1109/ACOMP53746.2021.00036

34. Medina-Mota, K.M., López-Morteo, G.A., Cruz-Flores, R.G.: Creation of a game-based augmented learning experience system (GL-XP AR). In: 2022 IEEE Mexican International Conference on Computer Science (ENC), pp. 1–8 (2022). https://doi.org/10.1109/ENC56672.2022.9882939
35. Tang, W., Ge, Q., Zhou, L., Zhu, Q.: Design of spot introduction and user interaction system based on AR augmented reality technology. In: 2018 17th International Symposium on Distributed Computing and Applications for Business Engineering and Science (DCABES), pp. 76–79 (2018). https://doi.org/10.1109/DCABES.2018.00029
36. Leon-Garza, H., Hagras, H., Pena-Rios, A., Owusu, G., Conway, A.: A fuzzy logic based system for cloud-based building information modelling rendering optimization in augmented reality. In: 2020 IEEE International Conference on Fuzzy Systems (FUZZ-IEEE). IEEE (2020). https://doi.org/10.1109/fuzz48607.2020.9177659
37. Papakostas, C., Troussas, C., Krouska, A., Sgouropoulou, C.: Modeling the knowledge of users in an augmented reality-based learning environment using fuzzy logic. In: Krouska, A., Troussas, C., Caro, J. (eds.) Novel & Intelligent Digital Systems: Proceedings of the 2nd International Conference (NiDS 2022), pp. 113–123, Springer, Cham (2023)
38. Pena-Rios, A., Hagras, H., Gardner, M., Owusu, G.: A fuzzy logic based system for geolocated augmented reality field service support. In: 2017 IEEE International Conference on Fuzzy Systems (FUZZ-IEEE). IEEE (2017). https://doi.org/10.1109/fuzz-ieee.2017.8015477
39. Strousopoulos, P., Papakostas, C., Troussas, C., Krouska, A., Mylonas, P., Sgouropoulou, C.: SculptMate: personalizing cultural heritage experience using fuzzy weights. In: Adjunct Proceedings of the 31st ACM Conference on User Modeling, Adaptation and Personalization. ACM (2023). https://doi.org/10.1145/3563359.3596667
40. Papakostas, C., Troussas, C., Sgouropoulou, C.: Review of the literature on AI-enhanced augmented reality in education. In: Papakostas, C., Troussas, C., Sgouropoulou, C. (eds.) Special Topics in Artificial Intelligence and Augmented Reality: The Case of Spatial Intelligence Enhancement, pp. 13–50. Springer, Cham (2024). https://doi.org/10.1007/978-3-031-52005-1_2
41. Papakostas, C., Troussas, C., Krouska, A., Sgouropoulou, C.: PARSAT: fuzzy logic for adaptive spatial ability training in an augmented reality system. Comput. Sci. Inf. Syst. **20**(4) (2023). https://doi.org/10.2298/CSIS230130043P
42. Papakostas, C., Troussas, C., Sgouropoulou, C.: Introduction and overview of AI-enhanced augmented reality in education. In: Papakostas, C., Troussas, C., Sgouropoulou, C. (eds.) Special Topics in Artificial Intelligence and Augmented Reality: The Case of Spatial Intelligence Enhancement, pp. 1–11. Springer, Cham (2024). https://doi.org/10.1007/978-3-031-52005-1_1
43. Papakostas, C., Troussas, C., Krouska, A., Sgouropoulou, C.: On the development of a personalized augmented reality spatial ability training mobile application. In: Frontiers in Artificial Intelligence and Applications, pp. V–VI. IOS Press (2021). https://doi.org/10.3233/FAIA210078

Author Index

M. Marchiori and F. García Peñalvo (Eds.): WEBIST 2023, LNBIP 543, pp. 267–268, 2025.
https://doi.org/10.1007/978-3-031-89621-7

GPSR Compliance

The European Union's (EU) General Product Safety Regulation (GPSR) is a set of rules that requires consumer products to be safe and our obligations to ensure this.

If you have any concerns about our products, you can contact us on ProductSafety@springernature.com

In case Publisher is established outside the EU, the EU authorized representative is:

Springer Nature Customer Service Center GmbH
Europaplatz 3
69115 Heidelberg, Germany

Batch number: 08544862

Printed by Printforce, the Netherlands